Student Solutions Manual

Differential Equations

FOURTH EDITION

Paul Blanchard
Boston University

Robert L. Devaney
Boston University

Glen R. Hall
Boston University

Prepared by

Paul Blanchard
Boston University

Robert L. Devaney
Boston University

Glen R. Hall
Boston University

BROOKS/COLE
CENGAGE Learning™

Australia • Brazil • Japan • Korea • Mexico • Singapore • Spain • United Kingdom • United States

ISBN-13: 978-0-495-82672-9

ISBN-10: 0-495-82672-3

For product information and technology assistance, contact us at
**Cengage Learning Customer & Sales Support,
1-800-354-9706**

For permission to use material from this text or product, submit all requests online at **www.cengage.com/permissions**
Further permissions questions can be emailed to
permissionrequest@cengage.com

Brooks/Cole
20 Davis Drive
Belmont, CA 94002
USA

Cengage Learning is a leading provider of customized learning solutions with office locations around the globe, including Singapore, the United Kingdom, Australia, Mexico, Brazil, and Japan. Locate your local office at **www.cengage.com/global**

Cengage Learning products are represented in Canada by Nelson Education, Ltd.

To learn more about Brooks/Cole, visit
www.cengage.com/brookscole

Purchase any of our products at your local college store or at our preferred online store **www.cengagebrain.com**

Printed in the United States of America
2 3 4 5 6   24 23 22 21 20

CONTENTS

Appendices   317

CHAPTER 1

First-Order Differential Equations

EXERCISES FOR SECTION 1.1

1. Note that $dy/dt = 0$ if and only if $y = -3$. Therefore, the constant function $y(t) = -3$ for all t is the only equilibrium solution.

3. **(a)** The equilibrium solutions correspond to the values of P for which $dP/dt = 0$ for all t. For this equation, $dP/dt = 0$ for all t if $P = 0$ or $P = 230$.

   (b) The population is increasing if $dP/dt > 0$. That is, $P(1 - P/230) > 0$. Hence, $0 < P < 230$.

   (c) The population is decreasing if $dP/dt < 0$. That is, $P(1 - P/230) < 0$. Hence, $P > 230$ or $P < 0$. Since this is a population model, $P < 0$ might be considered "nonphysical."

5. In order to answer the question, we first need to analyze the sign of the polynomial $y^3 - y^2 - 12y$. Factoring, we obtain

$$y^3 - y^2 - 12y = y(y^2 - y - 12) = y(y - 4)(y + 3).$$

   (a) The equilibrium solutions correspond to the values of y for which $dy/dt = 0$ for all t. For this equation, $dy/dt = 0$ for all t if $y = -3$, $y = 0$, or $y = 4$.

   (b) The solution $y(t)$ is increasing if $dy/dt > 0$. That is, $-3 < y < 0$ or $y > 4$.

   (c) The solution $y(t)$ is decreasing if $dy/dt < 0$. That is, $y < -3$ or $0 < y < 4$.

7. The general solution of the differential equation $dr/dt = -\lambda r$ is $r(t) = r_0 e^{-\lambda t}$ where $r(0) = r_0$ is the initial amount.

   (a) We have $r(t) = r_0 e^{-\lambda t}$ and $r(5230) = r_0/2$. Thus

$$\frac{r_0}{2} = r_0 e^{-\lambda \cdot 5230}$$

$$\frac{1}{2} = e^{-\lambda \cdot 5230}$$

$$\ln \frac{1}{2} = -\lambda \cdot 5230$$

$$-\ln 2 = -\lambda \cdot 5230$$

   because $\ln 1/2 = -\ln 2$. Thus,

$$\lambda = \frac{\ln 2}{5230} \approx 0.000132533.$$

   (b) We have $r(t) = r_0 e^{-\lambda t}$ and $r(8) = r_0/2$. By a computation similar to the one in part (a), we have

$$\lambda = \frac{\ln 2}{8} \approx 0.0866434.$$

   (c) If $r(t)$ is the number of atoms of C-14, then the units for dr/dt is number of atoms per year. Since $dr/dt = -\lambda r$, λ is "per year." Similarly, for I-131, λ is "per day." The unit of measurement of r does not matter.

   (d) We get the same answer because the original quantity, r_0, cancels from each side of the equation. We are only concerned with the proportion remaining (one-half of the original amount).

9. **(a)** The general solution of the exponential decay model $dr/dt = -\lambda r$ is $r(t) = r_0 e^{-\lambda t}$, where $r(0) = r_0$ is the initial amount. Since $r(\tau) = r_0/e$, we have

$$\frac{r_0}{e} = r_0 e^{-\lambda \tau}$$

$$e^{-1} = e^{-\lambda \tau}$$

$$-1 = -\lambda \tau$$

$$\tau = 1/\lambda.$$

(b) Let h be the half-life, that is, the amount of time it takes for a quantity to decay to one-half of its original amount. Since $\lambda = 1/\tau$, we get

$$\tfrac{1}{2} r_0 = r_0 e^{-\lambda h}$$

$$\tfrac{1}{2} r_0 = r_0 e^{-h/\tau}$$

$$\tfrac{1}{2} = e^{-h/\tau}$$

$$-\ln 2 = -h/\tau.$$

Thus,

$$\tau = \frac{h}{\ln 2}$$

(c) In Exercise 7, we stated that the half-life of Carbon 14 is 5230 years and that of Iodine 131 is 8 days. Therefore, the time constant for Carbon 14 is $5230/(\ln 2) \approx 7545$ years, and the time constant for Iodine 14 is $8/(\ln 2) \approx 11.5$ days.

(d) To determine the equation of the line passing through $(0, 1)$ and tangent to the curve $r(t)/r_0$, we need to determine the slope of $r(t)/r_0$ at $t = 0$. Since

$$\frac{d}{dt} \frac{r(t)}{r_0} = \frac{d}{dt} e^{-\lambda t} = -\lambda e^{-\lambda t}$$

the slope at $t = 0$ is $-\lambda e^0 = -\lambda$. Thus, the equation of the tangent line is

$$y = -\lambda t + 1.$$

The line crosses the t-axis when $-\lambda t + 1 = 0$. We obtain $t = 1/\lambda$, which is the time constant τ.

(e) An exponentially decaying function approaches zero asymptotically but is never actually equal to zero. Therefore, to say that an exponentially decaying function reaches its steady state in any amount of time is false. However, after five time constants, the original amount r_0 has decayed by a factor of $e^{-5} \approx 0.0067$. Therefore, less than one percent of the original quantity remains.

11. The solution of $dR/dt = kR$ with $R(0) = 4,000$ is

$$R(t) = 4,000\, e^{kt}.$$

Setting $t = 6$, we have $R(6) = 4{,}000\,e^{(k)(6)} = 130{,}000$. Solving for k, we obtain

$$k = \tfrac{1}{6}\ln\left(\tfrac{130{,}000}{4{,}000}\right) \approx 0.58.$$

Therefore, the rabbit population in the year 2010 would be $R(10) = 4{,}000\,e^{(0.58\cdot 10)} \approx 1{,}321{,}198$ rabbits.

13. The rate of learning is dL/dt. Thus, we want to know the values of L between 0 and 1 for which dL/dt is a maximum. As $k > 0$ and $dL/dt = k(1 - L)$, dL/dt attains it maximum value at $L = 0$.

15. **(a)** We have $L_B(0) = L_A(0) = 0$. So Aly's rate of learning at $t = 0$ is dL_A/dt evaluated at $t = 0$. At $t = 0$, we have

$$\frac{dL_A}{dt} = 2(1 - L_A) = 2.$$

Beth's rate of learning at $t = 0$ is

$$\frac{dL_B}{dt} = 3(1 - L_B)^2 = 3.$$

Hence Beth's rate is larger.

(b) In this case, $L_B(0) = L_A(0) = 1/2$. So Aly's rate of learning at $t = 0$ is

$$\frac{dL_A}{dt} = 2(1 - L_A) = 1$$

because $L_A = 1/2$ at $t = 0$. Beth's rate of learning at $t = 0$ is

$$\frac{dL_B}{dt} = 3(1 - L_B)^2 = \frac{3}{4}$$

because $L_B = 1/2$ at $t = 0$. Hence Aly's rate is larger.

(c) In this case, $L_B(0) = L_A(0) = 1/3$. So Aly's rate of learning at $t = 0$ is

$$\frac{dL_A}{dt} = 2(1 - L_A) = \frac{4}{3}.$$

Beth's rate of learning at $t = 0$ is

$$\frac{dL_B}{dt} = 3(1 - L_B)^2 = \frac{4}{3}.$$

They are both learning at the same rate when $t = 0$.

17. Let $P(t)$ be the population at time t, k be the growth-rate parameter, and N be the carrying capacity. The modified models are

(a) $dP/dt = k(1 - P/N)P - 100$
(b) $dP/dt = k(1 - P/N)P - P/3$
(c) $dP/dt = k(1 - P/N)P - a\sqrt{P}$, where a is a positive parameter.

19. Several different models are possible. Let $R(t)$ denote the rhinoceros population at time t. The basic assumption is that there is a minimum threshold that the population must exceed if it is to survive. In terms of the differential equation, this assumption means that dR/dt must be negative if R is close to zero. Three models that satisfy this assumption are:

- If k is a growth-rate parameter and M is a parameter measuring when the population is "too small", then

$$\frac{dR}{dt} = kR\left(\frac{R}{M} - 1\right).$$

- If k is a growth-rate parameter and b is a parameter that determines the level the population will start to decrease $(R < b/k)$, then

$$\frac{dR}{dt} = kR - b.$$

- If k is a growth-rate parameter and b is a parameter that determines the extinction threshold, then

$$\frac{dR}{dt} = kR - \frac{b}{R}.$$

In each case, if R is below a certain threshold, dR/dt is negative. Thus, the rhinos will eventually die out. The choice of which model to use depends on other assumptions. There are other equations that are also consistent with the basic assumption.

21. **(a)** The term governing the effect of the interaction of x and y on the rate of change of x is $+\beta xy$. Since this term is positive, the presence of y's helps the x population grow. Hence, x is the predator. Similarly, the term $-\delta xy$ in the dy/dt equation implies that when $x > 0$, y's grow more slowly, so y is the prey. If $y = 0$, then $dx/dt < 0$, so the predators will die out; thus, they must have insufficient alternative food sources. The prey has no limits on its growth other than the predator since, if $x = 0$, then $dy/dt > 0$ and the population increases exponentially.

(b) Since $-\beta xy$ is negative and $+\delta xy$ is positive, x suffers due to its interaction with y and y benefits from its interaction with x. Hence, x is the prey and y is the predator. The predator has other sources of food than the prey since $dy/dt > 0$ even if $x = 0$. Also, the prey has a limit on its growth due to the $-\alpha x^2/N$ term.

23. **(a)** The independent variable is t, and x and y are dependent variables. Since each xy-term is positive, the presence of either species increases the rate of change of the other. Hence, these species cooperate. The parameter α is the growth-rate parameter for x, and γ is the growth-rate parameter for y. The parameter N represents the carrying capacity for x, but y has no carrying capacity. The parameter β measures the benefit to x of the interaction of the two species, and δ measures the benefit to y of the interaction.

(b) The independent variable is t, and x and y are the dependent variables. Since both xy-terms are negative, these species compete. The parameter γ is the growth-rate coefficient for x, and α is the growth-rate parameter for y. Neither population has a carrying capacity. The parameter δ measures the harm to x caused by the interaction of the two species, and β measures the harm to y caused by the interaction.

EXERCISES FOR SECTION 1.2

1. **(a)** Let's check Bob's solution first. Since $dy/dt = 1$ and

$$\frac{y(t) + 1}{t + 1} = \frac{t + 1}{t + 1} = 1,$$

Bob's answer is correct.

Now let's check Glen's solution. Since $dy/dt = 2$ and

$$\frac{y(t) + 1}{t + 1} = \frac{2t + 2}{t + 1} = 2,$$

Glen's solution is also correct.

Finally let's check Paul's solution. We have $dy/dt = 2t$ on one hand and

$$\frac{y(t) + 1}{t + 1} = \frac{t^2 - 1}{t + 1} = t - 1$$

on the other. Paul is wrong.

(b) At first glance, they should have seen the equilibrium solution $y(t) = -1$ for all t because $dy/dt = 0$ for any constant function and $y = -1$ implies that

$$\frac{y + 1}{t + 1} = 0$$

independent of t.

Strictly speaking the differential equation is not defined for $t = -1$, and hence the solutions are not defined for $t = -1$.

3. In order to find one such $f(t, y)$, we compute the derivative of $y(t)$. We obtain

$$\frac{dy}{dt} = \frac{de^{t^3}}{dt} = 3t^2 e^{t^3}.$$

Now we replace e^{t^3} in the last expression by y and get the differential equation

$$\frac{dy}{dt} = 3t^2 y.$$

5. We separate variables and integrate to obtain

$$\int \frac{1}{y^2}\, dy = \int t^2\, dt$$

$$-\frac{1}{y} = \frac{t^3}{3} + c$$

$$y(t) = -\frac{1}{(t^3/3) + c},$$

where c is any real number. This function can also be written in the form

$$y(t) = -\frac{3}{t^3 + k}$$

where k is any constant. The constant function $y(t) = 0$ for all t is also a solution of this equation. It is the equilibrium solution at $y = 0$.

7. We separate variables and integrate to obtain

$$\int \frac{dy}{2y + 1} = \int dt.$$

We get

$$\frac{1}{2} \ln |2y + 1| = t + c$$

$$|2y + 1| = c_1 e^{2t},$$

where $c_1 = e^{2c}$. We can drop the absolute value signs by replacing $\pm c_1$ with a new constant k_1. Hence, we have

$$2y + 1 = k_1 e^{2t}$$

$$y = \frac{1}{2}\left(k_1 e^{2t} - 1\right),$$

and letting $k = k_1/2$, $y(t) = ke^{2t} - 1/2$. Note that, for $k = 0$, we get the equilibrium solution.

9. We separate variables and integrate to obtain

$$\int e^y \, dy = \int dt$$

$$e^y = t + c,$$

where c is any constant. We obtain $y(t) = \ln(t + c)$.

11. We separate variables and integrate to obtain

$$\int \frac{1}{y^2} \, dy = \int (2t + 3) \, dt$$

$$-\frac{1}{y} = t^2 + 3t + k$$

$$y(t) = -\frac{1}{t^2 + 3t + k},$$

where k is any constant. The constant function $y(t) = 0$ for all t is also a solution of this equation. It is the equilibrium solution at $y = 0$.

13. First note that the differential equation is not defined if $y = 0$.
In order to separate the variables, we write the equation as

$$\frac{dy}{dt} = \frac{t}{y(t^2 + 1)}$$

to obtain

$$\int y \, dy = \int \frac{t}{t^2 + 1} \, dt$$

$$\frac{y^2}{2} = \frac{1}{2} \ln(t^2 + 1) + c,$$

where c is any constant. So we get

$$y^2 = \ln\left(k(t^2 + 1)\right),$$

where $k = e^{2c}$ (hence any positive constant). We have

$$y(t) = \pm\sqrt{\ln\left(k(t^2 + 1)\right)},$$

where k is any positive constant and the sign is determined by the initial condition.

15. First note that the differential equation is not defined for $y = -1/2$. We separate variables and
integrate to obtain

$$\int (2y + 1) \, dy = \int dt$$

$$y^2 + y = t + k,$$

where k is any constant. So

$$y(t) = \frac{-1 \pm \sqrt{4t + 4k + 1}}{2} = \frac{-1 \pm \sqrt{4t + c}}{2},$$

where c is any constant and the $\pm$ sign is determined by the initial condition.
We can rewrite the answer in the more simple form

$$y(t) = -\frac{1}{2} \pm \sqrt{t + c_1}$$

where $c_1 = k + 1/4$. If k can be any possible constant, then c_1 can be as well.

17. First of all, the equilibrium solutions are $y = 0$ and $y = 1$. Now suppose $y \neq 0$ and $y \neq 1$. We
separate variables to obtain

$$\int \frac{1}{y(1 - y)} \, dy = \int dt = t + c,$$

where c is any constant. To integrate, we use partial fractions. Write

$$\frac{1}{y(1 - y)} = \frac{A}{y} + \frac{B}{1 - y}.$$

We must have $A = 1$ and $-A + B = 0$. Hence, $A = B = 1$ and

$$\frac{1}{y(1-y)} = \frac{1}{y} + \frac{1}{1-y}.$$

Consequently,

$$\int \frac{1}{y(1-y)} \, dy = \ln|y| - \ln|1-y| = \ln\left|\frac{y}{1-y}\right|.$$

After integration, we have

$$\ln\left|\frac{y}{1-y}\right| = t + c$$

$$\left|\frac{y}{1-y}\right| = c_1 e^t,$$

where $c_1 = e^c$ is any positive constant. To remove the absolute value signs, we replace the positive constant c_1 with a constant k that can be any real number and get

$$y(t) = \frac{ke^t}{1 + ke^t},$$

where $k = \pm c_1$. If $k = 0$, we get the first equilibrium solution. The formula $y(t) = ke^t/(1 + ke^t)$ yields all the solutions to the differential equation except for the equilibrium solution $y(t) = 1$.

19. The equation can be written in the form

$$\frac{dv}{dt} = (v+1)(t^2 - 2),$$

and we note that $v(t) = -1$ for all t is an equilibrium solution. Separating variables and integrating, we obtain

$$\int \frac{dv}{v+1} = \int t^2 - 2 \, dt$$

$$\ln|v+1| = \frac{t^3}{3} - 2t + c,$$

where c is any constant. Thus,

$$|v+1| = c_1 e^{-2t+t^3/3},$$

where $c_1 = e^c$. We can dispose of the absolute value signs by allowing the constant c_1 to be any real number. In other words,

$$v(t) = -1 + ke^{-2t+t^3/3},$$

where $k = \pm c_1$. Note that, if $k = 0$, we get the equilibrium solution.

21. The function $y(t) = 0$ for all t is an equilibrium solution.

Suppose $y \neq 0$ and separate variables. We get

$$\int y + \frac{1}{y} \, dy = \int e^t \, dt$$

$$\frac{y^2}{2} + \ln |y| = e^t + c,$$

where c is any real constant. We cannot solve this equation for y, so we leave the expression for y in this implicit form. Note that the equilibrium solution $y = 0$ cannot be obtained from this implicit equation.

23. The constant function $w(t) = 0$ is an equilibrium solution. Suppose $w \neq 0$ and separate variables. We get

$$\int \frac{dw}{w} = \int \frac{dt}{t}$$

$$\ln |w| = \ln |t| + c$$

$$= \ln c_1 |t|,$$

where c is any constant and $c_1 = e^c$. Therefore,

$$|w| = c_1 |t|.$$

We can eliminate the absolute value signs by allowing the constant to assume positive or negative values. We have

$$w = kt,$$

where $k = \pm c_1$. Moreover, if $k = 0$ we get the equilibrium solution.

25. Separating variables and integrating, we have

$$\int \frac{1}{x} \, dx = \int -t \, dt$$

$$\ln |x| = -\frac{t^2}{2} + c$$

$$|x| = k_1 e^{-t^2/2},$$

where $k_1 = e^c$. We can eliminate the absolute value signs by allowing the constant k_1 to be either positive or negative. Thus, the general solution is

$$x(t) = k e^{-t^2/2}$$

where $k = \pm k_1$. Using the initial condition to solve for k, we have

$$\frac{1}{\sqrt{\pi}} = x(0) = k e^0 = k.$$

Therefore,

$$x(t) = \frac{e^{-t^2/2}}{\sqrt{\pi}}.$$

27. Separating variables and integrating, we obtain

$$\int \frac{dy}{y^2} = -\int dt$$

$$-\frac{1}{y} = -t + c.$$

So we get

$$y = \frac{1}{t - c}.$$

Now we need to find the constant c so that $y(0) = 1/2$. To do this we solve

$$\frac{1}{2} = \frac{1}{0 - c}$$

and get $c = -2$. The solution of the initial-value problem is

$$y(t) = \frac{1}{t + 2}.$$

29. We do not need to do any computations to solve this initial-value problem. We know that the constant function $y(t) = 0$ for all t is an equilibrium solution, and it satisfies the initial condition.

31. From Exercise 7, we already know that the general solution is

$$y(t) = ke^{2t} - \tfrac{1}{2},$$

so we need only find the constant k for which $y(0) = 3$. We solve

$$3 = ke^0 - \tfrac{1}{2}$$

for k and obtain $k = 7/2$. The solution of the initial-value problem is

$$y(t) = \tfrac{7}{2}e^{2t} - \tfrac{1}{2}.$$

33. We write the equation in the form

$$\frac{dx}{dt} = \frac{t^2}{x(t^3 + 1)}$$

and separate variables to obtain

$$\int x \, dx = \int \frac{t^2}{t^3 + 1} \, dt$$

$$\frac{x^2}{2} = \frac{1}{3} \ln |t^3 + 1| + c,$$

where c is a constant. Hence,

$$x^2 = \frac{2}{3} \ln|t^3 + 1| + 2c.$$

The initial condition $x(0) = -2$ implies

$$4 = (-2)^2 = \frac{2}{3} \ln|1| + 2c.$$

Thus, $c = 2$. Solving for $x(t)$, we choose the negative square root because $x(0)$ is negative, and we drop the absolute value sign because $t^3 + 1 > 0$ for t near 0. The result is

$$x(t) = -\sqrt{\tfrac{2}{3} \ln(t^3 + 1) + 4}.$$

35. We separate variables to obtain

$$\int \frac{dy}{1 + y^2} = \int t\, dt$$

$$\arctan y = \frac{t^2}{2} + c,$$

where c is a constant. Hence the general solution is

$$y(t) = \tan\left(\frac{t^2}{2} + c\right).$$

Next we find c so that $y(0) = 1$. Solving

$$1 = \tan\left(\frac{0^2}{2} + c\right)$$

yields $c = \pi/4$, and the solution to the initial-value problem is

$$y(t) = \tan\left(\frac{t^2}{2} + \frac{\pi}{4}\right).$$

37. Separating variables and integrating, we have

$$\int \frac{1}{y^2}\, dy = \int 2t + 3t^2\, dt$$

$$-\frac{1}{y} = t^2 + t^3 + c$$

$$y = \frac{-1}{t^2 + t^3 + c}.$$

Using $y(1) = -1$ we have

$$-1 = y(1) = \frac{-1}{1 + 1 + c} = \frac{-1}{2 + c},$$

so $c = -1$. The solution to the initial-value problem is

$$y(t) = \frac{-1}{t^2 + t^3 - 1}.$$

39. Let $S(t)$ denote the amount of salt (in pounds) in the bucket at time t (in minutes). We derive a differential equation for S by considering the difference between the rate that salt is entering the bucket and the rate that salt is leaving the bucket. Salt is entering the bucket at the rate of 1/4 pounds per minute. The rate that salt is leaving the bucket is the product of the concentration of salt in the mixture and the rate that the mixture is leaving the bucket. The concentration is $S/5$, and the mixture is leaving the bucket at the rate of 1/2 gallons per minute. We obtain the differential equation

$$\frac{dS}{dt} = \frac{1}{4} - \frac{S}{5} \cdot \frac{1}{2},$$

which can be rewritten as

$$\frac{dS}{dt} = \frac{5 - 2S}{20}.$$

This differential equation is separable, and we can find the general solution by integrating

$$\int \frac{1}{5 - 2S} \, dS = \int \frac{1}{20} \, dt.$$

We have

$$-\frac{\ln|5 - 2S|}{2} = \frac{t}{20} + c$$

$$\ln|5 - 2S| = -\frac{t}{10} + c_1$$

$$|5 - 2S| = c_2 e^{-t/10}.$$

We can eliminate the absolute value signs and determine c_2 using the initial condition $S(0) = 0$ (the water is initially free of salt). We have $c_2 = 5$, and the solution is

$$S(t) = 2.5 - 2.5 e^{-t/10} = 2.5(1 - e^{-t/10}).$$

   (a) When $t = 1$, we have $S(1) = 2.5(1 - e^{-0.1}) \approx 0.238$ lbs.
   (b) When $t = 10$, we have $S(10) = 2.5(1 - e^{-1}) \approx 1.58$ lbs.
   (c) When $t = 60$, we have $S(60) = 2.5(1 - e^{-6}) \approx 2.49$ lbs.
   (d) When $t = 1000$, we have $S(1000) = 2.5(1 - e^{-100}) \approx 2.50$ lbs.
   (e) When t is very large, the $e^{-t/10}$ term is close to zero, so $S(t)$ is very close to 2.5 lbs. In this case, we can also reach the same conclusion by doing a qualitative analysis of the solutions of the equation. The constant solution $S(t) = 2.5$ is the only equilibrium solution for this equation, and by examining the sign of dS/dt, we see that all solutions approach $S = 2.5$ as t increases.

41. **(a)** If we let k denote the proportionality constant in Newton's law of cooling, the differential equation satisfied by the temperature T of the chocolate is

$$\frac{dT}{dt} = k(T - 70).$$

We also know that $T(0) = 170$ and that $dT/dt = -20$ at $t = 0$. Therefore, we obtain k by evaluating the differential equation at $t = 0$. We have

$$-20 = k(170 - 70),$$

so $k = -0.2$. The initial-value problem is

$$\frac{dT}{dt} = -0.2(T - 70), \quad T(0) = 170.$$

(b) We can solve the initial-value problem in part (a) by separating variables. We have

$$\int \frac{dT}{T - 70} = \int -0.2 \, dt$$

$$\ln |T - 70| = -0.2t + k$$

$$|T - 70| = ce^{-0.2t}.$$

Since the temperature of the chocolate cannot become lower than the temperature of the room, we can ignore the absolute value and conclude

$$T(t) = 70 + ce^{-0.2t}.$$

Now we use the initial condition $T(0) = 170$ to find the constant c because

$$170 = T(0) = 70 + ce^{-0.2(0)},$$

which implies that $c = 100$. The solution is

$$T = 70 + 100e^{-0.2t}.$$

In order to find t so that the temperature is $110°$ F, we solve

$$110 = 70 + 100e^{-0.2t}$$

for t obtaining

$$\tfrac{2}{5} = e^{-0.2t}$$

$$\ln \tfrac{2}{5} = -0.2t$$

so that

$$t = \frac{\ln(2/5)}{-0.2} \approx 4.6.$$

43. **(a)** We rewrite the differential equation as

$$\frac{dv}{dt} = g\left(1 - \frac{k}{mg}v^2\right).$$

Letting $\alpha = \sqrt{k/(mg)}$ and separating variables, we have

$$\int \frac{dv}{1 - \alpha^2 v^2} = \int g\,dt.$$

Now we use the partial fractions decomposition

$$\frac{1}{1 - \alpha^2 v^2} = \frac{1/2}{1 + \alpha v} + \frac{1/2}{1 - \alpha v}$$

to obtain

$$\int \frac{dv}{1 + \alpha v} + \int \frac{dv}{1 - \alpha v} = 2gt + c,$$

where c is an arbitrary constant. Integrating the left-hand side, we get

$$\frac{1}{\alpha}\left(\ln|1 + \alpha v| - \ln|1 - \alpha v|\right) = 2gt + c.$$

Multiplying through by α and using the properties of logarithms, we have

$$\ln\left|\frac{1 + \alpha v}{1 - \alpha v}\right| = 2\alpha gt + c.$$

Exponentiating and eliminating the absolute value signs yields

$$\frac{1 + \alpha v}{1 - \alpha v} = Ce^{2\alpha gt}.$$

Solving for v, we have

$$v = \frac{1}{\alpha}\frac{Ce^{2\alpha gt} - 1}{Ce^{2\alpha gt} + 1}.$$

Recalling that $\alpha = \sqrt{k/(mg)}$, we see that $\alpha g = \sqrt{kg/m}$, and we get

$$v(t) = \sqrt{\frac{mg}{k}}\left(\frac{Ce^{2\sqrt{(kg/m)}\,t} - 1}{Ce^{2\sqrt{(kg/m)}\,t} + 1}\right).$$

Note: If we assume that $v(0) = 0$, then $C = 1$. The solution to this initial-value problem is often expressed in terms of the hyperbolic tangent function as

$$v = \sqrt{\frac{mg}{k}}\tanh\left(\sqrt{\frac{kg}{m}}\,t\right).$$

(b) The fraction in the parentheses of the general solution

$$v(t) = \sqrt{\frac{mg}{k}}\left(\frac{Ce^{2\sqrt{(kg/m)}\,t} - 1}{Ce^{2\sqrt{(kg/m)}\,t} + 1}\right),$$

tends to 1 as $t \to \infty$, so the limit of $v(t)$ as $t \to \infty$ is $\sqrt{mg/k}$.

EXERCISES FOR SECTION 1.3

1.

3.

5.

7. **(a)**

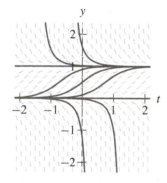

(b) The solution with $y(0) = 1/2$ approaches the equilibrium value $y = 1$ from below as t increases. It decreases toward $y = 0$ as t decreases.

9. **(a)**

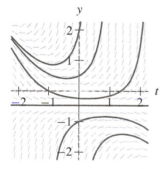

(b) The solution $y(t)$ with $y(0) = 1/2$ has $y(t) \to \infty$ both as t increases and as t decreases.

11. **(a)** On the line $y = 3$ in the ty-plane, all of the slope marks have slope -1.

(b) Because f is continuous, if y is close to 3, then $f(t, y) < 0$. So any solution close to $y = 3$ must be decreasing. Therefore, solutions $y(t)$ that satisfy $y(0) < 3$ can never be larger than 3 for $t > 0$, and consequently $y(t) < 3$ for all t.

13. The slope field in the ty-plane is constant along vertical lines.

15.

 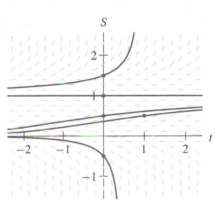

17. **(a)** Because the slope field is constant on vertical lines, the given information is enough to draw the entire slope field.

(b) The solution with initial condition $y(0) = 2$ is a vertical translation of the given solution. We only need change the "constant of integration" so that $y(0) = 2$.

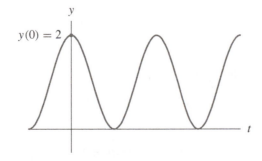

19. **(a)** Even though the question only asks for slope fields in this part, we superimpose the graphs of the equilibrium solutions on the fields to illustrate the equilibrium solutions (see part (b)).

$I_1 = -0.1$

$I_2 = 0.0$

$I_3 = 0.1$

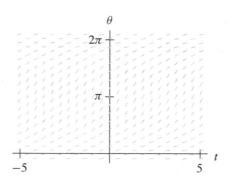

(b) For $I_1 = -0.1$, the equilibrium values satisfy the equation

$$1 - \cos\theta + (1 + \cos\theta)(-0.1) = 0.$$

We have

$$0.9 - 1.1\cos\theta = 0$$

$$\cos\theta = \frac{0.9}{1.1}$$

$$\theta \approx \pm 0.613.$$

Therefore, the equilibrium values are $\theta \approx 2\pi n \pm 0.613$ radians, where n is any integer. There are two equilibrium solutions with values $\theta \approx 0.613$ and $\theta \approx 5.670$ between 0 and 2π.

For $I_2 = 0.0$, similar calculations yield equilibrium values at even multiples of 2π, and for $I_3 = 0.1$, there are no equilibrium values.

(c) For $I_1 = -0.1$, the graphs of the equilibrium solutions divide the $t\theta$-plane into horizontal strips in which the signs of the slopes do not change. For example, if $0.613 < \theta < 5.670$ (approximately), then the slopes are positive. If $5.670 < \theta < 6.896$ (approximately), then the slopes are negative. Therefore, any solution $\theta(t)$ with an initial condition θ_0 that is between 0.613 and 6.896 (approximately) satisfies the limit $\theta(t) \to 5.670$ (approximately) as $t \to \infty$. Moreover, any solution $\theta(t)$ with an initial condition θ_0 that is between -0.613 and 5.670 (approximately) satisfies the limit $\theta(t) \to 0.613$ (approximately) as $t \to -\infty$.

For $I_2 = 0.0$, the graphs of the equilibrium solutions also divide the $t\theta$-plane into horizontal strips in which the signs of the slopes do not change. However, in this case, the slopes are always positive (or zero in the case of the equilibrium solutions). Therefore, for example, any solution $\theta(t)$ with an initial condition θ_0 that is between 0 and 2π satisfies the limits $\theta(t) \to 2\pi$ as $t \to \infty$ and $\theta(t) \to 0$ as $t \to -\infty$.

Lastly, if $I_3 = 0.1$, all of the slopes are positive, so all solutions are increasing for all t. The fact that $\theta(t) \to \infty$ as $t \to \infty$ requires an analytic estimate in addition to a qualitative analysis.

21. Separating variables, we obtain

$$\int \frac{dv_c}{K - v_c} = \int \frac{dt}{RC}.$$

Integrating both sides, we have

$$-\ln|K - v_c| = \frac{t}{RC} + c_1,$$

where c_1 is a constant. Thus,

$$|K - v_c| = c_2 e^{-t/RC}$$

where $c_2 = e^{-c_1}$. We can eliminate the absolute values by allowing c_2 to assume either positive or negative values. Therefore, we obtain the general solution

$$v_c(t) = K + c e^{-t/RC}$$

where c can be any constant.

To check that $v_c(t)$ is a solution, we calculate the left-hand side of the equation

$$\frac{dv_c}{dt} = -\frac{c}{RC} e^{-t/RC},$$

and the right-hand side of the equation

$$\frac{K - v_c}{RC} = \frac{K - \left(K + c e^{-t/RC}\right)}{RC} = -\frac{c}{RC} e^{-t/RC}.$$

Since they agree, $v_c(t)$ is a solution.

EXERCISES FOR SECTION 1.4

1.

Table 1.1
Results of Euler's method

k	t_k	y_k	m_k
0	0	3	7
1	0.5	6.5	14
2	1.0	13.5	28
3	1.5	27.5	56
4	2.0	55.5	

3.

Table 1.2
Results of Euler's method (shown rounded to two decimal places)

k	t_k	y_k	m_k
0	0	0.5	0.25
1	0.25	0.56	−0.68
2	0.50	0.39	−1.85
3	0.75	−0.07	−2.99
4	1.00	−0.82	−3.33
5	1.25	−1.65	−2.27
6	1.50	−2.22	−1.07
7	1.75	−2.49	−0.81
8	2.00	−2.69	

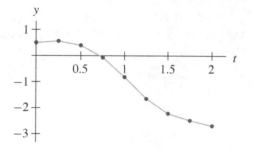

5.

Table 1.3
Results of Euler's method

k	t_k	w_k	m_k
0	0	4	−5
1	1	−1	0
2	2	−1	0
3	3	−1	0
4	4	−1	0
5	5	−1	

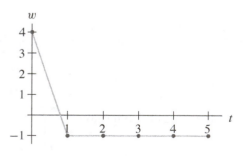

7.

Table 1.4
Results of Euler's method (shown rounded to two decimal places)

k	t_k	y_k	m_k
0	0	2	2.72
1	0.5	3.36	1.81
2	1.0	4.27	1.60
3	1.5	5.06	1.48
4	2.0	5.81	

9.

Table 1.5
Results of Euler's method (shown rounded to three decimal places)

k	t_k	y_k	m_k
0	0.0	0.2	0.032
1	0.1	0.203	0.033
2	0.2	0.206	0.034
3	0.3	0.210	0.035
⋮	⋮	⋮	⋮
99	9.9	0.990	0.010
100	10.0	0.991	

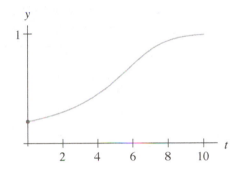

11. As the solution approaches the equilibrium solution corresponding to $w = 3$, its slope decreases. We do not expect the solution to "jump over" an equilibrium solution (see the Existence and Uniqueness Theorem in Section 1.5).

13. Because the differential equation is autonomous, the computation that determines y_{k+1} from y_k depends only on y_k and Δt and not on the actual value of t_k. Hence the approximate y-values that are obtained in both exercises are the same. It is useful to think about this fact in terms of the slope field of an autonomous equation.

15.

Table 1.6
Results of Euler's method with $\Delta t = 1.0$ (shown to two decimal places)

k	t_k	y_k	m_k
0	0	1	1
1	1	2	1.41
2	2	3.41	1.85
3	3	5.26	2.29
4	4	7.56	

Table 1.7
Results of Euler's method with $\Delta t = 0.5$ (shown to two decimal places)

k	t_k	y_k	m_k	k	t_k	y_k	m_k
0	0	1	1	5	2.5	4.64	2.15
1	0.5	1.5	1.22	6	3.0	5.72	2.39
2	1.0	2.11	1.45	7	3.5	6.91	2.63
3	1.5	2.84	1.68	8	4.0	8.23	
4	2.0	3.68	1.92				

Table 1.8
Results of Euler's method with $\Delta t = 0.25$ (shown to two decimal places)

k	t_k	y_k	m_k	k	t_k	y_k	m_k
0	0	1	1	9	2.25	4.32	2.08
1	0.25	1.25	1.12	10	2.50	4.84	2.20
2	0.50	1.53	1.24	11	2.75	5.39	2.32
3	0.75	1.84	1.36	12	3.0	5.97	2.44
4	1.0	2.18	1.48	13	3.25	6.58	2.56
5	1.25	2.55	1.60	14	3.50	7.23	2.69
6	1.50	2.94	1.72	15	3.75	7.90	2.81
7	1.75	3.37	1.84	16	4.0	8.60	
8	2.0	3.83	1.96				

The slopes in the slope field are positive and increasing. Hence, the graphs of all solutions are concave up. Since Euler's method uses line segments to approximate the graph of the actual solution, the approximate solutions will always be less than the actual solution. This error decreases as the step size decreases.

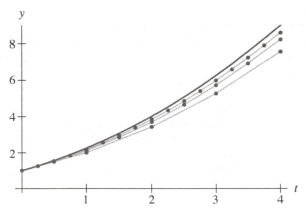

17. Assuming that $I(t) = 0.1$, the differential equation simplifies to

$$\frac{d\theta}{dt} = 0.9 - 1.1 \cos \theta.$$

Using Euler's method with $\Delta t = 0.1$, we obtain the results in the following table.

Table 1.9
Results of Euler's method (shown rounded to three decimal places)

k	t_k	y_k	m_k	k	t_k	y_k	m_k
0	0.0	1.0	0.306	23	2.3	3.376	1.970
1	0.1	1.031	0.334	24	2.4	3.573	1.899
2	0.2	1.064	0.366	25	2.5	3.763	1.794
$\vdots$	$\vdots$	$\vdots$	$\vdots$	$\vdots$	$\vdots$	$\vdots$	$\vdots$
21	2.1	2.978	1.985	49	4.9	5.452	0.159
22	2.2	3.176	1.999	50	5.0	5.467	

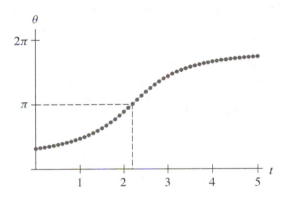

The graph of the results of Euler's method.

A neuron spikes when θ is equal to an odd multiple of π. Therefore, we need to determine when $\theta(t) = \pi$. From the results of Euler's method, we see that the neuron spikes when $t \approx 2.15$.

19.

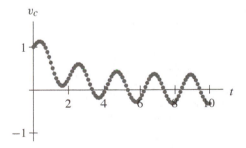

Graph of approximate solution obtained using Euler's method with $\Delta t = 0.1$.

21.

Graph of approximate solution obtained using Euler's method with $\Delta t = 0.1$.

EXERCISES FOR SECTION 1.5

1. Since the constant function $y_1(t) = 3$ for all t is a solution, then the graph of any other solution $y(t)$ with $y(0) < 3$ cannot cross the line $y = 3$ by the Uniqueness Theorem. So $y(t) < 3$ for all t in the domain of $y(t)$.

3. Because $y_2(0) < y(0) < y_1(0)$, we know that

$$-t^2 = y_2(t) < y(t) < y_1(t) = t + 2$$

for all t. This restricts how large positive or negative $y(t)$ can be for a given value of t (that is, between $-t^2$ and $t + 2$). As $t \to -\infty$, $y(t) \to -\infty$ between $-t^2$ and $t + 2$ ($y(t) \to -\infty$ as $t \to -\infty$ at least linearly, but no faster than quadratically).

5. The Existence Theorem implies that a solution with this initial condition exists, at least for a small t-interval about $t = 0$. This differential equation has equilibrium solutions $y_1(t) = 0$, $y_2(t) = 1$, and $y_3(t) = 3$ for all t. Since $y(0) = 4$, the Uniqueness Theorem implies that $y(t) > 3$ for all t in the domain of $y(t)$. Also, $dy/dt > 0$ for all $y > 3$, so the solution $y(t)$ is increasing for all t in its domain. Finally, $y(t) \to 3$ as $t \to -\infty$.

7. The Existence Theorem implies that a solution with this initial condition exists, at least for a small t-interval about $t = 0$. Because $1 < y(0) < 3$ and $y_1(t) = 1$ and $y_2(t) = 3$ are equilibrium solutions of the differential equation, we know that the solution exists for all t and that $1 < y(t) < 3$ for all t by the Uniqueness Theorem. Also, $dy/dt < 0$ for $1 < y < 3$, so dy/dt is always negative for this solution. Hence, $y(t) \to 1$ as $t \to \infty$, and $y(t) \to 3$ as $t \to -\infty$.

9. **(a)** To check that $y_1(t) = t^2$ is a solution, we compute

$$\frac{dy_1}{dt} = 2t$$

and

$$-y_1^2 + y_1 + 2y_1 t^2 + 2t - t^2 - t^4 = -(t^2)^2 + (t^2) + 2(t^2)t^2 + 2t - t^2 - t^4$$
$$= 2t.$$

To check that $y_2(t) = t^2 + 1$ is a solution, we compute

$$\frac{dy_2}{dt} = 2t$$

and

$$-y_2^2 + y_2 + 2y_2 t^2 + 2t - t^2 - t^4 = -(t^2 + 1)^2 + (t^2 + 1) + 2(t^2 + 1)t^2$$
$$+ 2t - t^2 - t^4$$
$$= 2t.$$

(b) The initial values of the two solutions are $y_1(0) = 0$ and $y_2(0) = 1$. Thus if $y(t)$ is a solution and $y_1(0) = 0 < y(0) < 1 = y_2(0)$, then we can apply the Uniqueness Theorem to obtain

$$y_1(t) = t^2 < y(t) < t^2 + 1 = y_2(t)$$

for all t. Note that since the differential equation satisfies the hypothesis of the Existence and Uniqueness Theorem over the entire ty-plane, we can continue to extend the solution as long as it does not escape to $\pm\infty$ in finite time. Since it is bounded above and below by solutions that exist for all time, $y(t)$ is defined for all time also.

(c)

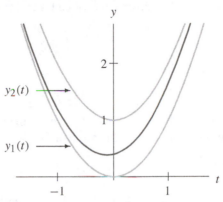

11. The key observation is that the differential equation is not defined when $t = 0$.

(a) Note that $dy_1/dt = 0$ and $y_1/t^2 = 0$, so $y_1(t)$ is a solution.

(b) Separating variables, we have

$$\int \frac{dy}{y} = \int \frac{dt}{t^2}.$$

Solving for y we obtain $y(t) = ce^{-1/t}$, where c is any constant. Thus, for any real number c, define the function $y_c(t)$ by

$$y_c(t) = \begin{cases} 0 & \text{for } t \le 0; \\ ce^{-1/t} & \text{for } t > 0. \end{cases}$$

For each c, $y_c(t)$ satisfies the differential equation for all $t \ne 0$.

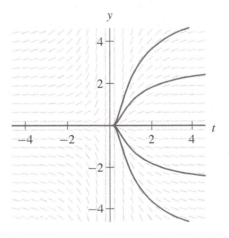

There are infinitely many solutions of the form $y_c(t)$ that agree with $y_1(t)$ for $t < 0$.

(c) Note that $f(t, y) = y/t^2$ is not defined at $t = 0$. Therefore, we *cannot* apply the Uniqueness Theorem for the initial condition $y(0) = 0$. The "solution" $y_c(t)$ given in part (b) actually represents two solutions, one for $t < 0$ and one for $t > 0$.

13. (a) The equation is separable. We separate the variables and compute

$$\int y^{-3} \, dy = \int dt.$$

Solving for y, we obtain

$$y(t) = \frac{1}{\sqrt{c - 2t}}$$

for any constant c. To find the desired solution, we use the initial condition $y(0) = 1$ and obtain $c = 1$. So the solution to the initial-value problem is

$$y(t) = \frac{1}{\sqrt{1 - 2t}}.$$

(b) This solution is defined when $-2t + 1 > 0$, which is equivalent to $t < 1/2$.

(c) As $t \to 1/2^-$, the denominator of $y(t)$ becomes a small positive number, so $y(t) \to \infty$. We only consider $t \to 1/2^-$ because the solution is defined only for $t < 1/2$. (The other "branch" of the function is also a solution, but the solution that includes $t = 0$ in its domain is not defined for $t \geq 1/2$.) As $t \to -\infty$, $y(t) \to 0$.

15. (a) The equation is separable. We separate, integrate

$$\int (y + 2)^2 \, dy = \int dt,$$

and solve for y to obtain the general solution

$$y(t) = (3t + c)^{1/3} - 2,$$

where c is any constant. To obtain the desired solution, we use the initial condition $y(0) = 1$ and solve

$$1 = (3 \cdot 0 + c)^{1/3} - 2$$

for c to obtain $c = 27$. So the solution to the given initial-value problem is

$$y(t) = (3t + 27)^{1/3} - 2.$$

(b) This function is defined for all t. However, $y(-9) = -2$, and the differential equation is not defined at $y = -2$. Strictly speaking, the solution exists only for $t > -9$.

(c) As $t \to \infty$, $y(t) \to \infty$. As $t \to -9^+$, $y(t) \to -2$.

17. This exercise shows that solutions of autonomous equations cannot have local maximums or minimums. Hence they must be either constant or monotonically increasing or monotonically decreasing. A useful corollary is that a function $y(t)$ that oscillates cannot be the solution of an autonomous differential equation.

(a) Note $dy_1/dt = 0$ at $t = t_0$ because $y_1(t)$ has a local maximum. Because $y_1(t)$ is a solution, we know that $dy_1/dt = f(y_1(t))$ for all t in the domain of $y_1(t)$. In particular,

$$0 = \left. \frac{dy_1}{dt} \right|_{t=t_0} = f(y_1(t_0)) = f(y_0),$$

so $f(y_0) = 0$.

(b) This differential equation is autonomous, so the slope marks along any given horizontal line are parallel. Hence, the slope marks along the line $y = y_0$ must all have zero slope.

(c) For all t,

$$\frac{dy_2}{dt} = \frac{d(y_0)}{dt} = 0$$

because the derivative of a constant function is zero, and for all t

$$f(y_2(t)) = f(y_0) = 0.$$

So $y_2(t)$ is a solution.

(d) By the Uniqueness Theorem, we know that two solutions that are in the same place at the same time are the same solution. We have $y_1(t_0) = y_0 = y_2(t_0)$. Moreover, $y_1(t)$ is assumed to be a solution, and we showed that $y_2(t)$ is a solution in parts (a) and (b) of this exercise. So $y_1(t) = y_2(t)$ for all t. In other words, $y_1(t) = y_0$ for all t.

(e) Follow the same four steps as before. We still have $dy_1/dt = 0$ at $t = t_0$ because y_1 has a local minimum at $t = t_0$.

EXERCISES FOR SECTION 1.6

1. The equilibrium points of $dy/dt = f(y)$ are the numbers y where $f(y) = 0$. For $f(y) = 3y(y - 2)$, the equilibrium points are $y = 0$ and $y = 2$. Since $f(y)$ is positive for $y < 0$, negative for $0 < y < 2$, and positive for $y > 2$, the equilibrium point $y = 0$ is a sink and the equilibrium point $y = 2$ is a source.

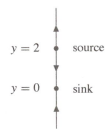

$y = 2$ source

$y = 0$ sink

3. The equilibrium points of $dy/dt = f(y)$ are the numbers y where $f(y) = 0$. For $f(y) = \cos y$, the equilibrium points are $y = \pi/2 + n\pi$, where $n = 0, \pm 1, \pm 2, \ldots$. Since $\cos y > 0$ for $-\pi/2 < y < \pi/2$ and $\cos y < 0$ for $\pi/2 < y < 3\pi/2$, we see that the equilibrium point at $y = \pi/2$ is a sink. Since the sign of $\cos y$ alternates between positive and negative in a period fashion, we see that the equilibrium points at $y = \pi/2 + 2n\pi$ are sinks and the equilibrium points at $y = 3\pi/2 + 2n\pi$ are sources.

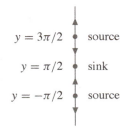

$y = 3\pi/2$ source

$y = \pi/2$ sink

$y = -\pi/2$ source

5. The equilibrium points of $dw/dt = f(w)$ are the numbers w where $f(w) = 0$. For $f(w) = (1 - w)\sin w$, the equilibrium points are $w = 1$ and $w = n\pi$, where $n = 0, \pm 1, \pm 2, \ldots$. The sign of $(1 - w)\sin w$ alternates between positive and negative at successive zeros. It is negative for $-\pi < w < 0$ and positive for $0 < w < 1$. Therefore, $w = 0$ is a source, and the equilibrium points alternate between sinks and sources.

7. The derivative dv/dt is always negative, so there are no equilibrium points, and all solutions are decreasing.

9. The equilibrium points of $dy/dt = f(y)$ are the numbers y where $f(y) = 0$. For $f(y) = 1 + \cos y$, the equilibrium points are $y = n\pi$, where $n = \pm 1, \pm 3, \ldots$. Since $f(y)$ is nonnegative for all values of y, all of the equilibrium points are nodes.

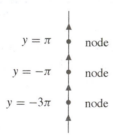

11. The equilibrium points of $dy/dt = f(y)$ are the numbers y where $f(y) = 0$. For $f(y) = y \ln|y|$, there are equilibrium points at $y = \pm 1$. In addition, although the function $f(y)$ is technically undefined at $y = 0$, the limit of $f(y)$ as $y \to 0$ is 0. Thus we can treat $y = 0$ as another equilibrium point. Since $f(y) < 0$ for $y < -1$ and $0 < y < 1$, and $f(y) > 0$ for $y > 1$ and $-1 < y < 0$, $y = -1$ is a source, $y = 0$ is a sink, and $y = 1$ is a source.

13.

15.

17.

19.

21.

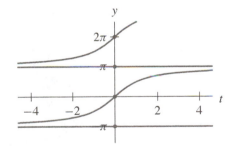

23. The initial value $y(0) = 2$ is between the equilibrium points $y = 2 - \sqrt{2}$ and $y = 2 + \sqrt{2}$. Also, $dy/dt < 0$ for $2 - \sqrt{2} < y < 2 + \sqrt{2}$. Hence the solution is decreasing and tends toward $y = 2 - \sqrt{2}$ as $t \to \infty$. It tends toward $y = 2 + \sqrt{2}$ as $t \to -\infty$.

25. The initial value $y(0) = -4$ is below both equilibrium points. Since $dy/dt > 0$ for $y < 2 - \sqrt{2}$, the solution is increasing for all t and tends to the equilibrium point $y = 2 - \sqrt{2}$ as $t \to \infty$. As t decreases, $y(t) \to -\infty$ in finite time.

27. The initial value $y(3) = 1$ is between the equilibrium points $y = 2 - \sqrt{2}$ and $y = 2 + \sqrt{2}$. Also, $dy/dt < 0$ for $2 - \sqrt{2} < y < 2 + \sqrt{2}$. Hence the solution is decreasing and tends toward the smaller equilibrium point $y = 2 - \sqrt{2}$ as $t \to \infty$. It tends toward the larger equilibrium point $y = 2 + \sqrt{2}$ as $t \to -\infty$.

29. The function $f(y)$ has two zeros $\pm y_0$, where y_0 is some positive number. So the differential equation $dy/dt = f(y)$ has two equilibrium solutions, one for each zero. Also, $f(y) < 0$ if $-y_0 < y < y_0$ and $f(y) > 0$ if $y < -y_0$ or if $y > y_0$. Hence y_0 is a source and $-y_0$ is a sink.

31. The function $f(y)$ has three zeros. We denote them as y_1, y_2, and y_3, where $y_1 < 0 < y_2 < y_3$. So the differential equation $dy/dt = f(y)$ has three equilibrium solutions, one for each zero. Also, $f(y) > 0$ if $y < y_1$, $f(y) < 0$ if $y_1 < y < y_2$, and $f(y) > 0$ if $y_2 < y < y_3$ or if $y > y_3$. Hence y_1 is a sink, y_2 is a source, and y_3 is a node.

33. Since there are two equilibrium points, the graph of $f(y)$ must touch the y-axis at two distinct numbers y_1 and y_2. Assume that $y_1 < y_2$. Since the arrows point up if $y < y_1$ and if $y > y_2$, we must have $f(y) > 0$ for $y < y_1$ and for $y > y_2$. Similarly, $f(y) < 0$ for $y_1 < y < y_2$.

The precise location of the equilibrium points is not given, and the direction of the arrows on the phase line is determined only by the sign (and not the magnitude) of $f(y)$. So the following graph is one of many possible answers.

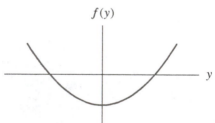

35. Since there are three equilibrium points (one appearing to be at $y = 0$), the graph of $f(y)$ must touch the y-axis at three numbers y_1, y_2, and y_3. We assume that $y_1 < y_2 = 0 < y_3$. Since the arrows point down for $y < y_1$ and $y_2 < y < y_3$, $f(y) < 0$ for $y < y_1$ and for $y_2 < y < y_3$. Similarly, $f(y) > 0$ if $y_1 < y < y_2$ and if $y > y_3$.

The precise location of the equilibrium points is not given, and the direction of the arrows on the phase line is determined only by the sign (and not the magnitude) of $f(y)$. So the following graph is one of many possible answers.

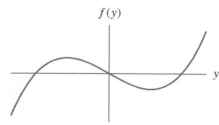

37. (a) This phase line has two equilibrium points, $y = 0$ and $y = 1$. Equations (ii), (iv), (vi), and (viii) have exactly these equilibria. There exists a node at $y = 0$. Only equations (iv) and (viii) have a node at $y = 0$. Moreover, for this phase line, $dy/dt < 0$ for $y > 1$. Only equation (viii) satisfies this property. Consequently, the phase line corresponds to equation (viii).

(b) This phase line has two equilibrium points, $y = 0$ and $y = 1$. Equations (ii), (iv), (vi) and (viii) have exactly these equilibria. Moreover, for this phase line, $dy/dt > 0$ for $y > 1$. Only equations (iv) and (vi) satisfy this property. Lastly, $dy/dt > 0$ for $y < 0$. Only equation (vi) satisfies this property. Consequently, the phase line corresponds to equation (vi).

(c) This phase line has an equilibrium point at $y = 3$. Only equations (i) and (v) have this equilibrium point. Moreover, this phase line has another equilibrium point at $y = 0$. Only equation (i) satisfies this property. Consequently, the phase line corresponds to equation (i).

(d) This phase line has an equilibrium point at $y = 2$. Only equations (iii) and (vii) have this equilibrium point. Moreover, there exists a node at $y = 0$. Only equation (vii) satisfies this property. Consequently, the phase line corresponds to equation (vii).

39. (a) In terms of the phase line with $P \geq 0$, there are three equilibrium points. If we assume that $f(P)$ is differentiable, then a decreasing population at $P = 100$ implies that $f(P) < 0$ for $P > 50$. An increasing population at $P = 25$ implies that $f(P) > 0$ for $10 < P < 50$. These assumptions leave two possible phase lines since the arrow between $P = 0$ and $P = 10$ is undetermined.

(b) Given the observations in part (a), we see that there are two basic types of graphs that go with the assumptions. However, there are many graphs that correspond to each possibility. The following two graphs are representative.

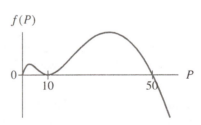

(c) The functions $f(P) = P(P - 10)(50 - P)$ and $f(P) = P(P - 10)^2(50 - P)$ respectively are two examples but there are many others.

41. The equilibrium points occur at solutions of $dy/dt = y^2 + a = 0$. For $a > 0$, there are no equilibrium points. For $a = 0$, there is one equilibrium point, $y = 0$. For $a < 0$, there are two equilibrium points, $y = \pm\sqrt{-a}$.

To draw the phase lines, note that:

- If $a > 0, dy/dt = y^2 + a > 0$, so the solutions are always increasing.
- If $a = 0, dy/dt > 0$ unless $y = 0$. Thus, $y = 0$ is a node.
- For $a < 0, dy/dt < 0$ for $-\sqrt{-a} < y < \sqrt{-a}$, and $dy/dt > 0$ for $y < -\sqrt{-a}$ and for $y > \sqrt{-a}$.

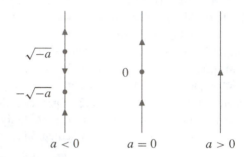

(a) The phase lines for $a < 0$ are qualitatively the same, and the phase lines for $a > 0$ are qualitatively the same.

(b) The phase line undergoes a qualitative change at $a = 0$.

43. (a) Because the first and second derivative are zero at y_0 and the third derivative is positive, Taylor's Theorem implies that the function $f(y)$ is approximately equal to

$$\frac{f'''(y_0)}{3!}(y - y_0)^3$$

for y near y_0. Since $f'''(y_0) > 0, f(y)$ is increasing near y_0. Hence, y_0 is a source.

(b) Just as in part (a), we see that $f(y)$ is decreasing near y_0, so y_0 is a sink.

(c) In this case, we can approximate $f(y)$ near y_0 by

$$\frac{f''(y_0)}{2!}(y - y_0)^2.$$

Since the second derivative of $f(y)$ at y_0 is assumed to be positive, $f(y)$ is positive on both sides of y_0 for y near y_0. Hence y_0 is a node.

45. One assumption of the model is that, if no people are present, then the time between trains decreases at a constant rate. Hence the term $-\alpha$ represents this assumption. The parameter α should be positive, so that $-\alpha$ makes a negative contribution to dx/dt.

The term βx represents the effect of the passengers. The parameter β should be positive so that βx contributes positively to dx/dt.

47. Note that the only equilibrium point is a source. If the initial gap between trains is too large, then x will increase without bound. If it is too small, x will decrease to zero. When $x = 0$, the two trains are next to each other, and they will stay together since $x < 0$ is not physically possible in this problem.

If the time between trains is exactly the equilibrium value ($x = \alpha/\beta$), then theoretically $x(t)$ is constant. However, any disruption to x causes the solution to tend away from the source. Since it is very likely that some stops will have fewer than the expected number of passengers and some stops will have more, it is unlikely that the time between trains will remain constant for long.

EXERCISES FOR SECTION 1.7

1. The equilibrium points occur at solutions of $dy/dt = y^2 + a = 0$. For $a > 0$, there are no equilibrium points. For $a = 0$, there is one equilibrium point, $y = 0$. For $a < 0$, there are two equilibrium points, $y = \pm\sqrt{-a}$. Thus, $a = 0$ is a bifurcation value.

To draw the phase lines, note that:

- If $a > 0$, $dy/dt = y^2 + a > 0$, so the solutions are always increasing.
- If $a = 0$, $dy/dt > 0$ unless $y = 0$. Thus, $y = 0$ is a node.
- For $a < 0$, $dy/dt < 0$ for $-\sqrt{-a} < y < \sqrt{-a}$, and $dy/dt > 0$ for $y < -\sqrt{-a}$ and for $y > \sqrt{-a}$.

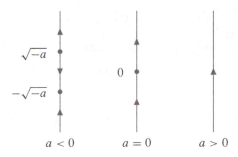

Phase lines for $a < 0$, $a = 0$, and $a > 0$.

3. The equilibrium points occur at solutions of $dy/dt = y^2 - ay + 1 = 0$. From the quadratic formula, we have

$$y = \frac{a \pm \sqrt{a^2 - 4}}{2}.$$

If $-2 < a < 2$, then $a^2 - 4 < 0$, and there are no equilibrium points. If $a > 2$ or $a < -2$, there are two equilibrium points. For $a = \pm 2$, there is one equilibrium point at $y = a/2$. The bifurcations occur at $a = \pm 2$.

To draw the phase lines, note that:

- For $-2 < a < 2$, $dy/dt = y^2 - ay + 1 > 0$, so the solutions are always increasing.
- For $a = 2$, $dy/dt = (y - 1)^2 \geq 0$, and $y = 1$ is a node.
- For $a = -2$, $dy/dt = (y + 1)^2 \geq 0$, and $y = -1$ is a node.
- For $a < -2$ or $a > 2$, let

$$y_1 = \frac{a - \sqrt{a^2 - 4}}{2} \quad \text{and} \quad y_2 = \frac{a + \sqrt{a^2 - 4}}{2}.$$

Then $dy/dt < 0$ if $y_1 < y < y_2$, and $dy/dt > 0$ if $y < y_1$ or $y > y_2$.

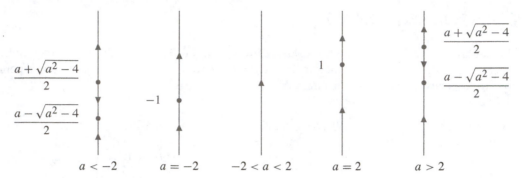

The five possible phase lines.

5. To find the equilibria we solve

$$(y^2 - \alpha)(y^2 - 4) = 0,$$

obtaining $y = \pm 2$ and $y = \pm\sqrt{\alpha}$ if $\alpha \geq 0$. Hence, there are two bifurcation values of α, $\alpha = 0$ and $\alpha = 4$.

For $\alpha < 0$, there are only two equilibria. The point $y = -2$ is a sink and $y = 2$ is a source. At $\alpha = 0$, there are three equilibria. There is a sink at $y = -2$, a source at $y = 2$, and a node at $y = 0$. For $0 < \alpha < 4$, there are four equilibria. The point $y = -2$ is still a sink, $y = -\sqrt{\alpha}$ is a source, $y = \sqrt{\alpha}$ is a sink, and $y = 2$ is still a source.

For $\alpha = 4$, there are only two equilibria, $y = \pm 2$. Both are nodes. For $\alpha > 4$, there are four equilibria again. The point $y = -\sqrt{\alpha}$ is a sink, $y = -2$ is now a source, $y = 2$ is now a sink, and $y = \sqrt{\alpha}$ is a source.

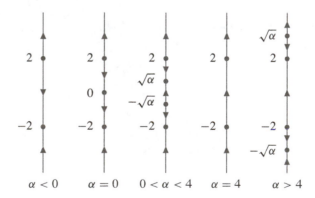

7. We have

$$\frac{dy}{dt} = y^4 + \alpha y^2 = y^2(y^2 + \alpha).$$

If $\alpha > 0$, there is one equilibrium point at $y = 0$, and $dy/dt > 0$ otherwise. Hence, $y = 0$ is a node.

If $\alpha < 0$, there are equilibria at $y = 0$ and $y = \pm\sqrt{-\alpha}$. From the sign of $y^4 + \alpha y^2$, we know that $y = 0$ is a node, $y = -\sqrt{-\alpha}$ is a sink, and $y = \sqrt{-\alpha}$ is a source.

The bifurcation value of α is $\alpha = 0$. As α increases through 0, a sink and a source come together with the node at $y = 0$, leaving only the node. For $\alpha < 0$, there are three equilibria, and for $\alpha \geq 0$, there is only one equilibrium.

9. The bifurcations occur at values of α for which the graph of $\sin y + \alpha$ is tangent to the y-axis. That is, $\alpha = -1$ and $\alpha = 1$.

   For $\alpha < -1$, there are no equilibria, and all solutions become unbounded in the negative direction as t increases.

   If $\alpha = -1$, there are equilibrium points at $y = \pi/2 \pm 2n\pi$ for every integer n. All equilibria are nodes, and as $t \to \infty$, all other solutions decrease toward the nearest equilibrium solution below the given initial condition.

   For $-1 < \alpha < 1$, there are infinitely many sinks and infinitely many sources, and they alternate along the phase line. Successive sinks differ by 2π. Similarly, successive sources are separated by 2π.

   As α increases from -1 to $+1$, nearby sink and source pairs move apart. This separation continues until α is close to 1 where each source is close to the next sink with larger value of y.

   At $\alpha = 1$, there are infinitely many nodes, and they are located at $y = 3\pi/2 \pm 2n\pi$ for every integer n. For $\alpha > 1$, there are no equilibria, and all solutions become unbounded in the positive direction as t increases.

11. For $\alpha = 0$, there are three equilibria. There is a sink to the left of $y = 0$, a source at $y = 0$, and a sink to the right of $y = 0$.

   As α decreases, the source and sink on the right move together. A bifurcation occurs at $\alpha \approx -2$. At this bifurcation value, there is a sink to the left of $y = 0$ and a node to the right of $y = 0$. For α below this bifurcation value, there is only the sink to the left of $y = 0$.

   As α increases from zero, the sink to the left of $y = 0$ and the source move together. There is a bifurcation at $\alpha \approx 2$ with a node to the left of $y = 0$ and a sink to the right of $y = 0$. For α above this bifurcation value, there is only the sink to the right of $y = 0$.

13. (a) Each phase line has an equilibrium point at $y = 0$. This corresponds to equations (i), (iii), and (vi). Since $y = 0$ is the only equilibrium point for $A < 0$, this only corresponds to equation (iii).

    (b) The phase line corresponding to $A = 0$ is the only phase line with $y = 0$ as an equilibrium point, which corresponds to equations (ii), (iv), and (v). For the phase lines corresponding to $A < 0$, there are no equilibrium points. Only equations (iv) and (v) satisfy this property. For the phase lines corresponding to $A > 0$, note that $dy/dt < 0$ for $-\sqrt{A} < y < \sqrt{A}$. Consequently, the bifurcation diagram corresponds to equation (v).

    (c) The phase line corresponding to $A = 0$ is the only phase line with $y = 0$ as an equilibrium point, which corresponds to equations (ii), (iv), and (v). For the phase lines corresponding to $A < 0$, there are no equilibrium points. Only equations (iv) and (v) satisfy this property. For the phase lines corresponding to $A > 0$, note that $dy/dt > 0$ for $-\sqrt{A} < y < \sqrt{A}$. Consequently, the bifurcation diagram corresponds to equation (iv).

    (d) Each phase line has an equilibrium point at $y = 0$. This corresponds to equations (i), (iii), and (vi). The phase lines corresponding to $A > 0$ only have two nonnegative equilibrium points. Consequently, the bifurcation diagram corresponds to equation (i).

15. The graph of f needs to cross the y-axis exactly four times so that there are exactly four equilibria if $\alpha = 0$. The function must be greater than -3 everywhere so that there are no equilibria if $\alpha \geq 3$. Finally, the graph of f must cross horizontal lines three or more units above the y-axis exactly twice so that there are exactly two equilibria for $\alpha \leq -3$. The following graph is an example of the graph of such a function.

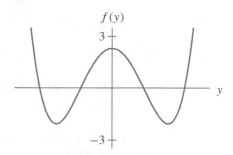

17. No such $f(y)$ exists. To see why, suppose that there is exactly one sink y_0 for $\alpha = 0$. Then, $f(y) > 0$ for $y < y_0$, and $f(y) < 0$ for $y > y_0$. Now consider the system $dy/dt = f(y) + 1$. Then $dy/dt \geq 1$ for $y < y_0$. If this system has an equilibrium point y_1 that is a source, then $y_1 > y_0$ and $dy/dt < 0$ for y slightly less than y_1. Since $f(y)$ is continuous and $dy/dt \geq 1$ for $y \leq y_0$, then dy/dt must have another zero between y_0 and y_1.

19.   (a) A model of the fish population that includes fishing is

$$\frac{dP}{dt} = 2P - \frac{P^2}{50} - 3L,$$

where L is the number of licenses issued. The coefficient of 3 represents the average catch of 3 fish per year. As L is increased, the two equilibrium points for $L = 0$ (at $P = 0$ and $P = 100$) will move together. If L is sufficiently large, there are no equilibrium points. Hence we wish to pick L as large as possible so that there is still an equilibrium point present. In other words, we want the bifurcation value of L. The bifurcation value of L occurs if the equation

$$\frac{dP}{dt} = 2P - \frac{P^2}{50} - 3L = 0$$

has just one solution for P in terms of L. Using the quadratic formula, we see that there is exactly one equilibrium point if $L = 50/3$. Since this value of L is not an integer, the largest number of licenses that should be allowed is 16.

(b) If we allow the fish population to come to equilibrium then the population will be at the carrying capacity, which is $P = 100$ if $L = 0$. If we then allow 16 licenses to be issued, we expect that the population is a solution to the new model with $L = 16$ and initial population $P = 100$. The model becomes

$$\frac{dP}{dt} = 2P - \frac{P^2}{50} - 48,$$

which has a source at $P = 40$ and a sink at $P = 60$.

Thus, any initial population greater than 40 when fishing begins tends to the equilibrium level $P = 60$. If the initial population of fish was less than 40 when fishing begins, then the model predicts that the population will decrease to zero in a finite amount of time.

(c) The maximum "number" of licenses is $16\frac{2}{3}$. With $L = 16\frac{2}{3}$, there is an equilibrium at $P = 50$. This equilibrium is a node, and if $P(0) > 50$, the population will approach 50 as t increases. However, it is dangerous to allow this many licenses since an unforeseen event might cause the death of a few extra fish. That event would push the number of fish below the equilibrium value of $P = 50$. In this case, $dP/dt < 0$, and the population decreases to extinction.

If, however, we restrict to $L = 16$ licenses, then there are two equilibria, a sink at $P = 60$ and source at $P = 40$. As long as $P(0) > 40$, the population will tend to 60 as t increases. In this case, we have a small margin of safety. If $P \approx 60$, then it would have to drop to less than 40 before the fish are in danger of extinction.

21. If $C < kN/4$, the differential equation has two equilibria

$$P_1 = \frac{N}{2} - \sqrt{\frac{N^2}{4} - \frac{CN}{k}} \quad \text{and} \quad P_2 = \frac{N}{2} + \sqrt{\frac{N^2}{4} - \frac{CN}{k}}.$$

The smaller one, P_1, is a source, and the larger one, P_2, is a sink. Note that they are equidistant from $N/2$. Also, note that any population below P_1 tends to extinction.

If C is near $kN/4$, then P_1 and P_2 are near $N/2$. Consequently, if the population is near zero, it will tend to extinction. As C is decreased, P_1 and P_2 move apart until they reach $P_1 = 0$ and $P_2 = N$ for $C = 0$.

Once P is near zero, the parameter C must be reset essentially to zero so that P will be greater than P_1. Simply reducing C slightly below $kN/4$ leaves P in the range where $dP/dt < 0$ and the population will still die out.

23. (a) If $a \leq 0$, there is a single equilibrium point at $y = 0$, and it is a sink. For $a > 0$, there are equilibrium points at $y = 0$ and $y = \pm\sqrt{a}$. The equilibrium point at $y = 0$ is a source, and the other two are sinks.

Phase lines for $dy/dt = ay - y^3$.

(b) Given the results in part (a), there is one bifurcation value, $a = 0$.

(c) The equilibrium points satisfy the cubic equation

$$r + ay - y^3 = 0.$$

Rather than solving it explicitly, we rely on `PhaseLines`.

If $r > 0$, there is a positive bifurcation value $a = a_0$. For $a < a_0$, the phase line has one equilibrium point, a positive sink. If $a > a_0$, there are two negative equilibria in addition to the positive sink. The larger of the two negative equilibria is a source and the smaller is a sink.

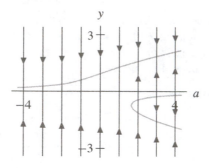

The bifurcation diagram for $r = 0.8$.

(d) If $r < 0$, there is a positive bifurcation value $a = a_0$. For $a < a_0$, the phase line has one equilibrium point, a negative sink. If $a > a_0$, there are two positive equilibria in addition to the negative sink. The larger of the two positive equilibria is a sink and the smaller is a source.

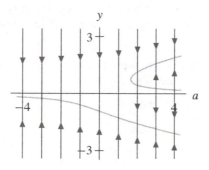

The bifurcation diagram for $r = -0.8$.

EXERCISES FOR SECTION 1.8

1. The general solution to the associated homogeneous equation is $y_h(t) = ke^{-4t}$. For a particular solution of the nonhomogeneous equation, we guess a solution of the form $y_p(t) = \alpha e^{-t}$. Then

$$\frac{dy_p}{dt} + 4y_p = -\alpha e^{-t} + 4\alpha e^{-t}$$
$$= 3\alpha e^{-t}.$$

Consequently, we must have $3\alpha = 9$ for $y_p(t)$ to be a solution. Hence, $\alpha = 3$, and the general solution to the nonhomogeneous equation is

$$y(t) = ke^{-4t} + 3e^{-t}.$$

3. The general solution to the associated homogeneous equation is $y_h(t) = ke^{-3t}$. For a particular solution of the nonhomogeneous equation, we guess a solution of the form $y_p(t) = \alpha \cos 2t + \beta \sin 2t$. Then

$$\frac{dy_p}{dt} + 3y_p = -2\alpha \sin 2t + 2\beta \cos 2t + (3\alpha \cos 2t + 3\beta \sin 2t)$$
$$= (3\alpha + 2\beta)\cos 2t + (3\beta - 2\alpha)\sin 2t$$

Consequently, we must have

$$(3\alpha + 2\beta)\cos 2t + (3\beta - 2\alpha)\sin 2t = 4\cos 2t$$

for $y_p(t)$ to be a solution. We must solve

$$\begin{cases} 3\alpha + 2\beta = 4 \\ 3\beta - 2\alpha = 0. \end{cases}$$

Hence, $\alpha = 12/13$ and $\beta = 8/13$. The general solution is

$$y(t) = ke^{-3t} + \tfrac{12}{13}\cos 2t + \tfrac{8}{13}\sin 2t.$$

5. The general solution to the associated homogeneous equation is $y_h(t) = ke^{3t}$. For a particular solution of the nonhomogeneous equation, we guess $y_p(t) = \alpha t e^{3t}$ rather than αe^{3t} because αe^{3t} is a solution of the homogeneous equation. Then

$$\frac{dy_p}{dt} - 3y_p = \alpha e^{3t} + 3\alpha t e^{3t} - 3\alpha t e^{3t}$$
$$= \alpha e^{3t}.$$

Consequently, we must have $\alpha = -4$ for $y_p(t)$ to be a solution. Hence, the general solution to the nonhomogeneous equation is

$$y(t) = ke^{3t} - 4te^{3t}.$$

7. The general solution to the associated homogeneous equation is $y_h(t) = ke^{-2t}$. For a particular solution of the nonhomogeneous equation, we guess a solution of the form $y_p(t) = \alpha e^{t/3}$. Then

$$\frac{dy_p}{dt} + 2y_p = \tfrac{1}{3}\alpha e^{t/3} + 2\alpha e^{t/3}$$
$$= \tfrac{7}{3}\alpha e^{t/3}.$$

Consequently, we must have $\tfrac{7}{3}\alpha = 1$ for $y_p(t)$ to be a solution. Hence, $\alpha = 3/7$, and the general solution to the nonhomogeneous equation is

$$y(t) = ke^{-2t} + \tfrac{3}{7}e^{t/3}.$$

Since $y(0) = 1$, we have

$$1 = k + \tfrac{3}{7},$$

so $k = 4/7$. The function $y(t) = \tfrac{4}{7}e^{-2t} + \tfrac{3}{7}e^{t/3}$ is the solution of the initial-value problem.

9. The general solution of the associated homogeneous equation is $y_h(t) = ke^{-t}$. For a particular solution of the nonhomogeneous equation, we guess a solution of the form $y_p(t) = \alpha \cos 2t + \beta \sin 2t$. Then

$$\frac{dy_p}{dt} + y_p = -2\alpha \sin 2t + 2\beta \cos 2t + \alpha \cos 2t + \beta \sin 2t$$

$$= (\alpha + 2\beta) \cos 2t + (-2\alpha + \beta) \sin 2t.$$

Consequently, we must have

$$(\alpha + 2\beta) \cos 2t + (-2\alpha + \beta) \sin 2t = \cos 2t$$

for $y_p(t)$ to be a solution. We must solve

$$\begin{cases} \alpha + 2\beta = 1 \\ -2\alpha + \beta = 0. \end{cases}$$

Hence, $\alpha = 1/5$ and $\beta = 2/5$. The general solution to the differential equation is

$$y(t) = ke^{-t} + \tfrac{1}{5} \cos 2t + \tfrac{2}{5} \sin 2t.$$

To find the solution of the given initial-value problem, we evaluate the general solution at $t = 0$ and obtain

$$y(0) = k + \tfrac{1}{5}.$$

Since the initial condition is $y(0) = 5$, we see that $k = 24/5$. The desired solution is

$$y(t) = \tfrac{24}{5} e^{-t} + \tfrac{1}{5} \cos 2t + \tfrac{2}{5} \sin 2t.$$

11. The general solution to the associated homogeneous equation is $y_h(t) = ke^{2t}$. For a particular solution of the nonhomogeneous equation, we guess $y_p(t) = \alpha t e^{2t}$ rather than αe^{2t} because αe^{2t} is a solution of the homogeneous equation. Then

$$\frac{dy_p}{dt} - 2y_p = \alpha e^{2t} + 2\alpha t e^{2t} - 2\alpha t e^{2t}$$

$$= \alpha e^{2t}.$$

Consequently, we must have $\alpha = 7$ for $y_p(t)$ to be a solution. Hence, the general solution to the nonhomogeneous equation is

$$y(t) = ke^{2t} + 7t e^{2t}.$$

Note that $y(0) = k = 3$, so the solution to the initial-value problem is

$$y(t) = 3e^{2t} + 7t e^{2t} = (3 + 7t)e^{2t}.$$

13. (a) For the guess $y_p(t) = \alpha \cos 3t$, we have $dy_p/dt = -3\alpha \sin 3t$, and substituting this guess into the differential equation, we get

$$-3\alpha \sin 3t + 2\alpha \cos 3t = \cos 3t.$$

If we evaluate this equation at $t = \pi/6$, we get $-3\alpha = 0$. Therefore, $\alpha = 0$. However, $\alpha = 0$ does not produce a solution to the differential equation. Consequently, there is no value of α for which $y_p(t) = \alpha \cos 3t$ is a solution.

(b) If we guess $y_p(t) = \alpha \cos 3t + \beta \sin 3t$, then the derivative

$$\frac{dy_p}{dt} = -3\alpha \sin 3t + 3\beta \cos 3t$$

is also a simple combination of terms involving $\cos 3t$ and $\sin 3t$. Substitution of this guess into the equation leads to two linear algebraic equations in two unknowns, and such systems of equations usually have a unique solution.

15. The Linearity Principle says that all nonzero solutions of a homogeneous linear equation are constant multiples of each other.

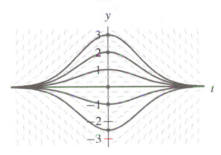

17. (a) We compute

$$\frac{dy_1}{dt} = \frac{1}{(1-t)^2} = (y_1(t))^2$$

to see that $y_1(t)$ is a solution.

(b) We compute

$$\frac{dy_2}{dt} = 2\frac{1}{(1-t)^2} \neq (y_2(t))^2$$

to see that $y_2(t)$ is not a solution.

(c) The equation $dy/dt = y^2$ is not linear. It contains y^2.

19. Let $y(t) = y_h(t) + y_1(t) + y_2(t)$. Then

$$\frac{dy}{dt} + a(t)y = \frac{dy_h}{dt} + \frac{dy_1}{dt} + \frac{dy_2}{dt} + a(t)y_h + a(t)y_1 + a(t)y_2$$

$$= \frac{dy_h}{dt} + a(t)y_h + \frac{dy_1}{dt} + a(t)y_1 + \frac{dy_2}{dt} + a(t)y_2$$

$$= 0 + b_1(t) + b_2(t).$$

This computation shows that $y_h(t) + y_1(t) + y_2(t)$ is a solution of the original differential equation.

21. To find the general solution, we use the technique suggested in Exercise 19. We calculate two particular solutions—one for the right-hand side $t^2 + 2t + 1$ and one for the right-hand side e^{4t}.

With the right-hand side $t^2 + 2t + 1$, we guess a solution of the form

$$y_{p_1}(t) = at^2 + bt + c.$$

Then

$$\frac{dy_{p_1}}{dt} + 2y_{p_1} = 2at + b + 2(at^2 + bt + c)$$
$$= 2at^2 + (2a + 2b)t + (b + 2c).$$

Then y_{p_1} is a solution if

$$\begin{cases} 2a = 1 \\ 2a + 2b = 2 \\ b + 2c = 1. \end{cases}$$

We get $a = 1/2, b = 1/2,$ and $c = 1/4$.

With the right-hand side e^{4t}, we guess a solution of the form

$$y_{p_2}(t) = \alpha e^{4t}.$$

Then

$$\frac{dy_{p_2}}{dt} + 2y_{p_2} = 4\alpha e^{4t} + 2\alpha e^{4t} = 6\alpha e^{4t},$$

and y_{p_2} is a solution if $\alpha = 1/6$.

The general solution of the associated homogeneous equation is $y_h(t) = ke^{-2t}$, so the general solution of the original equation is

$$ke^{-2t} + \tfrac{1}{2}t^2 + \tfrac{1}{2}t + \tfrac{1}{4} + \tfrac{1}{6}e^{4t}.$$

To find the solution that satisfies the initial condition $y(0) = 0$, we evaluate the general solution at $t = 0$ and obtain

$$k + \tfrac{1}{4} + \tfrac{1}{6} = 0.$$

Hence, $k = -5/12$.

23. To find the general solution, we use the technique suggested in Exercise 19. We calculate two particular solutions—one for the right-hand side $2t$ and one for the right-hand side $-e^{4t}$.

With the right-hand side $2t$, we guess a solution of the form

$$y_{p_1}(t) = at + b.$$

Then

$$\frac{dy_{p_1}}{dt} - 3y_{p_1} = a - 3(at + b)$$
$$= -3at + (a - 3b).$$

Then y_{p_1} is a solution if

$$\begin{cases} -3a = 2 \\ a - 3b = 0. \end{cases}$$

We get $a = -2/3$, and $b = -2/9$.

With the right-hand side $-e^{4t}$, we guess a solution of the form

$$y_{p_2}(t) = \alpha e^{4t}.$$

Then

$$\frac{dy_{p_2}}{dt} - 3y_{p_2} = 4\alpha e^{4t} - 3\alpha e^{4t} = \alpha e^{4t},$$

and y_{p_2} is a solution if $\alpha = -1$.

The general solution of the associated homogeneous equation is $y_h(t) = ke^{3t}$, so the general solution of the original equation is

$$y(t) = ke^{3t} - \tfrac{2}{3}t - \tfrac{2}{9} - e^{4t}.$$

To find the solution that satisfies the initial condition $y(0) = 0$, we evaluate the general solution at $t = 0$ and obtain

$$y(0) = k - \tfrac{2}{9} - 1.$$

Hence, $k = 11/9$ if $y(0) = 0$.

25. Since the general solution of the associated homogeneous equation is $y_h(t) = ke^{-2t}$ and since these $y_h(t) \to 0$ as $t \to \infty$, we only have to determine the long-term behavior of one solution to the nonhomogeneous equation. However, that is easier said than done.

Consider the slopes in the slope field for the equation. We rewrite the equation as

$$\frac{dy}{dt} = -2y + b(t).$$

Using the fact that $b(t) < 2$ for all t, we observe that $dy/dt < 0$ if $y > 1$ and, as y increases beyond $y = 1$, the slopes become more negative. Similarly, using the fact that $b(t) > -1$ for all t, we observe that $dy/dt > 0$ if $y < -1/2$ and, as y decreases below $y = -1/2$, the slopes become more positive. Thus, the graphs of all solutions must approach the strip $-1/2 \le y \le 1$ in the ty-plane as t increases. More precise information about the long-term behavior of solutions is difficult to obtain without specific knowledge of $b(t)$.

27. Since the general solution of the associated homogeneous equation is $y_h(t) = ke^{-t}$ and since these $y_h(t) \to 0$ as $t \to \infty$, we only have to determine the long-term behavior of one solution to the nonhomogeneous equation. However, that is easier said than done.

Consider the slopes in the slope field for the equation. We rewrite the equation as

$$\frac{dy}{dt} = -y + b(t).$$

For any number $T > 3$, let ϵ be a positive number less than $T - 3$, and fix t_0 such that $b(t) < T - \epsilon$ if $t > t_0$. If $t > t_0$ and $y(t) > T$, then

$$\frac{dy}{dt} < -T + T - \epsilon = -\epsilon.$$

Hence, no solution remains greater than T for all time. Since $T > 3$ is arbitrary, no solution remains greater than 3 (by a fixed amount) for all time.

The same idea works to show that no solution can remain less than 3 (by a fixed amount) for all time. Hence, every solution tends to 3 as $t \to \infty$.

29. **(a)** The differential equation modeling the problem is

$$\frac{dP}{dt} = .011P + 1{,}040,$$

where \$1,040 is the amount of money added to the account per year (assuming a "continuous deposit").

(b) To find the general solution, we first compute the general solution of the associated homogeneous equation. It is $P_h(t) = ke^{0.011t}$.

To find a particular solution of the nonhomogeneous equation, we observe that the equation is autonomous, and we calculate its equilibrium solution. It is $P(t) = -1{,}040/.011 \approx -94{,}545.46$ for all t. (This equilibrium solution is what we would have calculated if we had guessed a constant.)

Hence, the general solution is

$$P(t) = -94{,}545.46 + ke^{0.011t}.$$

Since the account initially has \$1,000 in it, the initial condition is $P(0) = 1{,}000$. Solving

$$1000 = -94{,}545.46 + ke^{0.011(0)}$$

yields $k = 95{,}545.46$. Therefore, our model is

$$P(t) = -94{,}545.46 + 95{,}545.46e^{0.011t}.$$

To find the amount on deposit after 5 years, we evaluate $P(5)$ and obtain

$$-94{,}545.46 + 95{,}545.46e^{0.011(5)} \approx 6{,}402.20.$$

31. Step 1: *Before retirement*

First we calculate how much money will be in her retirement fund after 30 years. The differential equation modeling the situation is

$$\frac{dy}{dt} = .07y + 5{,}000,$$

where $y(t)$ represents the fund's balance at time t.

The general solution of the homogeneous equation is $y_h(t) = ke^{0.07t}$.

To find a particular solution, we observe that the nonhomogeneous equation is autonomous and that it has an equilibrium solution at $y = -5{,}000/0.07 \approx -71{,}428.57$. We can use this equilibrium solution as the particular solution. (It is the solution we would have computed if we had guessed a constant solution). We obtain

$$y(t) = ke^{0.07t} - 71{,}428.57.$$

From the initial condition, we see that $k = 71{,}428.57$, and

$$y(t) = 71{,}428.57(e^{0.07t} - 1).$$

Letting $t = 30$, we compute that the fund contains $\approx \$511,869.27$ after 30 years.

Step 2: *After retirement*
We need a new model for the remaining years since the professor is withdrawing rather than depositing. Since she withdraws at a rate of \$3,000 per month (\$36,000 per year), we write

$$\frac{dy}{dt} = .07y - 36,000,$$

where we continue to measure time t in years.

Again, the solution of the homogeneous equation is $y_h(t) = ke^{0.07t}$.

To find a particular solution of the nonhomogeneous equation, we note that the equation is autonomous and that it has an equilibrium at $y = 36,000/0.07 \approx 514,285.71$. Hence, we may take the particular solution to be this equilibrium solution. (Again, this solution is what we would have computed if we had guessed a constant function for y_p.)

The general solution is

$$y(t) = ke^{0.07t} + 514,285.71.$$

In this case, we have the initial condition $y(0) = 511,869.27$ since now $y(t)$ is the amount in the fund t years after she retires. Solving $511,869.27 = k + 514,285.71$, we get $k = -2,416.44$. The solution in this case is

$$y(t) = -2,416.44e^{0.07t} + 514,285.71.$$

Finally, we wish to know when her money runs out. That is, at what time t is $y(t) = 0$? Solving

$$y(t) = -2,416.44e^{0.07t} + 514,285.71 = 0$$

yields $t \approx 76.58$ years (approximately 919 months).

33. **(a)** We know that

$$\frac{dy_h}{dt} = a(t)y_h \quad \text{and} \quad \frac{dy_p}{dt} = a(t)y_p + b(t).$$

Then

$$\frac{d(y_h + y_p)}{dt} = a(t)y_h + a(t)y_p + b(t)$$
$$= a(t)(y_h + y_p) + b(t).$$

(b) We know that

$$\frac{dy_p}{dt} = a(t)y_p + b(t) \quad \text{and} \quad \frac{dy_q}{dt} = a(t)y_q + b(t).$$

Then

$$\frac{d(y_p - y_q)}{dt} = (a(t)y_p + b(t)) - (a(t)y_q + b(t))$$
$$= a(t)(y_p - y_q).$$

EXERCISES FOR SECTION 1.9

1. We rewrite the equation in the form

$$\frac{dy}{dt} + \frac{y}{t} = 2$$

and note that the integrating factor is

$$\mu(t) = e^{\int (1/t)\, dt} = e^{\ln t} = t.$$

Multiplying both sides by $\mu(t)$, we obtain

$$t\frac{dy}{dt} + y = 2t.$$

Applying the Product Rule to the left-hand side, we see that this equation is the same as

$$\frac{d(ty)}{dt} = 2t,$$

and integrating both sides with respect to t, we obtain

$$ty = t^2 + c,$$

where c is an arbitrary constant. The general solution is

$$y(t) = \frac{1}{t}(t^2 + c) = t + \frac{c}{t}.$$

3. We rewrite the equation in the form

$$\frac{dy}{dt} + \frac{y}{1+t} = t^2$$

and note that the integrating factor is

$$\mu(t) = e^{\int (1/(1+t))\, dt} = e^{\ln(1+t)} = 1 + t.$$

Multiplying both sides by $\mu(t)$, we obtain

$$(1+t)\frac{dy}{dt} + y = (1+t)t^2.$$

Applying the Product Rule to the left-hand side, we see that this equation is the same as

$$\frac{d((1+t)y)}{dt} = t^3 + t^2,$$

and integrating both sides with respect to t, we obtain

$$(1+t)y = \frac{t^4}{4} + \frac{t^3}{3} + c,$$

where c is an arbitrary constant. The general solution is

$$y(t) = \frac{3t^4 + 4t^3 + 12c}{12(t+1)}.$$

5. Note that the integrating factor is

$$\mu(t) = e^{\int (-2t/(1+t^2))\, dt} = e^{-\ln(1+t^2)} = \left(e^{\ln(1+t^2)}\right)^{-1} = \frac{1}{1+t^2}.$$

Multiplying both sides by $\mu(t)$, we obtain

$$\frac{1}{1+t^2}\frac{dy}{dt} - \frac{2t}{(1+t^2)^2}\, y = \frac{3}{1+t^2}.$$

Applying the Product Rule to the left-hand side, we see that this equation is the same as

$$\frac{d}{dt}\left(\frac{y}{1+t^2}\right) = \frac{3}{1+t^2}.$$

Integrating both sides with respect to t, we obtain

$$\frac{y}{1+t^2} = 3\arctan(t) + c,$$

where c is an arbitrary constant. The general solution is

$$y(t) = (1+t^2)(3\arctan(t) + c).$$

7. We rewrite the equation in the form

$$\frac{dy}{dt} + \frac{y}{1+t} = 2$$

and note that the integrating factor is

$$\mu(t) = e^{\int (1/(1+t))\, dt} = e^{\ln(1+t)} = 1 + t.$$

Multiplying both sides by $\mu(t)$, we obtain

$$(1+t)\frac{dy}{dt} + y = 2(1+t).$$

Applying the Product Rule to the left-hand side, we see that this equation is the same as

$$\frac{d((1+t)y)}{dt} = 2(1+t),$$

and integrating both sides with respect to t, we obtain

$$(1+t)y = 2t + t^2 + c,$$

where c is an arbitrary constant. The general solution is

$$y(t) = \frac{t^2 + 2t + c}{1+t}.$$

To find the solution that satisfies the initial condition $y(0) = 3$, we evaluate the general solution at $t = 0$ and obtain

$$c = 3.$$

The desired solution is

$$y(t) = \frac{t^2 + 2t + 3}{1+t}.$$

9. In Exercise 1, we derived the general solution

$$y(t) = t + \frac{c}{t}.$$

To find the solution that satisfies the initial condition $y(1) = 3$, we evaluate the general solution at $t = 1$ and obtain $c = 2$. The desired solution is

$$y(t) = t + \frac{2}{t}.$$

11. Note that the integrating factor is

$$\mu(t) = e^{\int -(2/t)\,dt} = e^{-2\int (1/t)\,dt} = e^{-2\ln t} = e^{\ln(t^{-2})} = \frac{1}{t^2}.$$

Multiplying both sides by $\mu(t)$, we obtain

$$\frac{1}{t^2}\frac{dy}{dt} - \frac{2y}{t^3} = 2.$$

Applying the Product Rule to the left-hand side, we see that this equation is the same as

$$\frac{d}{dt}\left(\frac{y}{t^2}\right) = 2,$$

and integrating both sides with respect to t, we obtain

$$\frac{y}{t^2} = 2t + c,$$

where c is an arbitrary constant. The general solution is

$$y(t) = 2t^3 + ct^2.$$

To find the solution that satisfies the initial condition $y(-2) = 4$, we evaluate the general solution at $t = -2$ and obtain

$$-16 + 4c = 4.$$

Hence, $c = 5$, and the desired solution is

$$y(t) = 2t^3 + 5t^2.$$

13. We rewrite the equation in the form

$$\frac{dy}{dt} - (\sin t)y = 4$$

and note that the integrating factor is

$$\mu(t) = e^{\int (-\sin t)\,dt} = e^{\cos t}.$$

Multiplying both sides by $\mu(t)$, we obtain

$$e^{\cos t}\frac{dy}{dt} - e^{\cos t}(\sin t)y = 4e^{\cos t}.$$

Applying the Product Rule to the left-hand side, we see that this equation is the same as

$$\frac{d(e^{\cos t}y)}{dt} = 4e^{\cos t},$$

and integrating both sides with respect to t, we obtain

$$e^{\cos t}y = \int 4e^{\cos t}\,dt.$$

Since the integral on the right-hand side is impossible to express using elementary functions, we write the general solution as

$$y(t) = 4e^{-\cos t}\int e^{\cos t}\,dt.$$

15. We rewrite the equation in the form

$$\frac{dy}{dt} - \frac{y}{t^2} = 4\cos t$$

and note that the integrating factor is

$$\mu(t) = e^{\int(-1/t^2)\,dt} = e^{1/t}.$$

Multiplying both sides by $\mu(t)$, we obtain

$$e^{1/t}\frac{dy}{dt} - \frac{e^{1/t}}{t^2}y = 4e^{1/t}\cos t.$$

Applying the Product Rule to the left-hand side, we see that this equation is the same as

$$\frac{d(e^{1/t}y)}{dt} = 4e^{1/t}\cos t,$$

and integrating both sides with respect to t, we obtain

$$e^{1/t}y = \int 4e^{1/t}\cos t\,dt.$$

Since the integral on the right-hand side is impossible to express using elementary functions, we write the general solution as

$$y(t) = 4e^{-1/t}\int e^{1/t}\cos t\,dt.$$

17. We rewrite the equation in the form

$$\frac{dy}{dt} + e^{-t^2}y = \cos t$$

and note that the integrating factor is

$$\mu(t) = e^{\int e^{-t^2} dt}.$$

This integral is impossible to express in terms of elementary functions. Multiplying both sides by $\mu(t)$, we obtain

$$\left(e^{\int e^{-t^2} dt}\right) \frac{dy}{dt} + \left(e^{\int e^{-t^2} dt}\right) e^{-t^2} y = \left(e^{\int e^{-t^2} dt}\right) \cos t.$$

Applying the Product Rule to the left-hand side, we see that this equation is the same as

$$\frac{d\left(\left(e^{\int e^{-t^2} dt}\right) y\right)}{dt} = \left(e^{\int e^{-t^2} dt}\right) \cos t,$$

and integrating both sides with respect to t, we obtain

$$\left(e^{\int e^{-t^2} dt}\right) y = \int \left(e^{\int e^{-t^2} dt}\right) \cos t \, dt.$$

These integrals are also impossible to express in terms of elementary functions, so we write the general solution in the form

$$y(t) = \left(e^{-\int e^{-t^2} dt}\right) \int \left(e^{\int e^{-t^2} dt}\right) \cos t \, dt.$$

19. We rewrite the equation in the form

$$\frac{dy}{dt} - aty = 4e^{-t^2}$$

and note that the integrating factor is

$$\mu(t) = e^{\int (-at) dt} = e^{-at^2/2}.$$

Multiplying both sides by $\mu(t)$, we obtain

$$e^{-at^2/2} \frac{dy}{dt} - ate^{-at^2/2} y = 4e^{-t^2} e^{-at^2/2}.$$

Applying the Product Rule to the left-hand side and simplifying the right-hand side, we see that this equation is the same as

$$\frac{d(e^{-at^2/2} y)}{dt} = 4e^{-(1+a/2)t^2}.$$

Integrating both sides with respect to t, we obtain

$$e^{-at^2/2} y = \int 4e^{-(1+a/2)t^2} dt.$$

The integral on the right-hand side can be expressed in terms of elementary functions only if $1+a/2 = 0$ (that is, if the factor involving e^{t^2} really isn't there). Hence, the only value of a that yields an integral we can express in terms of elementary functions form is $a = -2$ (see Exercise 4).

21. **(a)** The integrating factor is

$$\mu(t) = e^{0.4t}.$$

Multiplying both sides of the differential equation by $\mu(t)$ and collecting terms, we obtain

$$\frac{d(e^{0.4t}v)}{dt} = 3e^{0.4t}\cos 2t.$$

Integrating both sides with respect to t yields

$$e^{0.4t}v = \int 3e^{0.4t}\cos 2t\, dt.$$

To calculate the integral on the right-hand side, we must integrate by parts twice.

For the first integration, we pick $u_1(t) = \cos 2t$ and $v_1(t) = e^{0.4t}$. Using the fact that $0.4 = 2/5$, we get

$$\int e^{0.4t}\cos 2t\, dt = \tfrac{5}{2}e^{0.4t}\cos 2t + 5\int e^{0.4t}\sin 2t\, dt.$$

For the second integration, we pick $u_2(t) = \sin 2t$ and $v_2(t) = e^{0.4t}$. We get

$$\int e^{0.4t}\sin 2t\, dt = \tfrac{5}{2}e^{0.4t}\sin 2t - 5\int e^{0.4t}\cos 2t\, dt.$$

Combining these results yields

$$\int e^{0.4t}\cos 2t\, dt = \tfrac{5}{2}e^{0.4t}\cos 2t + \tfrac{25}{2}e^{0.4t}\sin 2t - 25\int e^{0.4t}\cos 2t\, dt.$$

Solving for $\int e^{0.4t}\cos 2t\, dt$, we have

$$\int e^{0.4t}\cos 2t\, dt = \frac{5\,e^{0.4t}\cos 2t + 25\,e^{0.4t}\sin 2t}{52}.$$

To obtain the general solution, we multiply this integral by 3, add the constant of integration, and solve for v. We obtain the general solution

$$v(t) = ke^{-0.4t} + \tfrac{15}{52}\cos 2t + \tfrac{75}{52}\sin 2t.$$

(b) The solution of the associated homogeneous equation is

$$v_h(t) = e^{-0.4t}.$$

We guess

$$v_p(t) = \alpha\cos 2t + \beta\sin 2t$$

for the a solution to the nonhomogeneous equation and solve for α and β. Substituting this guess into the differential equation, we obtain

$$-2\alpha\sin 2t + 2\beta\cos 2t + 0.4\alpha\cos 2t + 0.4\beta\sin 2t = 3\cos 2t.$$

Collecting sine and cosine terms, we get the system of equations

$$\begin{cases} -2\alpha + 0.4\beta = 0 \\ 0.4\alpha + 2\beta = 3. \end{cases}$$

Using the fact that $0.4 = 2/5$, we solve this system of equations and obtain

$$\alpha = \tfrac{15}{52} \quad \text{and} \quad \beta = \tfrac{75}{52}.$$

The general solution of the original nonhomogeneous equation is

$$v(t) = ke^{-0.4t} + \tfrac{15}{52}\cos 2t + \tfrac{75}{52}\sin 2t.$$

Both methods require quite a bit of computation. If we use an integrating factor, we must do a complicated integral, and if we use the guessing technique, we have to be careful with our algebra.

23. The integrating factor is

$$\mu(t) = e^{\int 2\,dt} = e^{2t}.$$

Multiplying both sides by $\mu(t)$, we obtain

$$e^{2t}\frac{dy}{dt} + 2e^{2t}y = 3e^{2t}e^{-2t}$$

$$= 3.$$

Applying the Product Rule to the left-hand side, we see that this equation is the same as

$$\frac{d(e^{2t}y)}{dt} = 3,$$

and integrating both sides with respect to t, we obtain

$$e^{2t}y = 3t + k,$$

where k is an arbitrary constant. The general solution is

$$y(t) = (3t + k)e^{-2t}.$$

We know that ke^{-2t} is the general solution of the associated homogeneous equation, so $y_p(t) = 3te^{-2t}$ is a particular solution of the nonhomogeneous equation. Note that the factor of t arose after we multiplied the right-hand side of the equation by the integrating factor and ended up with the constant 3. After integrating, the constant produces a factor of t.

25. We will use the term "parts" as shorthand for the product of parts per billion of dioxin and the volume of water in the tank. Basically this product represents the total amount of dioxin in the tank. The tank initially contains 200 gallons at a concentration of 2 parts per billion, which results in 400 parts of dioxin.

Let $y(t)$ be the amount of dioxin in the tank at time t. Since water with 4 parts per billion of dioxin flows in at the rate of 5 gallons per minute, 20 parts of dioxin enter the tank each minute.

Also, the volume of water in the tank at time t is $200 + 2t$, so the concentration of dioxin in the tank is $y/(200 + 2t)$. Since well-mixed water leaves the tank at the rate of 2 gallons per minute, the differential equation that represents the change in the amount of dioxin in the tank is

$$\frac{dy}{dt} = 20 - 2\left(\frac{y}{200 + 2t}\right),$$

which can be simplified and rewritten as

$$\frac{dy}{dt} + \left(\frac{1}{100 + t}\right) y = 20.$$

The integrating factor is

$$\mu(t) = e^{\int (1/(100+t))\, dt} = e^{\ln(100+t)} = 100 + t.$$

Multiplying both sides by $\mu(t)$, we obtain

$$(100 + t)\frac{dy}{dt} + y = 20(100 + t),$$

which is equivalent to

$$\frac{d((100 + t)y)}{dt} = 20(100 + t)$$

by the Product Rule. Integrating both sides with respect to t, we obtain

$$(100 + t)y = 2000t + 10t^2 + c.$$

Since $y(0) = 400$, we see that $c = 40,000$. Therefore,

$$y(t) = \frac{10t^2 + 2000t + 40,000}{t + 100}.$$

The tank fills up at $t = 100$, and $y(100) = 1,700$. To express our answer in terms of concentration, we calculate $y(100)/400 = 4.25$ parts per billion.

27. (a) Let $y(t)$ be the amount of salt in the tank at time t. Since the tank is being filled at a total rate of 1 gallon per minute, the volume at time t is $V_0 + t$ and the concentration of salt in the tank is

$$\frac{y}{V_0 + t}.$$

The amount of salt entering the tank is the product of 2 gallons per minute and 0.25 pounds of salt per minute. The amount of salt leaving the tank is the product of the concentration of salt in the tank and the rate that brine is leaving. In this case, the rate is 1 gallon per minute, so the amount of salt leaving the tank is $y/(V_0 + t)$. The differential equation for $y(t)$ is

$$\frac{dy}{dt} = \frac{1}{2} - \frac{y}{V_0 + t}.$$

Since the water is initially clean, the initial condition is $y(0) = 0$.

(b) If $V_0 = 0$, the differential equation above becomes

$$\frac{dy}{dt} = \frac{1}{2} - \frac{y}{t}.$$

Note that this differential equation is undefined at $t = 0$. Thus, we *cannot* apply the Existence and Uniqueness Theorem to guarantee a unique solution at time $t = 0$. However, we can still solve the equation using our standard techniques assuming that $t \neq 0$.

Rewriting the equation as

$$\frac{dy}{dt} + \frac{y}{t} = \frac{1}{2},$$

we see that the integrating factor is

$$\mu(t) = e^{\int (1/t)\,dt} = e^{\ln t} = t.$$

Multiplying both sides of the differential equation by $\mu(t)$, we obtain

$$t\frac{dy}{dt} + y = \frac{t}{2},$$

which is equivalent to

$$\frac{d(ty)}{dt} = \frac{t}{2}.$$

Integrating both sides with respect to t, we get

$$ty = \frac{t^2}{4} + c,$$

so that the general solution is

$$y(t) = \frac{t}{4} + \frac{c}{t}.$$

Since the above expression is undefined at $t = 0$, we cannot make use of the initial condition $y(0) = 0$ to find the desired solution.

However, if the tank is initially empty, the concentration of salt in the tank remains constant over time at 0.25 pounds of salt per gallon. Therefore, we reconsider the equation

$$\frac{y}{t} = \frac{1}{4} + \frac{c}{t^2}.$$

If $c = 0$, we have $y/t = 1/4$. Hence, $c = 0$ yields the solution $y(t) = t/4$ which *is* a valid model for this situation.

It is useful to note that, if $V_0 = 0$, then we do not really need a differential equation to model the amount of the salt in the tank as a function of time. Clearly the concentration is constant as a function of time, and therefore the amount of salt in the tank is the product of the concentration and the volume of brine in the tank.

REVIEW EXERCISES FOR CHAPTER 1

1. The simplest differential equation with $y(t) = 2t$ as a solution is $dy/dt = 2$. The initial condition $y(0) = 3$ specifies the desired solution.

3. There are no values of y for which dy/dt is zero for all t. Hence, there are no equilibrium solutions.

5. The right-hand side is zero for all t only if $y = -1$. Consequently, the function $y(t) = -1$ for all t is the only equilibrium solution.

7. The equations $dy/dt = y$ and $dy/dt = 0$ are first-order, autonomous, separable, linear, and homogeneous.

9. The graph of $f(y)$ must cross the y-axis from negative to positive at $y = 0$. For example, the graph of the function $f(y) = y$ produces this phase line.

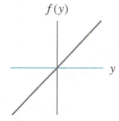

11. True. We have $dy/dt = e^{-t}$, which agrees with $|y(t)|$.

13. True. Autonomous equations have the form $dy/dt = f(y)$. Therefore, we can separate variables by dividing by $f(y)$. That is,

$$\frac{1}{f(y)}\frac{dy}{dt} = 1.$$

15. False. For example, $dy/dt = ty^2$ is separable but not linear.

17. True. Note that the function $y(t) = 3$ for all t is an equilibrium solution for the equation. The Uniqueness Theorem says that graphs of different solutions cannot touch. Hence, a solution with $y(0) > 3$ must have $y(t) > 3$ for all t.

19. False. By the Uniqueness Theorem, graphs of different solutions cannot touch. Hence, if one solution $y_1(t) \to \infty$ as t increases, any solution $y_2(t)$ with $y_2(0) > y_1(0)$ satisfies $y_2(t) > y_1(t)$ for all t. Therefore, $y_2(t) \to \infty$ as t increases.

21.   (a) The equation is autonomous, separable, and linear and nonhomogeneous.

   (b) The general solution to the associated homogeneous equation is $y_h(t) = ke^{-2t}$. For a particular solution of the nonhomogeneous equation, we guess a solution of the form $y_p(t) = \alpha$. Then

$$\frac{dy_p}{dt} + 2y_p = 2\alpha.$$

   Consequently, we must have $2\alpha = 3$ for $y_p(t)$ to be a solution. Hence, $\alpha = 3/2$, and the general solution to the nonhomogeneous equation is

$$y(t) = \tfrac{3}{2} + ke^{-2t}.$$

23. **(a)** The equation is linear and nonhomogeneous. (It is nonautonomous as well.)

(b) The general solution of the associated homogeneous equation is $y_h(t) = ke^{3t}$. For a particular solution of the nonhomogeneous equation, we guess a solution of the form $y_p(t) = \alpha e^{7t}$. Then

$$\frac{dy_p}{dt} - 3y_p = 7\alpha e^{7t} - 3\alpha e^{7t} = 4\alpha e^{7t}.$$

Consequently, we must have $4\alpha = 1$ for $y_p(t)$ to be a solution. Hence, $\alpha = 1/4$, and the general solution to the nonhomogeneous equation is

$$y(t) = ke^{3t} + \tfrac{1}{4}e^{7t}.$$

25. **(a)** This equation is linear and nonhomogeneous.

(b) To find the general solution, we first note that $y_h(t) = ke^{-5t}$ is the general solution of the associated homogeneous equation.

To get a particular solution of the nonhomogeneous equation, we guess

$$y_p(t) = \alpha \cos 3t + \beta \sin 3t.$$

Substituting this guess into the nonhomogeneous equation gives

$$\frac{dy_p}{dt} + 5y_p = -3\alpha \sin 3t + 3\beta \cos 3t + 5\alpha \cos 3t + 5\beta \sin 3t$$

$$= (5\alpha + 3\beta) \cos 3t + (5\beta - 3\alpha) \sin 3t.$$

In order for $y_p(t)$ to be a solution, we must solve the simultaneous equations

$$\begin{cases} 5\alpha + 3\beta = 0 \\ 5\beta - 3\alpha = 1. \end{cases}$$

From these equations, we get $\alpha = -3/34$ and $\beta = 5/34$. Hence, the general solution is

$$y(t) = ke^{-5t} - \tfrac{3}{34} \cos 3t + \tfrac{5}{34} \sin 3t.$$

27. **(a)** The equation is autonomous and separable.

(b) When we separate variables, we obtain

$$\int \frac{1}{3 + y^2}\, dy = \int dt.$$

Integrating, we get

$$\frac{1}{\sqrt{3}} \arctan \left(\frac{y}{\sqrt{3}} \right) = t + c,$$

and solving for $y(t)$ produces

$$y(t) = \sqrt{3} \tan \left(\sqrt{3}\, t + k \right).$$

29. (a) This equation is linear and nonhomogeneous.

(b) First we note that the general solution of the associated homogeneous equation is ke^{-3t}.

Next we use the technique suggested in Exercise 19 of Section 1.8. We could find particular solutions of the two nonhomogeneous equations

$$\frac{dy}{dt} = -3y + e^{-2t} \quad \text{and} \quad \frac{dy}{dt} = -3y + t^2$$

separately and add the results to obtain a particular solution for the original equation. However, these two steps can be combined by making a more complicated guess for the particular solution.

We guess $y_p(t) = ae^{-2t} + bt^2 + ct + d$, and we have

$$\frac{dy_p}{dt} + 3y_p = -2ae^{-2t} + 2bt + c + 3ae^{-2t} + 3bt^2 + 3ct + d$$

$$= ae^{-2t} + 3bt^2 + (2b + 3c)t + (c + 3d).$$

Hence, for $y_p(t)$ to be a solution we must have $a = 1, b = \frac{1}{3}, c = -\frac{2}{9}$, and $d = \frac{2}{27}$. Therefore, a particular solution is $y_p(t) = e^{-2t} + \frac{1}{3}t^2 - \frac{2}{9}t + \frac{2}{27}$. and the general solution is

$$y(t) = ke^{-3t} + e^{-2t} + \frac{1}{3}t^2 - \frac{2}{9}t + \frac{2}{27}.$$

31. (a) This equation is linear and nonhomogeneous. (It is nonautonomous as well.)

(b) The general solution of the associated homogeneous equaion is $y_h(t) = ke^{2t}$. To find a particular solution of the nonhomogeneous equation, we guess $y_p(t) = \alpha \cos 4t + \beta \sin 4t$. Then

$$\frac{dy_p}{dt} - 2y_p = -4\alpha \sin 4t + 4\beta \cos 4t - 2(\alpha \cos 4t + \beta \sin 4t)$$

$$= (-2\alpha + 4\beta) \cos 4t + (-4\alpha - 2\beta) \sin 4t.$$

Consequently, we must have

$$(-2\alpha + 4\beta) \cos 4t + (-4\alpha - 2\beta) \sin 4t = \cos 4t$$

for $y_p(t)$ to be a solution. We must solve

$$\begin{cases} -2\alpha + 4\beta = 1 \\ -4\alpha - 2\beta = 0. \end{cases}$$

Hence, $\alpha = -1/10$ and $\beta = 1/5$, and the general solution of the nonhomogeneous equation is

$$y(t) = ke^{2t} - \frac{1}{10} \cos 4t + \frac{1}{5} \sin 4t.$$

To find the solution of the given initial-value problem, we evaluate the general solution at $t = 0$ and obtain

$$y(0) = k - \frac{1}{10}.$$

Since the initial condition is $y(0) = 1$, we see that $k = 11/10$. The desired solution is

$$y(t) = \frac{11}{10}e^{2t} - \frac{1}{10} \cos 4t + \frac{1}{5} \sin 4t.$$

33. **(a)** The equation is separable because

$$\frac{dy}{dt} = (t^2 + 1)y^3.$$

(b) Separating variables and integrating, we have

$$\int y^{-3}\, dy = \int (t^2 + 1)\, dt$$

$$\frac{y^{-2}}{-2} = \frac{t^3}{3} + t + c$$

$$y^{-2} = -\frac{2}{3}t^3 - 2t + k.$$

Using the initial condition $y(0) = -1/2$, we get that $k = 4$. Therefore,

$$y^2 = \frac{1}{4 - 2t - \frac{2}{3}t^3}.$$

Taking the square root of both sides yields

$$y = \frac{\pm 1}{\sqrt{4 - 2t - \frac{2}{3}t^3}}.$$

In this case, we take the negative square root because $y(0) = -1/2$. The solution to the initial-value problem is

$$y(t) = \frac{-1}{\sqrt{4 - 2t - \frac{2}{3}t^3}}.$$

35. **(a)** This equation is linear and nonhomogeneous. (It is nonautonomous as well.)

(b) We rewrite the equation as

$$\frac{dy}{dt} - 2ty = 3te^{t^2}$$

and note that the integrating factor is

$$\mu(t) = e^{\int -2t\, dt} = e^{-t^2}.$$

Multiplying both sides by $\mu(t)$, we obtain

$$e^{-t^2}\frac{dy}{dt} - 2te^{-t^2}y = 3t.$$

Applying the Product Rule to the left-hand side, we see that this equation is the same as

$$\frac{d}{dt}\left(e^{-t^2}y\right) = 3t,$$

and integrating both sides with respect to t, we obtain $e^{-t^2}y = \frac{3}{2}t^2 + k$, where k is an arbitrary constant. The general solution is

$$y(t) = \left(\tfrac{3}{2}t^2 + k\right)e^{t^2}.$$

To find the solution that satisfies the initial condition $y(0) = 1$, we evaluate the general solution at $t = 0$ and obtain $k = 1$. The desired solution is

$$y(t) = \left(\tfrac{3}{2}t^2 + 1\right) e^{t^2}.$$

37. **(a)** This equation is separable.

(b) We separate variables and integrate to obtain

$$\int \frac{1}{y^2}\, dy = \int (2t + 3t^2)\, dt$$

$$-\frac{1}{y} = t^2 + t^3 + k$$

$$y = \frac{-1}{t^2 + t^3 + k}.$$

To find the solution of the initial-value problem, we evaluate the general solution at $t = 1$ and obtain

$$y(1) = \frac{-1}{2 + k}.$$

Since the initial condition is $y(1) = -1$, we see that $k = -1$. The solution to the initial-value problem is

$$y(t) = \frac{1}{1 - t^2 - t^3}.$$

39. **(a)** The differential equation is separable.

(b) We can write the equation in the form

$$\frac{dy}{dt} = \frac{t^2}{y(t^3 + 1)}$$

and separate variables to get

$$\int y\, dy = \int \frac{t^2}{t^3 + 1}\, dt$$

$$\frac{y^2}{2} = \frac{1}{3} \ln |t^3 + 1| + c,$$

where c is a constant. Hence,

$$y^2 = \frac{2}{3} \ln |t^3 + 1| + 2c.$$

The initial condition $y(0) = -2$ implies

$$(-2)^2 = \frac{2}{3} \ln |1| + 2c.$$

Thus, $c = 2$, and

$$y(t) = -\sqrt{\tfrac{2}{3} \ln |t^3 + 1| + 4}.$$

We choose the negative square root because $y(0)$ is negative.

41. (a)

(b)

43. The constant function $y(t) = 2$ for all t is an equilibrium solution. If $y > 2$, then $dy/dt > 0$. Moreover, solutions with initial conditions above $y = 2$ satisfy $y(t) \to \infty$ as t increases and $y(t) \to 2$ as $t \to -\infty$.

If $y < -2$, then $dy/dt > 0$, so solutions with initial conditions below $y = -2$ increase until they cross the line $y = -2$. If $0 < y < 2$, then $dy/dt < 0$, and solutions in this strip decrease until they cross the t-axis.

For all initial conditions on the y-axis below $y = 2$, the solutions tend toward a periodic solution of period 2π as t increases. This periodic solution crosses the y-axis at $y_0 \approx -0.1471$. If $y(0) < y_0$, then the solution satisfies $y(t) \to -\infty$ as t decreases. If $y_0 < y(0) < 2$, the $y(t) \to 2$ as $t \to -\infty$.

45. Note that

$$\frac{dy}{dt} = (1 + t^2)y + 1 + t^2 = (1 + t^2)(y + 1).$$

(a) Separating variables and integrating, we obtain

$$\int \frac{1}{y + 1}\, dy = \int (1 + t^2)\, dt$$

$$\ln|y + 1| = t + \frac{t^3}{3} + c,$$

where c is any constant. Thus, $|y + 1| = c_1 e^{t + t^3/3}$, where $c_1 = e^c$. We can dispose of the absolute value signs by allowing the constant c_1 to be any real number. In other words,

$$y(t) = -1 + k e^{t + t^3/3},$$

where $k = \pm c_1$. Note that, if $k = 0$, we have the equilibrium solution $y(t) = -1$ for all t.

(b) The associated homogeneous equation is $dy/dt = (1 + t^2)y$, and the Linearity Principle implies that

$$y(t) = k e^{\int (1 + t^2)\, dt}$$

$$= k e^{t + t^3/3}.$$

where k can be any real number (see page 113 in Section 1.8).

(c) When we write the differential equation as $dy/dt = (1 + t^2)(y + 1)$, we can immediately see that $y = -1$ corresponds to the equilibrium solution $y(t) = -1$ for all t.

(d) This equilibrium solution is a particular solution of the nonhomogeneous equation. Therefore, using the result of part (b), we get the general solution

$$y(t) = -1 + ke^{t+t^3/3}$$

of the nonhomogeneous equation using the Extended Linearity Principle. Note that this result agrees with the result of part (a).

47. (a) Using Euler's method, we obtain the values $y_0 = 0$, $y_1 = 1.5$, $y_2 = 1.875$, $y_3 = 1.617$, and $y_4 = 1.810$ (rounded to three decimal places).

(b)

(c) The phase line tells us that the solution with initial condition $y(0) = 0$ must be increasing. Moreover, its graph is below and asymptotic to the line $y = \sqrt{3}$ as $t \to \infty$. The oscillations obtained using Euler's method come from numerical error.

49. (a) Note that the slopes are constant along vertical lines—lines along which t is constant, so the right-hand side of the corresponding equation depends only on t. The only choices are equations (i) and (iv). Because the slopes are negative for $t > 1$ and positive for $t < 1$, this slope field corresponds to equation (iv).

(b) This slope field has an equilibrium solution corresponding to the line $y = 1$, as does equations (ii), (v), (vii), and (viii). Equations (ii), (v), and (viii) are autonomous, and this slope field is not constant along horizontal lines. Consequently, it corresponds to equation (vii).

(c) This slope field is constant along horizontal lines, so it corresponds to an autonomous equation. The autonomous equations are (ii), (v), and (viii). This field does not correspond to equation (v) because it has the equilibrium solution $y = -1$. The slopes are negative between $y = -1$ and $y = 1$. Consequently, this field corresponds to equation (viii).

(d) This slope field depends both on y and on t, so it can only correspond to equations (iii), (vi), or (vii). It does not correspond to (vii) because it does not have an equilibrium solution at $y = 1$. Also, the slopes are positive if $y > 0$. Therefore, it must correspond to equation (vi).

51. **(a)**

$y = b$ • sink

(b) As $t \to \infty$, $y(t) \to b$ for every solution $y(t)$.

(c) The equation is separable and linear. Hence, you can find the general solution by separating variables or by either of the methods for solving linear equations (undetermined coefficients or integrating factors).

(d) The associated homogeneous equation is $dy/dt = -(1/a)y$, and its general solution is $ke^{-t/a}$. One particular solution of the nonhomogeneous equation is the equilibrium solution $y(t) = b$ for all t. Therefore, the general solution of the nonhomogeneous equation is

$$y(t) = ke^{-t/a} + b.$$

(e) The authors love all the methods, just in different ways and for different reasons.

(f) Since $a > 0$, $e^{-t/a} \to 0$ as $t \to \infty$. Hence, $y(t) \to b$ as $t \to \infty$ independent of k.

53. **(a)** Let $C(t)$ be the volume of carbon monoxide at time t where t is measured in hours. Initially, the amount of the carbon monoxide is 3% by volume. Since the volume of the room is 1000 cubic feet, there are 30 cubic feet of carbon monoxide in the room at time $t = 0$. Carbon monoxide is being blown into the room at the rate of one cubic foot per hour. The concentration of carbon monoxide is $C/1000$, so carbon monoxide leaves the room at the rate of

$$100 \left(\frac{C}{1000} \right).$$

The initial-value problem that models this situation is

$$\frac{dC}{dt} = 1 - \frac{C}{10}, \qquad C(0) = 30.$$

(b) There is one equilibrium point, $C = 10$, and it is a sink. As t increases, $C(t)$ approaches 10, so the concentration approaches 1% carbon monoxide, the concentration of the air being blown into the room.

$C = 10$ • sink

(c) The differential equation is linear. It is also autonomous and, therefore, separable. We can solve the initial-value problem by separating variables, using integrating factors, or by the Extended Linearity Principle. Since we already know one solution to the equation, that is, the equilibrium solution, we use the Extended Linearity Principle.

The associated homogeneous equation is $dC/dt = -C/10$, and its general solution is $ke^{-0.1t}$. Therefore, the general solution of the nonhomogeneous equation is

$$C(t) = 10 + ke^{-0.1t}.$$

given $C(0) = 30$, $k = 20$.

To find the value of t for which $C(t) = 20$, we solve

$$10 + 20e^{-t/10} = 20$$

We get

$$20e^{-t/10} = 10$$
$$e^{-t/10} = \tfrac{1}{2}$$
$$-t/10 = \ln(\tfrac{1}{2})$$
$$t = 10\ln 2.$$

The air in the room is 2% carbon monoxide in approximately 6.93 hours.

CHAPTER 2

First-Order Systems

EXERCISES FOR SECTION 2.1

1. In the case where it takes many predators to eat one prey, the constant in the negative effect term of predators on the prey is small. Therefore, (ii) corresponds the system of large prey and small predators. On the other hand, one predator eats many prey for the system of large predators and small prey, and, therefore, the coefficient of negative effect term on predator-prey interaction on the prey is large. Hence, (i) corresponds to the system of small prey and large predators.

3. Substitution of $y = 0$ into the equation for dy/dt yields $dy/dt = 0$ for all t. Therefore, $y(t)$ is constant, and since $y(0) = 0$, $y(t) = 0$ for all t.
    Note that to verify this assertion rigorously, we need a uniqueness theorem (see Section 2.5).

5. Substitution of $x = 0$ into the equation for dx/dt yields $dx/dt = 0$ for all t. Therefore, $x(t)$ is constant, and since $x(0) = 0$, $x(t) = 0$ for all t.
    Note that to verify this assertion rigorously, we need a uniqueness theorem (see Section 2.5).

7. The population starts with a relatively large rabbit (R) and a relatively small fox (F) population. The rabbit population grows, then the fox population grows while the rabbit population decreases. Next the fox population decreases until both populations are close to zero. Then the rabbit population grows again and the cycle starts over. Each repeat of the cycle is less dramatic (smaller total oscillation) and both populations oscillate toward an equilibrium which is approximately $(R, F) = (1/2, 3/2)$.

9. By hunting, the number of prey decreases α units per unit of time. Therefore, the rate of change dR/dt of the number of prey has the term $-\alpha$. Only the equation for dR/dt needs modification.
    (i) $dR/dt = 2R - 1.2RF - \alpha$
    (ii) $dR/dt = R(2 - R) - 1.2RF - \alpha$

11. Since the second food source is unlimited, if $R = 0$ and k is the growth parameter for the predator population, F obeys an exponential growth model, $dF/dt = kF$. The only change we have to make is in the rate of F, dF/dt. For both (i) and (ii), $dF/dt = kF + 0.9RF$.

13. If $R - 5F > 0$, the number of predators increases and, if $R - 5F < 0$, the number of predators decreases. Since the condition on prey is same, we modify only the predator part of the system. the modified rate of change of the predator population is

$$\frac{dF}{dt} = -F + 0.9RF + k(R - 5F)$$

where $k > 0$ is the immigration parameter for the predator population.

15. Suppose $y = 1$. If we can find a value of x such that $dy/dt = 0$, then for this x and $y = 1$ the predator population is constant. (This point may not be an equilibrium point because we do not know if $dx/dt = 0$.) The required value of x is $x = 0.05$ in system (i) and $x = 20$ in system (ii). Survival for one unit of predators requires 0.05 units of prey in (i) and 20 units of prey in (ii). Therefore, (i) is a system of inefficient predators and (ii) is a system of efficient predators.

17.   (a) For the initial condition close to zero, the pest population increases much more rapidly than the predator. After a sufficient increase in the predator population, the pest population starts to decrease while the predator population keeps increasing. After a sufficient decrease in the pest population, the predator population starts to decrease. Then, the population comes back to the initial point.

(b) After applying the pest control, you may see the increase of the pest population due to the absence of the predator. So in the short run, this sort of pesticide can cause an explosion in the pest population.

19. (a) Substituting $y(t) = \sin t$ into the left-hand side of the differential equation gives

$$\frac{d^2y}{dt^2} + y = \frac{d^2(\sin t)}{dt^2} + \sin t$$

$$= -\sin t + \sin t$$

$$= 0,$$

so the left-hand side equals the right-hand side for all t.

(b)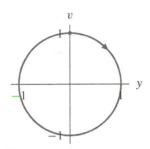

(c) These two solutions trace the same curve in the yv-plane—the unit circle.

(d) The difference in the two solution curves is in how they are parameterized. The solution in this problem is at $(0, 1)$ at time $t = 0$ and hence it lags behind the solution in the section by $\pi/2$. This information cannot be observed solely by looking at the solution curve in the phase plane.

21. Hooke's law tells us that the restoring force exerted by a spring is linearly proportional to the spring's displacement from its rest position. In this case, the displacement is 3 in. while the restoring force is 12 lbs. Therefore, 12 lbs. $= k \cdot 3$ in. or $k = 4$ lbs. per in. $= 48$ lbs. per ft.

23. An extra firm mattress does not deform when you lay on it. This means that it takes a great deal of force to compress the springs so the spring constant must be large.

25. Suppose $\alpha > 0$ is the reaction rate constant for A+B $\rightarrow$ C. The reaction rate is αab at time t, and after the reaction, a and b decrease by αab. We therefore obtain the system

$$\frac{da}{dt} = -\alpha ab$$

$$\frac{db}{dt} = -\alpha ab.$$

27. Suppose k_1 and k_2 are the rates of increase of A and B respectively. Since A and B are added to the solution at constant rates, k_1 and k_2 are added to da/dt and db/dt respectively. The system becomes

$$\frac{da}{dt} = k_1 - \alpha ab$$

$$\frac{db}{dt} = k_2 - \alpha ab.$$

29. Suppose γ is the reaction-rate coefficient for the reaction B + B $\rightarrow$ A. By the reaction, two B's react with each other to create one A. In other words, B decreases at the rate γb^2 and A increases at

the rate $\gamma b^2/2$. The resulting system of the differential equations is

$$\frac{da}{dt} = k_1 - \alpha ab + \frac{\gamma b^2}{2}$$
$$\frac{db}{dt} = k_2 - \alpha ab - \gamma b^2.$$

EXERCISES FOR SECTION 2.2

1. **(a)** $\mathbf{V}(x, y) = (1, 0)$ **(b)** See part (c).

    (c)   **(d)**

    (e) As t increases, solutions move along horizontal lines toward the right.

3. **(a)** $\mathbf{V}(y, v) = (-v, y)$ **(b)** See part (c).

    (c)   **(d)**

    (e) As t increases, solutions move on circles around $(0, 0)$ in the counter-clockwise direction.

5. **(a)** $\mathbf{V}(x, y) = (x, -y)$ **(b)** See part (c).

    (c)   **(d)**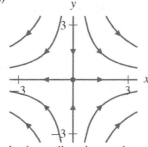

    (e) As t increases, solutions move toward the x-axis in the y-direction and away from the y-axis in the x-direction.

7. (a) Let $v = dy/dt$. Then

(b) See part (c).

$$\frac{dv}{dt} = \frac{d^2y}{dt^2} = y.$$

Thus the associated vector field is
$\mathbf{V}(y, v) = (v, y)$.

(c)

(d)

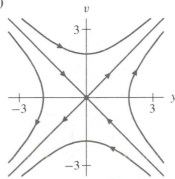

(e) As t increases, solutions in the 2nd and 4th quadrants move toward the origin and away from the line $y = -v$. Solutions in the 1st and 3rd quadrants move away from the origin and toward the line $y = v$.

9. (a)

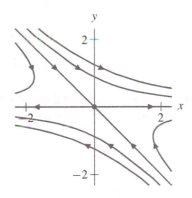

(b) The solution tends to the origin along the line $y = -x$ in the xy-phase plane. Therefore both $x(t)$ and $y(t)$ tend to zero as $t \to \infty$.

11. (a) There are equilibrium points at $(\pm 1, 0)$, so only systems (ii) and (vii) are possible. Since the direction field points toward the x-axis if $y \neq 0$, the equation $dy/dt = y$ does not match this field. Therefore, system (vii) is the system that generated this direction field.

(b) The origin is the only equilibrium point, so the possible systems are (iii), (iv), (v), and (viii). The direction field is not tangent to the y-axis, so it does not match either system (iv) or (v). Vectors point toward the origin on the line $y = x$, so $dy/dt = dx/dt$ if $y = x$. This condition is not satisfied by system (iii). Consequently, this direction field corresponds to system (viii).

(c) The origin is the only equilibrium point, so the possible systems are (iii), (iv), (v), and (viii). Vectors point directly away from the origin on the y-axis, so this direction field does not correspond to systems (iii) and (viii). Along the line $y = x$, the vectors are more vertical than horizontal. Therefore, this direction field corresponds to system (v) rather than system (iv).

(d) The only equilibrium point is $(1, 0)$, so the direction field must correspond to system (vi).

13. **(a)** To find the equilibrium points, we solve the system of equations

$$\begin{cases} 4x - 7y + 2 = 0 \\ 3x + 6y - 1 = 0. \end{cases}$$

These simultaneous equations have one solution, $(x, y) = (-1/9, 2/9)$.

(b)

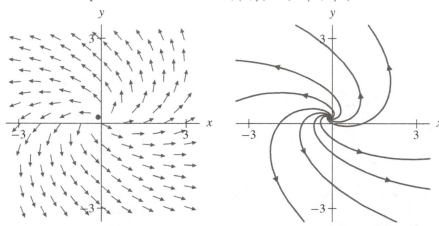

(c) As t increases, typical solutions spiral away from the origin in the counter-clockwise direction.

15. **(a)** To find the equilibrium points, we solve the system of equations

$$\begin{cases} \cos w = 0 \\ -z + w = 0. \end{cases}$$

The first equation implies that $w = \pi/2 + k\pi$ where k is any integer, and the second equation implies that $z = w$. The equilibrium points are $(\pi/2 + k\pi, \pi/2 + k\pi)$ for any integer k.

(b)

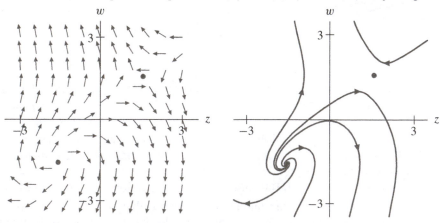

(c) As t increases, typical solutions move away from the line $z = w$, which contains the equilibrium points. The value of w is either increasing or decreasing without bound depending on the initial condition.

17. **(a)** To find the equilibrium points, we solve the system of equations

$$\begin{cases} y = 0 \\ -\cos x - y = 0. \end{cases}$$

We see that $y = 0$, and thus $\cos x = 0$. The equilibrium points are $(\pi/2 + k\pi, 0)$ for any integer k.

(b)

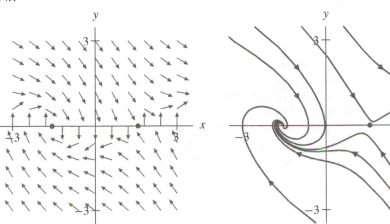

(c) As t increases, typical solutions spiral toward one of the equilibria on the x-axis. Which equilibrium point the solution approaches depends on the initial condition.

19. **(a)** Let $v = dx/dt$. Then

$$\frac{dv}{dt} = \frac{d^2x}{dt^2} = 3x - x^3 - 2v.$$

Thus the associated vector field is $\mathbf{V}(x, v) = (v, 3x - x^3 - 2v)$.

(b) Setting $\mathbf{V}(x, v) = (0, 0)$ and solving for (x, v), we get $v = 0$ and $3x - x^3 = 0$. Hence, the equilibria are $(x, v) = (0, 0)$ and $(x, v) = (\pm\sqrt{3}, 0)$.

(c)

(d)

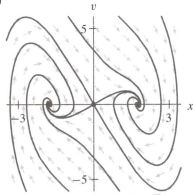

(e) As t increases, almost all solutions spiral to one of the two equilibria $(\pm\sqrt{3}, 0)$. There is a curve of initial conditions that divides these two phenomena. It consists of those initial conditions for which the corresponding solutions tend to the equilibrium point at $(0, 0)$.

21. **(a)** The $x(t)$- and $y(t)$-graphs are periodic, so they correspond to a solution curve that returns to its initial condition in the phase plane. In other words, its solution curve is a closed curve. Since the amplitude of the oscillation of $x(t)$ is relatively large, these graphs must correspond to the outermost closed solution curve.

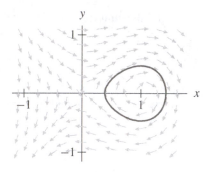

(b) The graphs are not periodic, so they cannot correspond to the two closed solution curves in the phase portrait. Both graphs cross the t-axis. The value of $x(t)$ is initially negative, then becomes positive and reaches a maximum, and finally becomes negative again. Therefore, the corresponding solution curve is the one that starts in the second quadrant, then travels through the first and fourth quadrants, and finally enters the third quadrant.

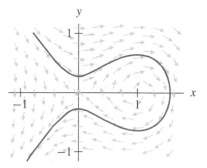

(c) The graphs are not periodic, so they cannot correspond to the two closed solution curves in the phase portrait. Only one graph crosses the t-axis. The other graph remains negative for all time. Note that the two graphs cross.

The corresponding solution curve is the one that starts in the second quadrant and crosses the x-axis and the line $y = x$ as it moves through the third quadrant.

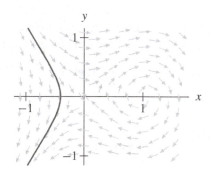

(d) The $x(t)$- and $y(t)$-graphs are periodic, so they correspond to a solution curve that returns to its initial condition in the phase plane. In other words, its solution curve is a closed curve. Since the amplitude of the oscillation of $x(t)$ is relatively small, these graphs must correspond to the intermost closed solution curve.

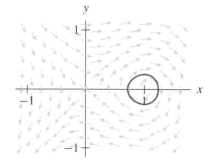

23. Since the solution curve spirals into the origin, the corresponding $x(t)$- and $y(t)$-graphs must oscillate about the t-axis with the decreasing amplitudes.

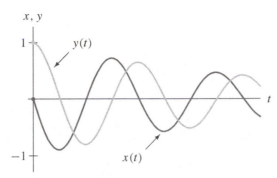

25. The $x(t)$-graph satisfies $-2 < x(0) < -1$ and increases as t increases. The $y(t)$-graph satisfies $1 < y(0) < 2$. Initially it decreases until it reaches its minimum value of $y = 1$ when $x = 0$. Then it increases as t increases.

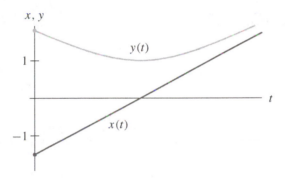

27. From the graphs, we see that $y(0) = 0$ and $x(0)$ is slightly positive. Initially both graphs increase. Then they cross, and slightly later $x(t)$ attains its maximum value. Continuing along we see that $y(t)$ attains its maximum at the same time as $x(t)$ crosses the t-axis.

In the xy-phase plane these graphs correspond to a solution curve that starts on the positive x-axis, enters the first quadrant, crosses the line $y = x$, and eventually crosses the y-axis into the second quadrant exactly when $y(t)$ assumes its maximum value. For this portion of the curve, $y(t)$ is increasing while $x(t)$ assumes a maximum and starts decreasing.

We see that once $y(t)$ attains its maximum, it decreases for a prolonged period of time until it assumes its minimum value. Throughout this interval, $x(t)$ remains negative although it assumes its minimum value twice and a local maximum value once. In the phase plane, the solution curve enters the second quadrant and then crosses into the third quadrant when $y(t) = 0$. The $x(t)$- and $y(t)$-graphs cross precisely when the solution curve crosses the line $y = x$ in the third quadrant.

Finally the $y(t)$-graph is increasing again while the $x(t)$-graph becomes positive and assumes its maximum value once more. The two graphs return to their initial values. In the phase plane this behavior corresponds to the solution curve moving from the third quadrant through the fourth quadrant and back to the original starting point.

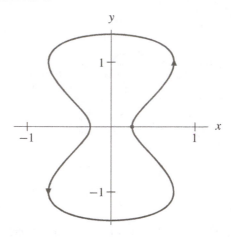

EXERCISES FOR SECTION 2.3

1. **(a)** See part (c).

   (b) We guess that there are solutions of the form $y(t) = e^{st}$ for some choice of the constant s. To determine these values of s, we substitute $y(t) = e^{st}$ into the left-hand side of the differential equation, obtaining

   $$\frac{d^2y}{dt^2} + 7\frac{dy}{dt} + 10y = \frac{d^2(e^{st})}{dt^2} + 7\frac{d(e^{st})}{dt} + 10(e^{st})$$

   $$= s^2e^{st} + 7se^{st} + 10e^{st}$$

   $$= (s^2 + 7s + 10)e^{st}$$

   In order for $y(t) = e^{st}$ to be a solution, this expression must be 0 for all t. In other words,

   $$s^2 + 7s + 10 = 0.$$

   This equation is satisfied only if $s = -2$ or $s = -5$. We obtain two solutions, $y_1(t) = e^{-2t}$ and $y_2(t) = e^{-5t}$, of this equation.

   (c)

   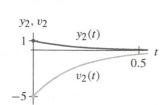

3. **(a)** See part (c).

    (b) We guess that there are solutions of the form $y(t) = e^{st}$ for some choice of the constant s. To determine these values of s, we substitute $y(t) = e^{st}$ into the left-hand side of the differential equation, obtaining

$$\frac{d^2y}{dt^2} + 4\frac{dy}{dt} + y = \frac{d^2(e^{st})}{dt^2} + 4\frac{d(e^{st})}{dt} + e^{st}$$

$$= s^2 e^{st} + 4s e^{st} + e^{st}$$

$$= (s^2 + 4s + 1)e^{st}$$

In order for $y(t) = e^{st}$ to be a solution, this expression must be 0 for all t. In other words,

$$s^2 + 4s + 1 = 0.$$

Applying the quadratic formula, we obtain the roots $s = -2 \pm \sqrt{3}$ and the two solutions, $y_1(t) = e^{(-2-\sqrt{3})t}$ and $y_2(t) = e^{(-2+\sqrt{3})t}$, of this equation.

    (c)

5. **(a)** See part (c).

    (b) We guess that there are solutions of the form $y(t) = e^{st}$ for some choice of the constant s. To determine these values of s, we substitute $y(t) = e^{st}$ into the left-hand side of the differential equation, obtaining

$$\frac{d^2y}{dt^2} + 3\frac{dy}{dt} - 10y = \frac{d^2(e^{st})}{dt^2} + 3\frac{d(e^{st})}{dt} - 10(e^{st})$$

$$= s^2 e^{st} + 3s e^{st} - 10e^{st}$$

$$= (s^2 + 3s - 10)e^{st}$$

In order for $y(t) = e^{st}$ to be a solution, this expression must be 0 for all t. In other words,

$$s^2 + 3s - 10 = 0.$$

This equation is satisfied only if $s = -5$ or $s = 2$. We obtain two solutions, $y_1(t) = e^{-5t}$ and $y_2(t) = e^{2t}$, of this equation.

(c)

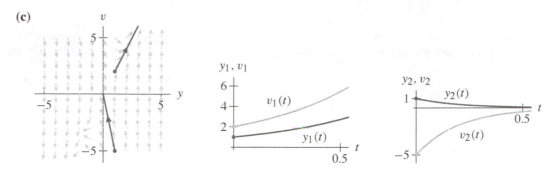

7. **(a)** Let $y_p(t)$ be any solution of the damped harmonic oscillator equation and $y_g(t) = \alpha y_p(t)$ where α is a constant. We substitute $y_g(t)$ into the left-hand side of the damped harmonic oscillator equation, obtaining

$$m\frac{d^2 y}{dt^2} + b\frac{dy}{dt} + ky = m\frac{d^2 y_g}{dt^2} + b\frac{dy_g}{dt} + ky_g$$

$$= m\alpha\frac{d^2 y_p}{dt^2} + b\alpha\frac{dy_p}{dt} + \alpha k y_p$$

$$= \alpha\left(m\frac{d^2 y_p}{dt^2} + b\frac{dy_p}{dt} + ky_p\right)$$

Since $y_p(t)$ is a solution, we know that the expression in the parentheses is zero. Therefore, $y_g(t) = \alpha y_p(t)$ is a solution of the damped harmonic oscillator equation.

(b) Substituting $y(t) = \alpha e^{-t}$ into the left-hand side of the damped harmonic oscillator equation, we obtain

$$\frac{d^2 y}{dt^2} + 3\frac{dy}{dt} + 2y = \frac{d^2(\alpha e^{-t})}{dt^2} + 3\frac{d(\alpha e^{-t})}{dt} + 2(\alpha e^{-t})$$

$$= \alpha e^{-t} - 3\alpha e^{-t} + 2\alpha e^{-t}$$

$$= (\alpha - 3\alpha + 2\alpha)e^{-t}$$

$$= 0.$$

We also get zero if we substitute $y(t) = \alpha e^{-2t}$ into the equation.

(c) If we obtain one nonzero solution to the equation with the guess-and-test method, then we obtain an infinite number of solutions because there are infinitely many constants α.

9. We choose the left wall to be the position $x = 0$ with $x > 0$ indicating positions to the right. Each spring exerts a force on the mass. If the position of the mass is x, then the left spring is stretched by the amount $x - L_1$. Therefore, the force F_1 exerted by this spring is

$$F_1 = k_1\,(L_1 - x).$$

Similarly, the right spring is stretched by the amount $(1 - x) - L_2$. However, the restoring force F_2 of the right spring acts in the direction of increasing values of x. Therefore, we have

$$F_2 = k_2\,((1 - x) - L_2).$$

Using Newton's second law, we have

$$m\frac{d^2x}{dt^2} = k_1\left(L_1 - x\right) + k_2\left((1 - x) - L_2\right) - b\frac{dx}{dt},$$

where the term involving dx/dt represents the force due to damping. After a little algebra, we obtain

$$m\frac{d^2x}{dt^2} + b\frac{dx}{dt} + (k_1 + k_2)x = k_1 L_1 - k_2 L_2 + k_2.$$

EXERCISES FOR SECTION 2.4

1. To check that $dx/dt = 2x + 2y$, we compute both

$$\frac{dx}{dt} = 2e^t$$

and

$$2x + 2y = 4e^t - 2e^t = 2e^t.$$

To check that $dy/dt = x + 3y$, we compute both

$$\frac{dy}{dt} = -e^t,$$

and

$$x + 3y = 2e^t - 3e^t = -e^t.$$

Both equations are satisfied for all t. Hence $(x(t), y(t))$ is a solution.

3. To check that $dx/dt = 2x + 2y$, we compute both

$$\frac{dx}{dt} = 2e^t - 4e^{4t}$$

and

$$2x + 2y = 4e^t - 2e^{4t} - 2e^t + 2e^{4t} = 2e^t.$$

Since the results of these two calculations do not agree, the first equation in the system is not satisfied, and $(x(t), y(t))$ is not a solution.

5. The second equation in the system is $dy/dt = -y$, and from Section 1.1, we know that $y(t)$ must be a function of the form $y_0 e^{-t}$, where y_0 is the initial value.

7. From the second equation, we know that $y(t) = k_1 e^{-t}$ for some constant k_1. Using this observation, the first equation in the system can be rewritten as

$$\frac{dx}{dt} = 2x + k_1 e^{-t}.$$

This equation is a first-order linear equation, and we can derive the general solution using the Extended Linearity Principle from Section 1.8 or integrating factors from Section 1.9.

Using the Extended Linearity Principle, we note that the general solution of the associated homogeneous equation is $x_h(t) = k_2 e^{2t}$.

To find one solution to the nonhomogeneous equation, we guess $x_p(t) = \alpha e^{-t}$. Then

$$\frac{dx_p}{dt} - 2x_p = -\alpha e^{-t} - 2\alpha e^{-t}$$
$$= -3\alpha e^{-t}.$$

Therefore, $x_p(t)$ is a solution if $\alpha = -k_1/3$.

The general solution for $x(t)$ is

$$x(t) = k_2 e^{2t} - \frac{k_1}{3} e^{-t}.$$

9. **(a)** Given the general solution

$$\left(k_2 e^{2t} - \frac{k_1}{3} e^{-t}, k_1 e^{-t} \right),$$

we see that $k_1 = 0$, and therefore $k_2 = 1$. We obtain $\mathbf{Y}(t) = (x(t), y(t)) = (e^{2t}, 0)$.

(b)

(c)

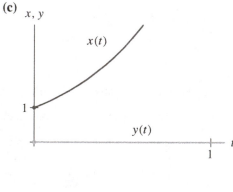

11. **(a)** Given the general solution

$$\left(k_2 e^{2t} - \frac{k_1}{3} e^{-t}, k_1 e^{-t} \right),$$

we see that $k_1 = 1$, and therefore $k_2 = 1/3$. We obtain

$$\mathbf{Y}(t) = (x(t), y(t)) = \left(\frac{1}{3} e^{2t} - \frac{1}{3} e^{-t}, e^{-t} \right).$$

(b)

(c)

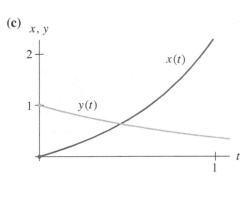

13. **(a)** For this system, we note that the equation for dy/dt is a homogeneous linear equation. Its general solution is

$$y(t) = k_2 e^{-3t}.$$

Substituting $y = k_2 e^{-3t}$ into the equation for dx/dt, we have

$$\frac{dx}{dt} = 2x - 8(k_2 e^{-3t})^2$$

$$= 2x - 8k_2^2 e^{-6t}$$

This equation is a linear and nonhomogeneous. The general solution of the associated homogeneous equation is $x_h(t) = k_1 e^{2t}$. To find one particular solution of the nonhomogeneous equation, we guess

$$x_p(t) = \alpha e^{-6t}.$$

With this guess, we have

$$\frac{dx_p}{dt} - 2x_p = -6\alpha e^{-6t} - 2\alpha e^{-6t}$$

$$= -8\alpha e^{-6t}.$$

Therefore, $x_p(t)$ is a solution if $\alpha = k_2^2$. The general solution for $x(t)$ is $k_1 e^{2t} + k_2^2 e^{-6t}$, and the general solution for the system is

$$(x(t), y(t)) = (k_1 e^{2t} + k_2^2 e^{-6t}, k_2 e^{-3t}).$$

(b) Setting $dy/dt = 0$, we obtain $y = 0$. From $dx/dt = 2x - 8y^2 = 0$, we see that $x = 0$ as well. Therefore, this system has exactly one equilibrium point, $(x, y) = (0, 0)$.

(c) If $(x(0), y(0)) = (0, 1)$, then $k_2 = 1$. We evaluate the expression for $x(t)$ at $t = 0$ and obtain $k_1 + 1 = 0$. Consequently, $k_1 = -1$, and the solution to the initial-value problem is

$$(x(t), y(t)) = (e^{-6t} - e^{2t}, e^{-3t}).$$

(d)

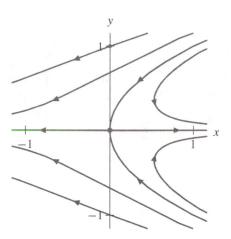

EXERCISES FOR SECTION 2.5

1. **(a)** We compute

$$\frac{dx}{dt} = \frac{d(\cos t)}{dt} = -\sin t = -y \quad \text{and} \quad \frac{dy}{dt} = \frac{d(\sin t)}{dt} = \cos t = x,$$

so $(\cos t, \sin t)$ is a solution.

(b)

Table 2.1

t	Euler's approx.	actual	distance
0	$(1, 0)$	$(1, 0)$	
4	$(-2.06, -1.31)$	$(-0.65, -0.76)$	1.51
6	$(2.87, -2.51)$	$(0.96, -0.28)$	2.94
10	$(-9.21, 1.41)$	$(-0.84, -0.54)$	8.59

(c)

Table 2.2

t	Euler's approx.	actual	distance
0	$(1, 0)$	$(1, 0)$	
4	$(-.81, -.91)$	$(-0.65, -0.76)$	0.22
6	$(1.29, -.40)$	$(0.96, -0.28)$	0.35
10	$(-1.41, -.85)$	$(-0.84, -.54)$	0.65

(d) The solution curves for this system are all circles centered at the origin. Since Euler's method uses tangent lines to approximate the solution curve and the tangent line to any point on a circle is entirely outside the circle (except at the point of tangency), each step of the Euler approximation takes the approximate solution farther from the origin. So the Euler approximations always spiral away from the origin for this system.

3.   **(a)** Euler approximation yields $(x_5, y_5) \approx (0.65, -0.59)$.

(b)

(c)

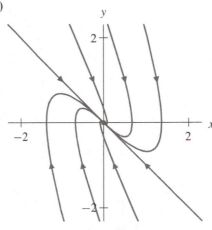

5.   **(a)** Euler approximation yields $(x_5, y_5) \approx (1.94, -0.72)$.

(b)

(c)

7. In order to be able to apply Euler's method to this second-order equation, we reduce the equation to a first-order system using $v = dy/dt$. We obtain

$$\frac{dy}{dt} = v$$

$$\frac{dv}{dt} = -2y - \frac{v}{2}.$$

The choice of Δt has an important effect on the long-term behavior of the approximate solution curve. The approximate solution curve for $\Delta t = 0.25$ seems almost periodic. If $(y_0, v_0) = (2, 0)$, then we obtain $(y_5, v_5) \approx (-0.06, -2.81)$, $(y_{10}, v_{10}) \approx (-1.98, 1.15)$, $(y_{15}, v_{15}) \approx (0.87, 2.34), \ldots$

However, the approximate solution curve for $\Delta t = 0.1$ spirals toward the origin. If $(y_0, v_0) = (2, 0)$, then we obtain $(y_5, v_5) \approx (1.62, -1.73)$, $(y_{10}, v_{10}) \approx (0.57, -2.44)$, $(y_{15}, v_{15}) \approx (-0.60, -1.94), \ldots$

The following figure illustrates the results of Euler's method with $\Delta t = 0.1$.

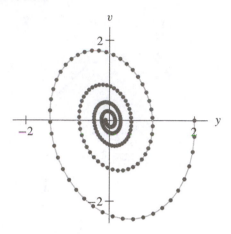

EXERCISES FOR SECTION 2.6

1. **(a)** If $y = 0$, the system is

$$\frac{dx}{dt} = -x$$

$$\frac{dy}{dt} = 0.$$

Therefore, any solution that lies on the x-axis tends toward the origin. Solutions on negative half of the x-axis approach the origin from the left, and solutions on the positive half of the x-axis approach from the right. The third solution curve is the equilibrium point at the origin.

(b)

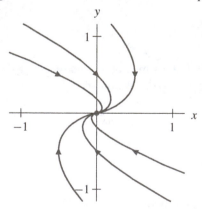

Since $dy/dt = -y$, we know that $y(t) = k_2 e^{-t}$ where k_2 can be any constant. Therefore, all solution curves not on the x-axis approach the x-axis but never touch it. Using the general solution for $y(t)$, the equation for dx/dt becomes $dx/dt = -x + k_2 e^{-t}$. This equation is a nonhomogeneous, linear equation, and there are many ways that we can solve it. The solution is $x(t) = k_1 e^{-t} + k_2 t e^{-t}$. We see that $(x(t), y(t)) \to (0, 0)$ as $t \to \infty$, but $(x(t), y(t))$ never equals $(0, 0)$ unless the initial condition is $(0, 0)$.

3. With $x(t) = e^{-t} \sin(3t)$ and $y(t) = e^{-t} \cos(3t)$, we have

$$
\begin{aligned}
\frac{dx}{dt} &= -e^{-t} \sin(3t) + 3e^{-t} \cos(3t) \\
&= -x + 3y \\
\frac{dy}{dt} &= -3e^{-t} \sin(3t) - e^{-t} \cos(3t) \\
&= -3x - y
\end{aligned}
$$

Therefore, $\mathbf{Y}_1(t)$ is a solution.

5.

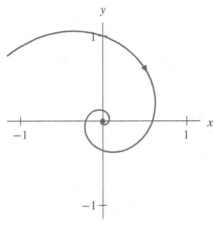

The solution curve swept out by $\mathbf{Y}_2(t)$ is identical to the solution curve swept out by $\mathbf{Y}_1(t)$ because $\mathbf{Y}_2(t)$ has $t - 1$ wherever $\mathbf{Y}_1(t)$ has a t. Whenever $\mathbf{Y}_1(t)$ occupies a point in the phase plane, $\mathbf{Y}_2(t)$ occupies that same point exactly one unit of time later. Since these curves never occupy the same point at the same time, they do not violate the Uniqueness Theorem.

Although the exercise does not ask for a verification that these curves spiral into the origin, we can show that they do spiral by expressing the solution curve for $\mathbf{Y}_1(t)$ in terms of polar coordinates (r, θ). Since $r^2 = x^2 + y^2$, we obtain $r = e^{-t}$, and

$$
\frac{x(t)}{y(t)} = \frac{e^{-t} \sin 3t}{e^{-t} \cos 3t} = \tan 3t.
$$

Also,

$$
\frac{x(t)}{y(t)} = \tan \phi,
$$

where $\phi = \pi/2 - \theta$. Therefore, $\tan 3t = \tan \phi$, and $3t = \pi/2 - \theta$. In other words, the angle θ changes according to the relationship $\theta = \pi/2 - 3t$.

These two computations imply that the solution curves for $\mathbf{Y}_1(t)$ and $\mathbf{Y}_2(t)$ spiral into the origin in a clockwise direction.

7. Assume the vector field satisfies the hypotheses of the Uniqueness Theorem. Since the vector field does not change with time, Gib will follow the same path as Harry, only one time unit behind.

9. From Exercise 8 we know that $\mathbf{Y}_1(t - 1)$ is a solution of the system and $\mathbf{Y}_1(1 - 1) = \mathbf{Y}_1(0) = \mathbf{Y}_2(1)$, so both $\mathbf{Y}_2(t)$ and $\mathbf{Y}_1(t - 1)$ occupy the point $\mathbf{Y}_1(0)$ at time $t = 1$. Hence, by the Uniqueness Theorem, they are the same solution. So $\mathbf{Y}_2(t)$ is a reparameterization by a constant time shift of $\mathbf{Y}_1(t)$.

11. As long as $y(t)$ is defined, we have $y(t) \geq 1$ if $t \geq 0$ because dy/dt is nonnegative. Using this observation, we have

$$\frac{dx}{dt} \geq x^2 + 1$$

for all $t \geq 0$ in the domain of $x(t)$. Since $x(t) = \tan t$ satisfies the initial-value problem $dx/dt = x^2 + 1, x(0) = 0$, we see that the $x(t)$-function for the solution to our system must satisfy

$$x(t) \geq \tan t.$$

Therefore, since $\tan t \to \infty$ as $t \to \pi/2^-, x(t) \to \infty$ as $t \to t_*,$ where $0 \leq t_* \leq \pi/2$.

EXERCISES FOR SECTION 2.7

1. The system of differential equations is

$$\frac{dS}{dt} = -\alpha SI$$

$$\frac{dI}{dt} = \alpha SI - \beta I$$

$$\frac{dR}{dt} = \beta I.$$

Note that

$$\frac{dS}{dt} + \frac{dI}{dt} + \frac{dR}{dt} = -\alpha SI + (\alpha SI - \beta I) + \beta I = 0.$$

Hence, the sum $S(t) + I(t) + R(t)$ is constant for all t. Since the model assumes that the total population is divided into these three groups at $t = 0$, $S(0) + I(0) + R(0) = 1$. Therefore, $S(t) + I(t) + R(t) = 1$ for all t.

3. **(a)** To guarantee that $dI/dt < 0$, we must have $\alpha SI - \beta I < 0$. Factoring, we obtain

$$(\alpha S - \beta)I < 0,$$

and since I is positive, we have $\alpha S - \beta < 0$. In other words,

$$S < \frac{\beta}{\alpha}.$$

Including initial conditions for which $S(0) = \beta/\alpha$ is debatable since $S(0) = \beta/\alpha$ implies that $I(t)$ is decreasing for $t \geq 0$.

(b) If $S(0) < \beta/\alpha$, then $dI/dt < 0$. In that case, any initial influx of infecteds will decrease toward zero, and the epidemic will die out. The fraction vaccinated must be at least $1 - \beta/\alpha$.

5. **(a)**

(b)

(c) As ρ increases, the limit of $S(t)$ as $t \to \infty$ approaches 1. Therefore, as ρ increases, the fraction of the population that contract the disease approaches zero.

7. **(a)** If $I = 0$, both equations are zero, so the S-axis consists entirely of equilibrium points. If $I \neq 0$, then S would have to be zero. However, in that case, the second equation reduces to $dI/dt = -\beta I$, which cannot be zero by assumption. Therefore, all equilibrium points must lie on the S-axis.

(b) We have $dI/dt > 0$ if and only if $\alpha S\sqrt{I} - \beta I > 0$. Factoring out $\sqrt{I}$, we obtain

$$(\alpha S - \beta\sqrt{I})\sqrt{I} > 0.$$

Since $\sqrt{I} \geq 0$, we have

$$\alpha S - \beta\sqrt{I} > 0$$

$$-\beta\sqrt{I} > \alpha S$$

$$\sqrt{I} < -\frac{\alpha}{\beta}S$$

$$I < \left(\frac{\alpha}{\beta}\right)^2 S^2.$$

The resulting region is bounded by the S-axis and the parabola

$$I = \left(\frac{\alpha S}{\beta}\right)^2,$$

and lies in the half-plane $I > 0$.

(c) The model predicts that the entire population will become infected. That is, $R(t) \to 1$ as $t \to \infty$.

9. **(a)** $\beta = 0.44$.

(b) As $t \to \infty$, $S(t) \approx 19$. Therefore, the total number of infected students is 744.

(c) Since β determines how quickly students move from being infected to recovered, a small value of β relative to α indicates that it will take a long time for the infected students to recover.

EXERCISES FOR SECTION 2.8

1. **(a)** Substitution of $(0, 0, 0)$ into the given system of differential equations yields $dx/dt = dy/dt = dz/dt = 0$. Similarly, for the case of $(\pm 6\sqrt{2}, \pm 6\sqrt{2}, 27)$, we obtain

    $$\frac{dx}{dt} = 10(\pm 6\sqrt{2} - (\pm 6\sqrt{2}))$$

    $$\frac{dy}{dt} = 28(\pm 6\sqrt{2}) - (\pm 6\sqrt{2}) - 27(\pm 6\sqrt{2})$$

    $$\frac{dz}{dt} = -\frac{8}{3}(27) - (\pm 6\sqrt{2})^2.$$

    Therefore, $dx/dt = dy/dt = dz/dt = 0$, and these three points are equilibrium points.

    (b) For equilibrium points, we must have $dx/dt = dy/dt = dz/dt = 0$. We therefore obtain the three simultaneous equations

    $$\begin{cases} 10(y - x) = 0 \\ 28x - y - xz = 0 \\ -\frac{8}{3}z + xy = 0. \end{cases}$$

    From the first equation, $x = y$. Eliminating y, we obtain

    $$\begin{cases} x(27 - z) = 0 \\ -\frac{8}{3}z + x^2 = 0 \end{cases}$$

    Then, $x = 0$ or $z = 27$. With $x = 0$, $z = 0$. With $z = 27$, $x^2 = 72$, hence $y = x = \pm 6\sqrt{2}$.

3. **(a)** We have

    $$\frac{dx}{dt} = 10(y - x) = 0 \quad \text{and} \quad \frac{dy}{dt} = 28x - y = 0,$$

    so $x(t) = y(t) = 0$ for all t if $x(0) = y(0) = 0$.

    (b) We have

    $$\frac{dz}{dt} = -\frac{8}{3}z,$$

    so $z(t) = ce^{-8t/3}$. Since $z(0) = 1$, it follows that $c = 1$, and the solution is $x(t) = 0$, $y(t) = 0$, and $z(t) = e^{-8t/3}$.

    (c) If $z(0) = z_0$, it follows that $c = z_0$, so the solution is $x(t) = 0$, $y(t) = 0$, and $z(t) = z_0 e^{-8t/3}$.

5. (a)

(b)

(c)

REVIEW EXERCISES FOR CHAPTER 2

1. The simplest solution is an equilibrium solution, and the origin is an equilibrium point for this system. Hence, the equilibrium solution $(x(t), y(t)) = (0, 0)$ for all t is a solution.

3. Let $v = dy/dt$. Then $dv/dt = d^2y/dt^2$, and we obtain the system

$$\frac{dy}{dt} = v$$
$$\frac{dv}{dt} = 1.$$

5. The equation for dx/dt gives $y = 0$. If $y = 0$, then $\sin(xy) = 0$, so $dy/dt = 0$. Hence, every point on the x-axis is an equilibrium point.

7. First, we check to see if $dx/dt = 2x - 2y^2$ is satisfied. We compute

$$\frac{dx}{dt} = -6e^{-6t} \quad \text{and} \quad 2x - 2y^2 = 2e^{-6t} - 8e^{-6t} = -6e^{-6t}.$$

Second, we check to see if $dy/dt = -3y$. We compute

$$\frac{dy}{dt} = -6e^{-3t} \quad \text{and} \quad -3y = -3(2e^{-3t}) = -6e^{-3t}.$$

Since both equations are satisfied, $(x(t), y(t))$ is a solution.

9. From the equation for dx/dt, we know that $x(t) = k_1 e^{2t}$, where k_1 is an arbitrary constant, and from the equation for dy/dt, we have $y(t) = k_2 e^{-3t}$, where k_2 is another arbitrary constant. The general solution is $(x(t), y(t)) = (k_1 e^{2t}, k_2 e^{-3t})$.

11. There are many examples. One is

$$\frac{dx}{dt} = (x^2 - 1)(x^2 - 4)(x^2 - 9)(x^2 - 16)(x^2 - 25)$$
$$\frac{dy}{dt} = y.$$

This system has equilibria at $(\pm 1, 0)$, $(\pm 2, 0)$, $(\pm 3, 0)$, $(\pm 4, 0)$, and $(\pm 5, 0)$.

13. The point $(1, 1)$ is on the line $y = x$. Along this line, the vector field for the system points toward the origin. Therefore, the solution curve consists of the half-line $y = x$ in the first quadrant. Note that the point $(0, 0)$ is not on this curve.

15. True. First, we check the equation for dx/dt. We have

$$\frac{dx}{dt} = \frac{d(e^{-6t})}{dt} = -6e^{-6t},$$

and

$$2x - 2y^2 = 2(e^{-6t}) - 2(2e^{-3t})^2 = 2e^{-6t} - 8e^{-6t} = -6e^{-6t}.$$

Since that equation holds, we check the equation for dy/dt. We have

$$\frac{dy}{dt} = \frac{d(2e^{-3t})}{dt} = -6e^{-3t},$$

and

$$-3y = -3(2e^{-3t}) = -6e^{-3t}.$$

Since the equations for both dx/dt and dy/dt hold, the function $(x(t), y(t)) = (e^{-6t}, 2e^{-3t})$ is a solution of this system.

17. False. The components of the vector field are the right-hand sides of the equations of the system.

19. False. Note that $(x(0), y(0)) = (x(\pi), y(\pi)) = (0, 0)$. However, $(dx/dt, dy/dt) = (1, 1)$ at $t = 0$, and $(dx/dt, dy/dt) = (-1, -1)$ at $t = \pi$. For an autonomous system, the vector in the vector field at any given point does not vary as t varies. This function cannot be a solution of any autonomous system. (This function parameterizes a line segment in the xy-plane from $(1, 1)$ to $(-1, -1)$. In fact, it sweeps out the segment twice for $0 \leq t \leq 2\pi$.)

21. True. Note that $\cos(t + \pi/2) = -\sin t$ and $\sin(t + \pi/2) = \cos t$. Consequently,

$$(-\sin t, \cos t) = (\cos(t + \pi/2), \sin(t + \pi/2)),$$

which is a time-translate of the solution $(\cos t, \sin t)$. Since the system is autonomous, a time-translate of a solution is another solution.

23. False. The point $(0, 0)$ is an equilibrium point, so the Uniqueness Theorem guarantees that it is not on the solution curve corresponding to $(1, 0)$.

25. False. These solutions are different because they have different values at $t = 0$. However, they do trace out the same curve in the phase plane.

27. False. The function $y(t)$ decreases monotonically, but $x(t)$ increases until it reaches its maximum at $x = -1$. It decreases monotonically after that.

29. **(a)** The equilibrium points satisfy the equations $x = 2y$ and $\cos 2y = 0$. From the second equation, we conclude that

$$2y = \frac{\pi}{2} + k\pi,$$

where $k = 0, \pm 1, \pm 2, \dots$. Since $2y = x$, we see that the equilibria are

$$(x, y) = \dots, (-3\pi/2, -3\pi/4), (-\pi/2, -\pi/4), (\pi/2, \pi/4), (3\pi/2, 3\pi/4), (5\pi/2, 5\pi/4), \dots$$

(b)

(c) Most solutions become unbounded in y as t increases. However, there appears to be a "curve" of solutions that tend toward the equilibria $\dots, (-\pi/2, -\pi/4), (3\pi/2, 3\pi/4), \dots$ as t increases.

31.

33.

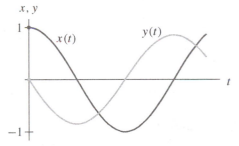

35. **(a)** First, we note that dy/dt depends only on y. In fact, the general solution of $dy/dt = 3y$ is $y(t) = k_2 e^{3t}$, where k_2 can be any constant.

Substituting this expression for y into the equation for dx/dt, we obtain

$$\frac{dx}{dt} = x + 2k_2 e^{3t} + 1.$$

The general solution of the associated homogeneous equation is $x_h(t) = k_1 e^t$. To find a particular solution of the nonhomogeneous equation, we guess $x_p(t) = ae^{3t} + b$. Substituting this guess into the equation gives

$$3ae^{3t} = ae^{3t} + b + 2k_2 e^{3t} + 1,$$

so if $x_p(t)$ is a solution, we must have $3a = a + 2k_2$ and $b + 1 = 0$. Hence, $a = k_2$ and $b = -1$, and the function $x_p(t) = k_2 e^{3t} - 1$ is a solution of the nonhomogeneous equation.
Therefore, the general solution of the system is

$$x(t) = k_1 e^t + k_2 e^{3t} - 1$$
$$y(t) = k_2 e^{3t}.$$

(b) To find the equilibrium points, we solve the system of equations

$$\begin{cases} x + 2y + 1 = 0 \\ 3y = 0, \end{cases}$$

so $(x, y) = (-1, 0)$ is the only equilibrium point.

(c) To find the solution with initial condition $(-1, 3)$, we set

$$-1 = x(0) = k_1 + k_2 - 1$$
$$3 = y(0) = k_2,$$

so $k_2 = 3$ and $k_1 = -3$. The solution with the desired initial condition is

$$(x(t), y(t)) = (-3e^t + 3e^{3t} - 1, 3e^{3t}).$$

(d)

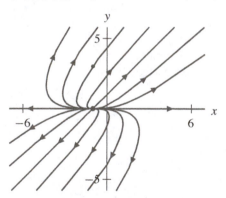

37. **(a)** Since θ represents an angle in this model, we restrict θ to the interval $-\pi < \theta < \pi$.
The equilibria must satisfy the equations $\cos \theta = s^2$ and $\sin \theta = -Ds^2$. Therefore,

$$\tan \theta = \frac{\sin \theta}{\cos \theta} = \frac{-Ds^2}{s^2} = -D,$$

and consequently, $\theta = -\arctan D$.
To find s, we note that $s^2 = \cos(-\arctan D)$. From trigonometry, we know that

$$\cos(-\arctan D) = \frac{1}{\sqrt{1 + D^2}}.$$

If $-\pi < \theta < \pi$, there is a single equilibrium point for each value of the parameter D. It is

$$(\theta, s) = \left(-\arctan D, \frac{1}{\sqrt[4]{1 + D^2}} \right).$$

(b) The equilibrium point represents motion along a line at a given angle from the horizon with a constant speed.

CHAPTER 3

Linear Systems

EXERCISES FOR SECTION 3.1

1. Since $a > 0$, Paul's making a profit ($x > 0$) has a beneficial effect on Paul's profits in the future because the ax term makes a positive contribution to dx/dt. However, since $b < 0$, Bob's making a profit ($y > 0$) hinders Paul's ability to make profit because the by term contributes negatively to dx/dt. Roughly speaking, business is good for Paul if his store is profitable and Bob's is not. In fact, since $dx/dt = x - y$, Paul's profits will increase whenever his store is more profitable than Bob's.

Even though $dx/dt = dy/dt = x - y$ for this choice of parameters, the interpretation of the equation is exactly the opposite from Bob's point of view. Since $d < 0$, Bob's future profits are hurt whenever he is profitable because $dy < 0$. But Bob's profits are helped whenever Paul is profitable since $cx > 0$. Once again, since $dy/dt = x - y$, Bob's profits will increase whenever Paul's store is more profitable than his.

Finally, note that both x and y change by identical amounts since dx/dt and dy/dt are always equal.

3. Since $a = 1$ and $b = 0$, we have $dx/dt = x$. Hence, if Paul is making a profit ($x > 0$), then those profits will increase since dx/dt is positive. However, Bob's profits have no effect on Paul's profits. (Note that $dx/dt = x$ is the standard exponential growth model.)

Since $c = 2$ and $d = 1$, profits from both stores have a positive effect on Bob's profits. In some sense, Paul's profits have twice the impact of Bob's profits on dy/dt.

5. $\mathbf{Y} = \begin{pmatrix} x \\ y \end{pmatrix}, \quad \dfrac{d\mathbf{Y}}{dt} = \begin{pmatrix} 2 & 1 \\ 1 & 1 \end{pmatrix} \mathbf{Y}$

7. $\mathbf{Y} = \begin{pmatrix} p \\ q \\ r \end{pmatrix}, \quad \dfrac{d\mathbf{Y}}{dt} = \begin{pmatrix} 3 & -2 & -7 \\ -2 & 0 & 6 \\ 0 & 7.3 & 2 \end{pmatrix} \mathbf{Y}$

9. $\dfrac{dx}{dt} = \beta y$

$\dfrac{dy}{dt} = \gamma x - y$

11.   (a)   **(b)**   **(c)**

13. **(a)**   **(b)**   **(c)**

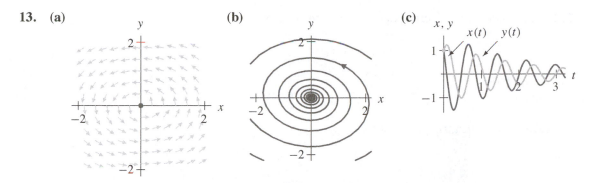

15. The vector field at a point (x_0, y_0) is (ax_0+by_0, cx_0+dy_0), so in order for a point to be an equilibrium point, it must be a solution to the system of simultaneous linear equations

$$\begin{cases} ax_0 + by_0 = 0 \\ cx_0 + dy_0 = 0. \end{cases}$$

If $a \neq 0$, we know that the first equation is satisfied if and only if

$$x_0 = -\frac{b}{a} y_0.$$

Now we see that any point that lies on this line $x_0 = (-b/a)y_0$ also satisfies the second linear equation $cx_0 + dy_0 = 0$. In fact, if we substitute a point of this form into the second component of the vector field, we have

$$cx_0 + dy_0 = c\left(-\frac{b}{a}\right)y_0 + dy_0$$

$$= \left(-\frac{bc}{a} + d\right)y_0$$

$$= \left(\frac{ad - bc}{a}\right)y_0$$

$$= \frac{\det \mathbf{A}}{a}y_0$$

$$= 0,$$

since we are assuming that $\det \mathbf{A} = 0$. Hence, the line $x_0 = (-b/a)y_0$ consists entirely of equilibrium points.

If $a = 0$ and $b \neq 0$, then the determinant condition $\det \mathbf{A} = ad - bc = 0$ implies that $c = 0$. Consequently, the vector field at the point (x_0, y_0) is (by_0, dy_0). Since $b \neq 0$, we see that we get equilibrium points if and only if $y_0 = 0$. In other words, the set of equilibrium points is exactly the x-axis.

Finally, if $a = b = 0$, then the vector field at the point (x_0, y_0) is $(0, cx_0 + dy_0)$. In this case, we see that a point (x_0, y_0) is an equilibrium point if and only if $cx_0 + dy_0 = 0$. Since at least one of c or d is nonzero, the set of points (x_0, y_0) that satisfy $cx_0 + dy_0 = 0$ is precisely a line through the origin.

17. The first-order system corresponding to this equation is

$$\frac{dy}{dt} = v$$

$$\frac{dv}{dt} = -qy - pv.$$

(a) If $q = 0$, then the system becomes

$$\frac{dy}{dt} = v$$

$$\frac{dv}{dt} = -pv,$$

and the equilibrium points are the solutions of the system of equations

$$\begin{cases} v = 0 \\ -pv = 0. \end{cases}$$

Thus, the point (y, v) is an equilibrium point if and only if $v = 0$. In other words, the set of all equilibria agrees with the horizontal axis in the yv-plane.

(b) If $p = q = 0$, then the system becomes

$$\frac{dy}{dt} = v$$

$$\frac{dv}{dt} = 0$$

but the equilibrium points are again the points with $v = 0$.

19. Letting $v = dy/dt$ and $w = d^2y/dt^2$ we can write this equation as the system

$$\frac{dy}{dt} = v$$

$$\frac{dv}{dt} = \frac{d^2y}{dt^2} = w$$

$$\frac{dw}{dt} = \frac{d^3y}{dt^3} = -ry - qv - pw.$$

In matrix notation, this system is

$$\frac{d\mathbf{Y}}{dt} = \mathbf{AY}$$

where

$$\mathbf{Y} = \begin{pmatrix} y \\ v \\ w \end{pmatrix} \quad \text{and} \quad \mathbf{A} = \begin{pmatrix} 0 & 1 & 0 \\ 0 & 0 & 1 \\ -r & -q & -p \end{pmatrix}.$$

21. If there are fewer than the usual number of buyers, then $b < 0$. If this level of b has a negative effect on the number of sellers, we expect the γb-term in ds/dt to be negative. If $\gamma b < 0$ and $b < 0$, then we must have $\gamma > 0$.

23. In the model, $ds/dt = \gamma b + \delta s$. If $s > 0$, then the number of sellers is greater than usual and house prices should decline. Since declining prices should have a negative effect on the number of sellers, we expect the δs-term to be negative. If $\delta s < 0$ when $s > 0$, we should have $\delta < 0$.

25. **(a)** Note that substituting $\mathbf{Y}(t)$ into the left-hand side of the differential equation, we get

$$\frac{d\mathbf{Y}}{dt} = \begin{pmatrix} e^{2t} + 2te^{2t} \\ -e^{2t} - 2(t+1)e^{2t} \end{pmatrix}$$

$$= \begin{pmatrix} e^{2t} + 2te^{2t} \\ -3e^{2t} - 2te^{2t} \end{pmatrix}.$$

Substituting $\mathbf{Y}(t)$ into the right-hand side, we get

$$\begin{pmatrix} 1 & -1 \\ 1 & 3 \end{pmatrix} \begin{pmatrix} te^{2t} \\ -(t+1)e^{2t} \end{pmatrix} = \begin{pmatrix} te^{2t} + (t+1)e^{2t} \\ te^{2t} - 3(t+1)e^{2t} \end{pmatrix}$$

$$= \begin{pmatrix} e^{2t} + 2te^{2t} \\ -3e^{2t} - 2te^{2t} \end{pmatrix}.$$

Since the left-hand side of the differential equation equals the right-hand side, the function $\mathbf{Y}(t)$ is a solution.

(b) At $t = 0$, $\mathbf{Y}(0) = (0, -1)$. By the Linearity Principle, any constant multiple of the solution $\mathbf{Y}(t)$ is also a solution. Since the function $-2\mathbf{Y}(t)$ has the desired initial condition, we know that

$$-2\mathbf{Y}(t) = \begin{pmatrix} -2te^{2t} \\ (2t+2)e^{2t} \end{pmatrix}$$

is the desired solution. By the Uniqueness Theorem, this is the only solution with this initial condition. (Given the formula for $-2\mathbf{Y}(t)$ directly above, note that we can directly check our assertion that this function solves the initial-value problem without appealing to the Linearity Principle.)

27. **(a)** Substitute $\mathbf{Y}_1(t)$ into the differential equation and compare the left-hand side to the right-hand side. On the left-hand side, we have

$$\frac{d\mathbf{Y}_1}{dt} = \begin{pmatrix} -3e^{-3t} + 8e^{-4t} \\ -3e^{-3t} + 16e^{-4t} \end{pmatrix},$$

and on the right-hand side, we have

$$\mathbf{AY}_1(t) = \begin{pmatrix} -2 & -1 \\ 2 & -5 \end{pmatrix} \begin{pmatrix} e^{-3t} - 2e^{-4t} \\ e^{-3t} - 4e^{-4t} \end{pmatrix} = \begin{pmatrix} -3e^{-3t} + 8e^{-4t} \\ -3e^{-3t} + 16e^{-4t} \end{pmatrix}.$$

Since the two sides agree, we know that $\mathbf{Y}_1(t)$ is a solution.

For $\mathbf{Y}_2(t)$,

$$\frac{d\mathbf{Y}_2}{dt} = \begin{pmatrix} -6e^{-3t} - 4e^{-4t} \\ -6e^{-3t} - 8e^{-4t} \end{pmatrix},$$

and

$$\mathbf{AY}_2(t) = \begin{pmatrix} -2 & -1 \\ 2 & -5 \end{pmatrix} \begin{pmatrix} 2e^{-3t} + e - 4t \\ 2e^{-3t} + 2e^{-4t} \end{pmatrix} = \begin{pmatrix} -6e^{-3t} - 4e^{-4t} \\ -6e^{-3t} - 8e^{-4t} \end{pmatrix}.$$

Since the two sides agree, the function $\mathbf{Y}_2(t)$ is also a solution.

Both $\mathbf{Y}_1(t)$ and $\mathbf{Y}_2(t)$ are solutions, and we proceed to the next part of the exercise.

(b) Note that $\mathbf{Y}_1(0) = (-1, -3)$ and $\mathbf{Y}_2(0) = (3, 4)$. These vectors are not on the same line through the origin, so the initial conditions are linearly independent. If the initial conditions are linearly independent, then the solutions must also be linearly independent. Since the two solutions are linearly independent, we proceed to part (c) of the exercise.

(c) We must find constants k_1 and k_2 such that

$$k_1 \mathbf{Y}_1(0) + k_2 \mathbf{Y}_2(0) = k_1 \begin{pmatrix} -1 \\ -3 \end{pmatrix} + k_2 \begin{pmatrix} 3 \\ 4 \end{pmatrix} = \begin{pmatrix} 2 \\ 3 \end{pmatrix}.$$

In other words, the constants k_1 and k_2 must satisfy the simultaneous system of linear equations

$$\begin{cases} -k_1 + 3k_2 = 2 \\ -3k_1 + 4k_2 = 3. \end{cases}$$

It follows that $k_1 = -1/5$ and $k_2 = 3/5$. Hence, the required solution is

$$-\frac{1}{5}\mathbf{Y}_1(t) + \frac{3}{5}\mathbf{Y}_2(t) = \begin{pmatrix} e^{-3t} + e^{-4t} \\ e^{-3t} + 2e^{-4t} \end{pmatrix}.$$

29. **(a)** First, we check to see if $\mathbf{Y}_1(t)$ is a solution. The left-hand side of the differential equation is

$$\frac{d\mathbf{Y}_1}{dt} = \begin{pmatrix} e^{-t} + 36e^{3t} \\ -e^{-t} + 12e^{3t} \end{pmatrix},$$

and the right-hand side is

$$\mathbf{AY}_1(t) = \begin{pmatrix} 2 & 3 \\ 1 & 0 \end{pmatrix} \begin{pmatrix} -e^{-t} + 12e^{3t} \\ e^{-t} + 4e^{3t} \end{pmatrix} = \begin{pmatrix} e^{-t}36e^{3t} \\ -e^{-t} + 12e^{3t} \end{pmatrix}.$$

Consequently, $\mathbf{Y}_1(t)$ is a solution. However,

$$\frac{d\mathbf{Y}_2}{dt} = \begin{pmatrix} e^{-t} \\ -2e^{-t} \end{pmatrix},$$

and

$$\mathbf{AY}_2(t) = \begin{pmatrix} 2 & 3 \\ 1 & 0 \end{pmatrix} \begin{pmatrix} -e^{-t} \\ 2e^{-t} \end{pmatrix} = \begin{pmatrix} 4e^{-t} \\ -e^{-t} \end{pmatrix}.$$

Consequently, the function $\mathbf{Y}_2(t)$ is not a solution. In this case, we are not able to solve the given initial-value problem, so we stop here.

31. (a) If $(x_1, y_1) = (0, 0)$, then (x_1, y_1) and (x_2, y_2) are on the same line through the origin because (x_1, y_1) is the origin. So (x_1, y_1) and (x_2, y_2) are linearly dependent.

(b) If $(x_1, y_1) = \lambda(x_2, y_2)$ for some λ, then (x_1, y_1) and (x_2, y_2) are on the same line through the origin. To see why, suppose that $x_2 \neq 0$ and $\lambda \neq 0$. (The $\lambda = 0$ case was handled in part (a) above.) In this case, $x_1 \neq 0$ as well. Then the slope of the line through the origin and (x_1, y_1) is y_1/x_1, and the slope of the line through the origin and (x_2, y_2) is y_2/x_2. However, because $(x_1, y_1) = \lambda(x_2, y_2)$, we have

$$\frac{y_1}{x_1} = \frac{\lambda y_2}{\lambda x_2} = \frac{y_2}{x_2}.$$

Since these two lines have the same slope and both contain the origin, they are the same line. (The special case where $x_2 = 0$ reduces to considering vertical lines through the origin.)

(c) If $x_1 y_2 - x_2 y_1 = 0$, then $x_1 y_2 = x_2 y_1$. Once again, this condition implies that (x_1, y_1) and (x_2, y_2) are on the same line through the origin. For example, suppose that $x_1 \neq 0$, then

$$y_2 = \frac{x_2 y_1}{x_1} = \frac{x_2}{x_1} y_1.$$

But we already know that

$$x_2 = \frac{x_2}{x_1} x_1,$$

so we have

$$(x_2, y_2) = \frac{x_2}{x_1}(x_1, y_1).$$

By part (b) above (where $\lambda = x_2/x_1$), the two vectors are linearly dependent.

If $x_1 = 0$ but $y_1 \neq 0$, it follows that $x_2 y_1 = 0$, and thus $x_2 = 0$. Thus, both (x_1, y_1) and (x_2, y_2) are on the vertical line through the origin.

Finally, if $x_1 = 0$ and $y_1 = 0$, we can use part (a) to show that the two vectors are linearly dependent.

33. The initial position of $\mathbf{Y}_1(t)$ is $\mathbf{Y}_1(0) = (-1, 1)$. By the Linearity Principle, we know that $k\mathbf{Y}_1(t)$ is also a solution of the system for any constant k. Hence, for any initial condition of the form $(-k, k)$, the solution is $k\mathbf{Y}_1(t)$.

(a) The curve $2\mathbf{Y}_1(t) = (-2e^{-t}, 2e^{-t})$ is the solution with this initial condition.

(b) We cannot find the solution for this initial condition using only $\mathbf{Y}_1(t)$.

(c) The constant function $0\mathbf{Y}_1(t) = (0, 0)$ (represented by the equilibrium point at the origin) is the solution with this initial condition.

(d) The curve $-3\mathbf{Y}_1(t) = (3e^{-t}, -3e^{-t})$ is the solution with this initial condition.

35. (a) Using the Product Rule we compute

$$\frac{dW}{dt} = \frac{dx_1}{dt}y_2 + x_1\frac{dy_2}{dt} - \frac{dx_2}{dt}y_1 - x_2\frac{dy_1}{dt}.$$

(b) Since $(x_1(t), y_1(t))$ and $(x_2(t), y_2(t))$ are solutions, we know that

$$\frac{dx_1}{dt} = ax_1 + by_1$$

$$\frac{dy_1}{dt} = cx_1 + dy_1$$

and that

$$\frac{dx_2}{dt} = ax_2 + by_2$$

$$\frac{dy_2}{dt} = cx_2 + dy_2.$$

Substituting these equations into the expression for dW/dt, we obtain

$$\frac{dW}{dt} = (ax_1 + by_1)y_2 + x_1(cx_2 + dy_2) - (ax_2 + by_2)y_1 - x_2(cx_1 + dy_1).$$

After we collect terms, we have

$$\frac{dW}{dt} = (a+d)W.$$

(c) This equation is a homogeneous, linear, first-order equation (as such it is also separable—see Sections 1.1, 1.2, and 1.8). Therefore, we know that the general solution is

$$W(t) = Ce^{(a+d)t}$$

where C is any constant (but note that $C = W(0)$).

(d) From Exercises 31 and 32, we know that $\mathbf{Y}_1(t)$ and $\mathbf{Y}_2(t)$ are linearly independent if and only if $W(t) \neq 0$. But, $W(t) = Ce^{(a+d)t}$, so $W(t) = 0$ if and only if $C = W(0) = 0$. Hence, $W(t) = 0$ is zero for some t if and only if $C = W(0) = 0$.

EXERCISES FOR SECTION 3.2

1. **(a)** The characteristic polynomial is

$$(3 - \lambda)(-2 - \lambda) = 0,$$

and therefore the eigenvalues are $\lambda_1 = -2$ and $\lambda_2 = 3$.

(b) To obtain the eigenvectors (x_1, y_1) for the eigenvalue $\lambda_1 = -2$, we solve the system of equations

$$\begin{cases} 3x_1 + 2y_1 = -2x_1 \\ -2y_1 = -2y_1 \end{cases}$$

and obtain $5x_1 = -2y_1$.

Using the same procedure, we see that the eigenvectors (x_2, y_2) for $\lambda_2 = 3$ must satisfy the equation $y_2 = 0$.

(c)

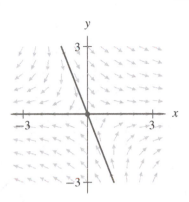

(d) One eigenvector $\mathbf{V}_1$ for λ_1 is $\mathbf{V}_1 = (-2, 5)$, and one eigenvector $\mathbf{V}_2$ for λ_2 is $\mathbf{V}_2 = (1, 0)$.

Given the eigenvalues and these eigenvectors, we have the two linearly independent solutions

$$\mathbf{Y}_1(t) = e^{-2t} \begin{pmatrix} -2 \\ 5 \end{pmatrix} \quad \text{and} \quad \mathbf{Y}_2(t) = e^{3t} \begin{pmatrix} 1 \\ 0 \end{pmatrix}.$$

The $x(t)$- and $y(t)$-graphs for $\mathbf{Y}_1(t)$.   The $x(t)$- and $y(t)$-graphs for $\mathbf{Y}_2(t)$.

(e) The general solution to this linear system is

$$\mathbf{Y}(t) = k_1 e^{-2t} \begin{pmatrix} -2 \\ 5 \end{pmatrix} + k_2 e^{3t} \begin{pmatrix} 1 \\ 0 \end{pmatrix}.$$

3. **(a)** The eigenvalues are the roots of the characteristic polynomial, so they are the solutions of

$$(-5 - \lambda)(-4 - \lambda) - 2 = \lambda^2 + 9\lambda + 18 = 0.$$

Therefore, the eigenvalues are $\lambda_1 = -3$ and $\lambda_2 = -6$.

(b) To obtain the eigenvectors (x_1, y_1) for the eigenvalue $\lambda_1 = -3$, we solve the system of equations

$$\begin{cases} -5x_1 - 2y_1 = -3x_1 \\ -x_1 - 4y_1 = -3y_1 \end{cases}$$

and obtain $y_1 = -x_1$.

Using the same procedure, we obtain the eigenvalues (x_2, y_2) where $x_2 = 2y_2$ for $\lambda_2 = -6$.

(c)

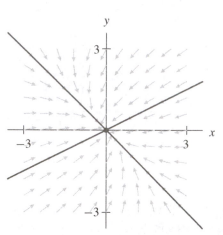

(d) One eigenvector $\mathbf{V}_1$ for $\lambda_1 = -3$ is $\mathbf{V}_1 = (1, -1)$, and one eigenvector $\mathbf{V}_2$ for $\lambda_2 = -6$ is $\mathbf{V}_2 = (2, 1)$.

Given the eigenvalues and these eigenvectors, we have two linearly independent solutions

$$\mathbf{Y}_1(t) = e^{-3t} \begin{pmatrix} 1 \\ -1 \end{pmatrix} \quad \text{and} \quad \mathbf{Y}_2(t) = e^{-6t} \begin{pmatrix} 2 \\ 1 \end{pmatrix}.$$

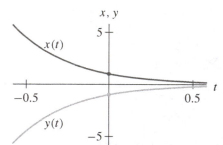

The $x(t)$- and $y(t)$-graphs for $\mathbf{Y}_1(t)$.   the $x(t)$- and $y(t)$-graphs for $\mathbf{Y}_2(t)$.

(e) The general solution to this linear system is

$$\mathbf{Y}(t) = k_1 e^{-3t} \begin{pmatrix} 1 \\ -1 \end{pmatrix} + k_2 e^{-6t} \begin{pmatrix} 2 \\ 1 \end{pmatrix}.$$

5. **(a)** The characteristic polynomial is

$$\left(-\tfrac{1}{2} - \lambda \right)^2 = 0,$$

and therefore there is only one eigenvalue, $\lambda = -1/2$.

(b) To obtain the eigenvectors (x_1, y_1) for the eigenvalue $\lambda = -1/2$, we solve the system of equations

$$\begin{cases} -\tfrac{1}{2}x_1 = -\tfrac{1}{2}x_1 \\ x_1 - \tfrac{1}{2}y_1 = -\tfrac{1}{2}y_1 \end{cases}$$

and obtain $x_1 = 0$.

(c)

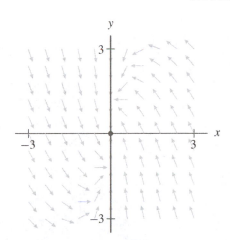

(d) Given the eigenvalue $\lambda = -1/2$ and the eigenvector $\mathbf{V} = (0, 1)$, we have the solution

$$\mathbf{Y}(t) = e^{-t/2} \begin{pmatrix} 0 \\ 1 \end{pmatrix}.$$

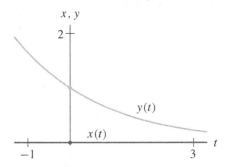

The $x(t)$- and $y(t)$-graphs for $\mathbf{Y}(t)$.

(e) Since the method of eigenvalues and eigenvectors does not give us a second solution that is linearly independent from $\mathbf{Y}(t)$, we cannot form the general solution.

7. **(a)** The characteristic polynomial is

$$(3 - \lambda)(-\lambda) - 4 = \lambda^2 - 3\lambda - 4 = (\lambda - 4)(\lambda + 1) = 0,$$

and therefore the eigenvalues are $\lambda_1 = -1$ and $\lambda_2 = 4$.

(b) To obtain the eigenvectors (x_1, y_1) for the eigenvalue $\lambda_1 = -1$, we solve the system of equations

$$\begin{cases} 3x_1 + 4y_1 = -x_1 \\ \\ x_1 = -y_1 \end{cases}$$

and obtain $y_1 = -x_1$.

Using the same procedure, we obtain the eigenvectors (x_2, y_2) where $x_2 = 4y_2$ for $\lambda_2 = 4$.

(c)

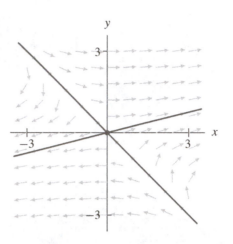

(d) One eigenvector $\mathbf{V}_1$ for λ_1 is $\mathbf{V}_1 = (1, -1)$, and one eigenvector $\mathbf{V}_2$ for λ_2 is $\mathbf{V}_2 = (4, 1)$. Given the eigenvalues and these eigenvectors, we have two linearly independent solutions

$$\mathbf{Y}_1(t) = e^{-t} \begin{pmatrix} 1 \\ -1 \end{pmatrix} \quad \text{and} \quad \mathbf{Y}_2(t) = e^{4t} \begin{pmatrix} 4 \\ 1 \end{pmatrix}.$$

 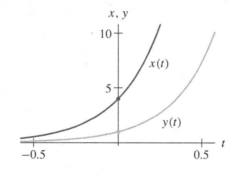

The $x(t)$- and $y(t)$-graphs for $\mathbf{Y}_1(t)$. The $x(t)$- and $y(t)$-graphs for $\mathbf{Y}_2(t)$.

(e) The general solution to this linear system is

$$\mathbf{Y}(t) = k_1 e^{-t} \begin{pmatrix} 1 \\ -1 \end{pmatrix} + k_2 e^{4t} \begin{pmatrix} 4 \\ 1 \end{pmatrix}$$

9. **(a)** The characteristic polynomial is

$$(2 - \lambda)(1 - \lambda) - 1 = \lambda^2 - 3\lambda + 1 = 0,$$

and therefore the eigenvalues are

$$\lambda_1 = \frac{3 + \sqrt{5}}{2} \quad \text{and} \quad \lambda_2 = \frac{3 - \sqrt{5}}{2}.$$

(b) To obtain the eigenvectors (x_1, y_1) for the eigenvalue $\lambda_1 = (3 + \sqrt{5})/2$, we solve the system of equations

$$\begin{cases} 2x_1 + y_1 = \dfrac{3 + \sqrt{5}}{2} x_1 \\[2mm] x_1 + y_1 = \dfrac{3 + \sqrt{5}}{2} y_1 \end{cases}$$

and obtain

$$y_1 = \frac{-1 + \sqrt{5}}{2} x_1,$$

which is equivalent to the equation $2y_1 = (-1 + \sqrt{5})x_1$.

Using the same procedure, we obtain the eigenvectors (x_2, y_2) where $2y_2 = (-1 - \sqrt{5})x_2$ for $\lambda_2 = (3 - \sqrt{5})/2$.

(c)

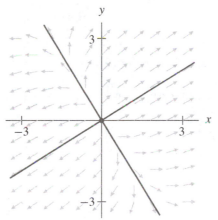

(d) One eigenvector $\mathbf{V}_1$ for the eigenvalue λ_1 is $\mathbf{V}_1 = (2, -1 + \sqrt{5})$, and one eigenvector $\mathbf{V}_2$ for the eigenvalue λ_2 is $\mathbf{V}_2 = (-2, 1 + \sqrt{5})$.

Given the eigenvalues and these eigenvectors, we have two linearly independent solutions

$$\mathbf{Y}_1(t) = e^{(3+\sqrt{5})t/2} \begin{pmatrix} 2 \\ -1 + \sqrt{5} \end{pmatrix} \quad \text{and} \quad \mathbf{Y}_2(t) = e^{(3-\sqrt{5})t/2} \begin{pmatrix} -2 \\ 1 + \sqrt{5} \end{pmatrix}.$$

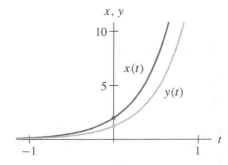

The $x(t)$- and $y(t)$-graphs for $\mathbf{Y}_1(t)$.

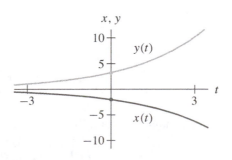

The $x(t)$- and $y(t)$-graphs for $\mathbf{Y}_2(t)$.

(e) The general solution to this linear system is

$$\mathbf{Y}(t) = k_1 e^{(3+\sqrt{5})t/2} \begin{pmatrix} 2 \\ -1 + \sqrt{5} \end{pmatrix} + k_2 e^{(3-\sqrt{5})t/2} \begin{pmatrix} -2 \\ 1 + \sqrt{5} \end{pmatrix}.$$

11. The eigenvalues are the roots of the characteristic polynomial, so they are solutions of

$$(-2 - \lambda)(1 - \lambda) - 4 = \lambda^2 + \lambda - 6 = 0.$$

Hence, $\lambda_1 = 2$ and $\lambda_2 = -3$ are the eigenvalues.

To find the eigenvectors for the eigenvalue $\lambda_1 = 2$, we solve

$$\begin{cases} -2x_1 - 2y_1 = 2x_1 \\ -2x_1 + y_1 = 2y_1, \end{cases}$$

so $y_1 = -2x_1$ is the line of eigenvectors. In particular, $(1, -2)$ is an eigenvector for $\lambda_1 = 2$.

Similarly, the line of eigenvectors for $\lambda_2 = -3$ is given by $x_1 = 2y_1$. In particular, $(2, 1)$ is an eigenvector for $\lambda_2 = -3$.

Given the eigenvalues and these eigenvectors, we have the two linearly independent solutions

$$\mathbf{Y}_1(t) = e^{2t} \begin{pmatrix} 1 \\ -2 \end{pmatrix} \quad \text{and} \quad \mathbf{Y}_2(t) = e^{-3t} \begin{pmatrix} 2 \\ 1 \end{pmatrix}.$$

The general solution is

$$\mathbf{Y}(t) = k_1 e^{2t} \begin{pmatrix} 1 \\ -2 \end{pmatrix} + k_2 e^{-3t} \begin{pmatrix} 2 \\ 1 \end{pmatrix}.$$

(a) Given the initial condition $\mathbf{Y}(0) = (1, 0)$, we must solve

$$\begin{pmatrix} 1 \\ 0 \end{pmatrix} = \mathbf{Y}(0) = k_1 \begin{pmatrix} 1 \\ -2 \end{pmatrix} + k_2 \begin{pmatrix} 2 \\ 1 \end{pmatrix}$$

for k_1 and k_2. This vector equation is equivalent to the two scalar equations

$$\begin{cases} k_1 + 2k_2 = 1 \\ -2k_1 + k_2 = 0. \end{cases}$$

Solving these equations, we obtain $k_1 = 1/5$ and $k_2 = 2/5$. Thus, the particular solution is

$$\mathbf{Y}(t) = \tfrac{1}{5} e^{2t} \begin{pmatrix} 1 \\ -2 \end{pmatrix} + \tfrac{2}{5} e^{-3t} \begin{pmatrix} 2 \\ 1 \end{pmatrix}.$$

(b) Given the initial condition $\mathbf{Y}(0) = (0, 1)$ we must solve

$$\begin{pmatrix} 0 \\ 1 \end{pmatrix} = \mathbf{Y}(0) = k_1 \begin{pmatrix} 1 \\ -2 \end{pmatrix} + k_2 \begin{pmatrix} 2 \\ 1 \end{pmatrix}$$

for k_1 and k_2. This vector equation is equivalent to the two scalar equations

$$\begin{cases} k_1 + 2k_2 = 0 \\ -2k_1 + k_2 = 1. \end{cases}$$

Solving these equations, we obtain $k_1 = -2/5$ and $k_2 = 1/5$. Thus, the particular solution is

$$\mathbf{Y}(t) = -\tfrac{2}{5}e^{2t} \begin{pmatrix} 1 \\ -2 \end{pmatrix} + \tfrac{1}{5}e^{-3t} \begin{pmatrix} 2 \\ 1 \end{pmatrix}.$$

(c) The initial condition $\mathbf{Y}(0) = (1, -2)$ is an eigenvector for the eigenvalue $\lambda_1 = 2$. Hence, the solution with this initial condition is

$$\mathbf{Y}(t) = e^{2t} \begin{pmatrix} 1 \\ -2 \end{pmatrix}.$$

13. The characteristic polynomial is

$$(-4 - \lambda)(-3 - \lambda) - 2 = \lambda^2 + 7\lambda + 10 = 0,$$

and therefore the eigenvalues are $\lambda_1 = -5$ and $\lambda_2 = -2$.

To obtain the eigenvectors (x_1, y_1) for the eigenvalue $\lambda_1 = -5$, we solve the system of equations

$$\begin{cases} -4x_1 + y_1 = -5x_1 \\ 2x_1 - 3y_1 = -5y_1 \end{cases}$$

and obtain

$$y_1 = -x_1.$$

Therefore, an eigenvector for the eigenvalue $\lambda_1 = -5$ is $\mathbf{V}_1 = (1, -1)$.

Using the same procedure, we obtain the eigenvector $\mathbf{V}_2 = (1, 2)$ for $\lambda_2 = -2$.

Given the eigenvalues and these eigenvectors, we have two linearly independent solutions

$$\mathbf{Y}_1(t) = e^{-5t} \begin{pmatrix} 1 \\ -1 \end{pmatrix} \quad \text{and} \quad \mathbf{Y}_2(t) = e^{-2t} \begin{pmatrix} 1 \\ 2 \end{pmatrix}.$$

The general solution to this linear system is

$$\mathbf{Y}(t) = k_1 e^{-5t} \begin{pmatrix} 1 \\ -1 \end{pmatrix} + k_2 e^{-2t} \begin{pmatrix} 1 \\ 2 \end{pmatrix}.$$

(a) We have $\mathbf{Y}(0) = (1, 0)$, so we must find k_1 and k_2 so that

$$\begin{pmatrix} 1 \\ 0 \end{pmatrix} = \mathbf{Y}(0) = k_1 \begin{pmatrix} 1 \\ -1 \end{pmatrix} + k_2 \begin{pmatrix} 1 \\ 2 \end{pmatrix}.$$

This vector equation is equivalent to the simultaneous system of linear equations

$$\begin{cases} k_1 + k_2 = 1 \\ -k_1 + 2k_2 = 0. \end{cases}$$

Solving these equations, we obtain $k_1 = 2/3$ and $k_2 = 1/3$. Thus, the particular solution is

$$\mathbf{Y}(t) = \tfrac{2}{3} e^{-5t} \begin{pmatrix} 1 \\ -1 \end{pmatrix} + \tfrac{1}{3} e^{-2t} \begin{pmatrix} 1 \\ 2 \end{pmatrix}.$$

(b) We have $\mathbf{Y}(0) = (2, 1)$, so we must find k_1 and k_2 so that

$$\begin{pmatrix} 2 \\ 1 \end{pmatrix} = \mathbf{Y}(0) = k_1 \begin{pmatrix} 1 \\ -1 \end{pmatrix} + k_2 \begin{pmatrix} 1 \\ 2 \end{pmatrix}.$$

This vector equation is equivalent to the simultaneous system of linear equations

$$\begin{cases} k_1 + k_2 = 2 \\ -k_1 + 2k_2 = 1. \end{cases}$$

Solving these equations, we obtain $k_1 = 1$ and $k_2 = 1$. Thus, the particular solution is

$$\mathbf{Y}(t) = e^{-5t} \begin{pmatrix} 1 \\ -1 \end{pmatrix} + e^{-2t} \begin{pmatrix} 1 \\ 2 \end{pmatrix}.$$

(c) We have $\mathbf{Y}(0) = (-1, -2)$. Since this initial condition is an eigenvector associated to the $\lambda = -2$ eigenvalue, we do not need to do any additional calculation. The desired solution to the initial-value problem is

$$\mathbf{Y}(t) = -e^{-2t} \begin{pmatrix} 1 \\ 2 \end{pmatrix}.$$

15. Given any vector $\mathbf{Y}_0 = (x_0, y_0)$, we have

$$\mathbf{AY}_0 = \begin{pmatrix} a & 0 \\ 0 & a \end{pmatrix} \begin{pmatrix} x_0 \\ y_0 \end{pmatrix} = \begin{pmatrix} ax_0 \\ ay_0 \end{pmatrix} = a \begin{pmatrix} x_0 \\ y_0 \end{pmatrix} = a\mathbf{Y}_0.$$

Therefore, every nonzero vector is an eigenvector associated to the eigenvalue a.

17. The characteristic polynomial of $\mathbf{B}$ is

$$\lambda^2 - (a+d)\lambda + ad - b^2.$$

The roots of this polynomial are

$$\begin{aligned} \lambda &= \frac{a + d \pm \sqrt{(a+d)^2 - 4(ad - b^2)}}{2} \\ &= \frac{a + d \pm \sqrt{a^2 + 2ad + d^2 - 4ad + 4b^2}}{2} \\ &= \frac{a + d \pm \sqrt{(a-d)^2 + 4b^2}}{2}. \end{aligned}$$

Since the discriminant $D = (a-d)^2 + 4b^2$ is always nonnegative, the roots λ are real. Therefore, the matrix $\mathbf{B}$ has real eigenvalues. If $b \neq 0$, then D is positive and hence $\mathbf{B}$ has two distinct eigenvalues. (The only way to have only one eigenvalue is for $D = 0$).

19. (a) To form the system, we introduce the new dependent variable $v = dy/dt$. Then

$$\frac{dv}{dt} = \frac{d^2y}{dt^2} = -p\frac{dy}{dt} - qy = -pv - qy.$$

Written in matrix form this system where $\mathbf{Y} = (y, v)$, we have

$$\frac{d\mathbf{Y}}{dt} = \begin{pmatrix} 0 & 1 \\ -q & -p \end{pmatrix} \mathbf{Y}.$$

(b) The characteristic polynomial is

$$(0 - \lambda)(-p - \lambda) + q = \lambda^2 + p\lambda + q.$$

(c) The roots of this polynomial (the eigenvalues) are

$$\frac{-p \pm \sqrt{p^2 - 4q}}{2}.$$

(d) The roots are distinct real numbers if the discriminant $D = p^2 - 4q$ is positive. In other words, the roots are distinct real numbers if $p^2 > 4q$.

(e) Since q is positive, $p^2 - 4q < p^2$, so we know that $\sqrt{p^2 - 4q} < \sqrt{p^2} = p$. Since the numerator in the expression for the eigenvalues is $-p \pm \sqrt{p^2 - 4q}$, we see that it must be negative. Since the denominator is positive, the eigenvalues must be negative.

21. (a) Given $v = dy/dt$, the corresponding system is

$$\frac{dy}{dt} = v$$

$$\frac{dv}{dt} = -10y - 7v.$$

(b) The characteristic polynomial is $\lambda^2 + 7\lambda + 10 = (\lambda + 5)(\lambda + 2) = 0$, and the eigenvalues are $\lambda_1 = -5$ and $\lambda_2 = -2$.

To find the eigenvectors $\mathbf{V}_1 = (y_1, v_1)$ associated to the eigenvalue $\lambda_1 = -5$, we solve the system of equations

$$\begin{cases} v_1 = -5y_1 \\ -10y_1 - 7v_1 = -5v_1 \end{cases}$$

and obtain $v_1 = -5y_1$.

By the same procedure, we can find the eigenvectors $\mathbf{V}_2 = (y_2, v_2)$ for the eigenvalue $\lambda_2 = -2$. They consist of all vectors that satisfy the equation $v_2 = -2y_2$.

(c) From part (b) we see that one eigenvector for $\lambda_1 = -5$ is $\mathbf{V}_1 = (1, -5)$. Therefore the solution $\mathbf{Y}_1(t)$ that satisfies $\mathbf{Y}_1(0) = \mathbf{V}_1$ is

$$\mathbf{Y}_1(t) = e^{-5t} \begin{pmatrix} 1 \\ -5 \end{pmatrix}.$$

One eigenvector for $\lambda_2 = -2$ is $\mathbf{V}_2 = (1, -2)$, and the solution $\mathbf{Y}_2(t)$ that satisfies $\mathbf{Y}_2(0) = \mathbf{V}_2$ is

$$\mathbf{Y}_2(t) = e^{-2t} \begin{pmatrix} 1 \\ -2 \end{pmatrix}.$$

(d) Note that the solutions obtained here are vector-valued functions of the form $\mathbf{Y}(t) = (y(t), v(t))$. In Section 2.4 we obtained $y_1(t) = e^{-2t}$ and $y_2(t) = e^{-5t}$. Using the fact that $v = dy/dt$, we can obtain $\mathbf{Y}_1(t)$ and $\mathbf{Y}_2(t)$ from $y_1(t)$ and $y_2(t)$.

23. **(a)** Given $v = dy/dt$, the corresponding system is

$$\frac{dy}{dt} = v$$
$$\frac{dv}{dt} = -y - 4v.$$

(b) The characteristic polynomial is $\lambda^2 + 4\lambda + 1 = 0$. Using the quadratic formula, we obtain the eigenvalues $\lambda_1 = -2 + \sqrt{3}$ and $\lambda_2 = -2 - \sqrt{3}$.

To find the eigenvectors $\mathbf{V}_1 = (y_1, v_1)$ associated to the eigenvalue $\lambda_1 = -2 + \sqrt{3}$, we solve the system of equations

$$\begin{cases} v_1 = (-2 + \sqrt{3})y_1 \\ -y_1 - 4v_1 = (-2 + \sqrt{3})v_1 \end{cases}$$

and obtain $v_1 = (-2 + \sqrt{3})y_1$.

By the same procedure, we can find the eigenvectors $\mathbf{V}_2 = (y_2, v_2)$ for the eigenvalue $\lambda_2 = -2 - \sqrt{3}$. They consist of all vectors that satisfy the equation $v_2 = (-2 - \sqrt{3})y_2$.

(c) From part (b) we see that one eigenvector for $\lambda_1 = -2 + \sqrt{3}$ is $\mathbf{V}_1 = (1, -2 + \sqrt{3})$. Therefore the solution $\mathbf{Y}_1(t)$ that satisfies $\mathbf{Y}_1(0) = \mathbf{V}_1$ is

$$\mathbf{Y}_1(t) = e^{(-2+\sqrt{3})t} \begin{pmatrix} 1 \\ -2 + \sqrt{3} \end{pmatrix}.$$

One eigenvector for $\lambda_2 = -2 - \sqrt{3}$ is $\mathbf{V}_2 = (1, -2 - \sqrt{3})$, and the solution $\mathbf{Y}_2(t)$ that satisfies $\mathbf{Y}_2(0) = \mathbf{V}_2$ is

$$\mathbf{Y}_2(t) = e^{(-2-\sqrt{3})t} \begin{pmatrix} 1 \\ -2 - \sqrt{3} \end{pmatrix}.$$

(d) Note that the solutions obtained here are vector-valued functions of the form $\mathbf{Y}(t) = (y(t), v(t))$. In Section 2.4 we obtained $y_1(t) = e^{(-2+\sqrt{3})t}$ and $y_2(t) = e^{(-2-\sqrt{3})t}$. Using the fact that $v = dy/dt$, we can obtain $\mathbf{Y}_1(t)$ and $\mathbf{Y}_2(t)$ from $y_1(t)$ and $y_2(t)$.

25. With $m = 1, k = 4$, and $b = 1$, the system is

$$\frac{dy}{dt} = v$$
$$\frac{dv}{dt} = -4y - v.$$

The characteristic polynomial is $\lambda^2 + \lambda + 4$, and its roots are the complex numbers $(-1 \pm \sqrt{15}\,i)/2$. Therefore there are no straight-line solutions. According to the direction field, the solution curves seem to spiral around the origin.

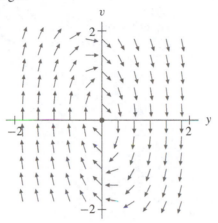

EXERCISES FOR SECTION 3.3

1. As we computed in Exercise 1 of Section 3.2, the eigenvalues are $\lambda_1 = -2$ and $\lambda_2 = 3$. The eigenvectors (x_1, y_1) for the eigenvalue $\lambda_1 = -2$ satisfy $5x_1 = -2y_1$, and the eigenvectors (x_2, y_2) for $\lambda_2 = 3$ satisfy the equation $y_2 = 0$. The equilibrium point at the origin is a saddle.

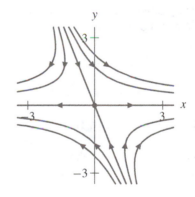

3. As we computed in Exercise 3 of Section 3.2, the eigenvalues are $\lambda_1 = -3$ and $\lambda_2 = -6$. The eigenvectors (x_1, y_1) for the eigenvalue $\lambda_1 = -3$ satisfy $y_1 = -x_1$, and the eigenvectors for $\lambda_2 = -6$ satisfy $x_2 = 2y_2$. The equilibrium point at the origin is a sink.

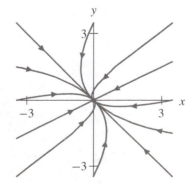

5. As we computed in Exercise 7 of Section 3.2, the eigenvalues are $\lambda_1 = -1$ and $\lambda_2 = 4$. The eigenvectors (x_1, y_1) for the eigenvalue $\lambda_1 = -1$ satisfy $y_1 = -x_1$, and the eigenvectors (x_2, y_2) for $\lambda_2 = 4$ satisfy $x_2 = 4y_2$. The equilibrium point at the origin is a saddle.

7. As we computed in Exercise 9 of Section 3.2, the eigenvalues are

$$\lambda_1 = \frac{3 + \sqrt{5}}{2} \quad \text{and} \quad \lambda_2 = \frac{3 - \sqrt{5}}{2}.$$

The eigenvectors (x_1, y_1) for the eigenvalue λ_1 satisfy $y_1 = (-1 + \sqrt{5})x_1/2$, and the eigenvectors (x_2, y_2) for λ_2 satisfy $y_2 = (-1 - \sqrt{5})x_2/2$. The equilibrium point at the origin is a source.

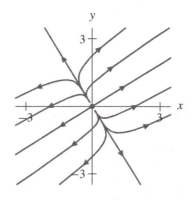

9. As we computed in Exercise 11 of Section 3.2, the eigenvalues are $\lambda_1 = 2$ and $\lambda_2 = -3$. The eigenvectors (x_1, y_1) for the eigenvalue $\lambda_1 = 2$ satisfy $y_1 = -2x_1$, and the eigenvectors for the eigenvalue $\lambda_2 = -3$ satisfy $x_1 = 2y_1$. The equilibrium point at the origin is a saddle. The solution curves in the phase plane for the initial conditions $(1, 0), (0, 1)$, and $(1, -2)$ are shown in the figure on the right.

(a) The solution with initial condition $(1, 0)$ is asymptotic to the line $y = -2x$ in the fourth quadrant as $t \to \infty$ and to the line $x = 2y$ in the first quadrant as $t \to -\infty$.

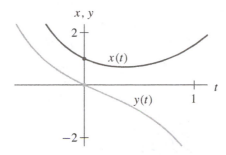

(b) The solution curve with initial condition $(1, 0)$ is asymptotic to the line $y = -2x$ in the second quadrant as $t \to \infty$ and to the line $x = 2y$ in the first quadrant as $t \to -\infty$.

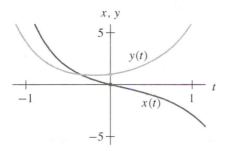

(c) The solution curve with initial condition $(1, -2)$ is on the line of eigenvectors for the eigenvalue $\lambda_1 = 2$. Hence, this solution curve stays on the line $y = -2x$. It approaches the origin as $t \to -\infty$, and it tends to ∞ in the fourth quadrant as $t \to \infty$.

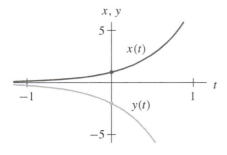

11. As we computed in Exercise 13 of Section 3.2, the eigenvalues are $\lambda_1 = -5$ and $\lambda_2 = -2$. The eigenvectors (x_1, y_1) for the eigenvalue $\lambda_1 = -5$ satisfy $y_1 = -x_1$, and the eigenvectors (x_2, y_2) for the eigenvalue $\lambda_2 = -2$ satisfy $y_2 = 2x_2$. The equilibrium point at the origin is a sink. The solution curves in the phase plane for the initial conditions $(1, 0)$, $(2, 1)$, and $(-1, -2)$ are shown in the following figure.

(a) The solution curve with initial condition $(1, 0)$ approaches the origin tangent to the line $y = 2x$.

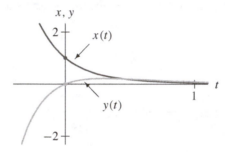

(b) The solution curve with initial condition $(2, 1)$ approaches the origin tangent to the line $y = 2x$.

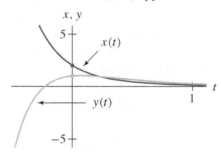

(c) The initial condition $(-1, -2)$ is an eigenvector associated to the eigenvalue $\lambda_2 = -2$. The corresponding solution curve approaches the origin along the line $y = 2x$ as $t \to \infty$.

13. As we computed in Exercise 21 of Section 3.2, the eigenvalues are $\lambda_1 = -5$ and $\lambda_2 = -2$. The eigenvectors (y_1, v_1) associated to the eigenvalue $\lambda_1 = -5$ satisfy $v_1 = -5y_1$, and the eigenvectors (y_2, v_2) for the eigenvalue $\lambda_2 = -2$ satisfy the equation $v_2 = -2y_2$. The equilibrium point at the origin is a sink.

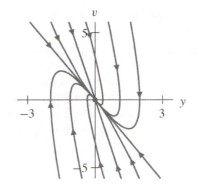

15. As we computed in Exercise 23 of Section 3.2, the eigenvalues are $\lambda_1 = -2 + \sqrt{3}$ and $\lambda_2 = -2 - \sqrt{3}$. The eigenvectors (y_1, v_1) associated to the eigenvalue $\lambda_1 = -2 + \sqrt{3}$ satisfy the equation $v_1 = (-2 + \sqrt{3})y_1$, and the eigenvectors (y_2, v_2) for the eigenvalue $\lambda_2 = -2 - \sqrt{3}$ satisfy the equation $v_2 = (-2 - \sqrt{3})y_2$. The equilibrium point at the origin is a sink.

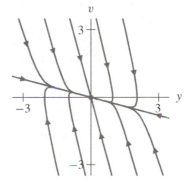

17. The characteristic equation is

$$(2 - \lambda)(-1 - \lambda) = 0,$$

and therefore, the eigenvalues are $\lambda_1 = 2$ and $\lambda_2 = -1$. The equilibrium point at the origin is a saddle.

To compute the eigenvectors associated to $\lambda_1 = 2$, we must solve the simultaneous equations

$$\begin{cases} 2x + y = 2x \\ -y = 2y. \end{cases}$$

Therefore, any vector of the form $(x, 0)$ is an eigenvector associated to the eigenvalue $\lambda_1 = 2$.

Similarly, for $\lambda_1 = -1$, the eigenvectors (x, y) associated to the eigenvalue $\lambda_2 = -1$ must satisfy the equation $y = -3x$. Therefore, we know that the phase portrait is

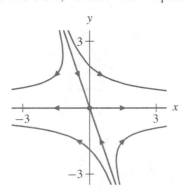

Given an initial condition on the line $y = -3x$, the corresponding solution is a straight-line solution that is asymptotic to the origin. For any other initial condition, the corresponding solution is asymptotic to the x-axis. Therefore, for any initial condition, Bob's profits, $y(t)$, eventually tend to 0 as $t \to \infty$.

To see what happens to Paul's profits, we must locate the initial condition relative to the line $y = -3x$. As stated above, if the initial condition (x_0, y_0) lies on the line $y = -3x$, then Paul's profits will also tend to 0 eventually. However, if (x_0, y_0) lies to the left of the line $y = -3x$, Paul goes broke. On the other hand, if (x_0, y_0) lies to the right of the line $y = -3x$, then Paul makes a fortune.

19. (a) The characteristic equation is
$$(-2 - \lambda)(-1 - \lambda) = 0,$$
so the eigenvalues are $\lambda_1 = -2$ and $\lambda_2 = -1$. Therefore, the equilibrium point at the origin is a sink.

(b) To find all the straight-line solutions, we must calculate the eigenvectors. For the eigenvalue $\lambda_1 = -2$, we have the simultaneous equations
$$\begin{cases} -2x_1 + \frac{1}{2}y_1 = -2x_1 \\ -y_1 = -2y_1. \end{cases}$$

The second equation implies that $y_1 = 0$. In other words, all vectors on the x-axis are eigenvectors for λ_1. Therefore, any solution of the form $e^{-2t}(x_1, 0)$ for any x_1 is a straight-line solution corresponding to the eigenvalue $\lambda_1 = -2$.

To calculate the eigenvectors associated to the eigenvalue $\lambda_2 = -1$, we must solve the equations
$$\begin{cases} -2x_2 + \frac{1}{2}y_2 = -x_2 \\ -y_2 = -y_2. \end{cases}$$

From the first equation, we see that $y_2 = 2x_2$. Therefore, any solution of the form $e^{-t}(x_2, 2x_2)$ for any x_2 is a straight-line solution corresponding to the eigenvalue $\lambda_2 = -1$.

(c) In the phase plane, all solution curves approach the origin as $t \to \infty$. If the initial condition is on the x-axis, it yields a straight-line solution that remains on the x-axis as $t \to \infty$. For any other initial condition, the solution approaches the origin tangent to the line $y = 2x$.

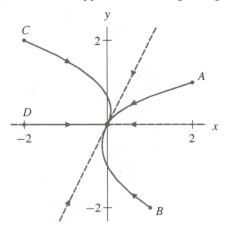

For the initial condition $A = (2, 1)$, the solution curve remains in the first quadrant. Since it approaches the origin tangent to the line $y = 2x$, it must cross the line $y = x$ at some time. Therefore, the $x(t)$- and $y(t)$-graphs are positive for all t, but they cross at some time $t > 0$.

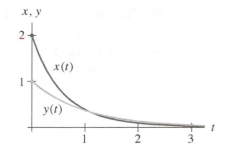

For the initial condition $B = (1, -2)$, we see that $y(t)$ is increasing but negative for all t. We also see that $x(t)$ is decreasing initially. It becomes negative, reaches a minimum, and then increases as it approaches 0. Note that $x(t) \neq y(t)$ for all t.

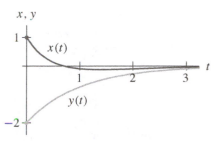

For the initial condition $C = (-2, 2)$, we see that $y(t)$ is decreasing but positive for all t. We also see that $x(t)$ is increasing initially. It becomes positive, reaches a maximum, and then decreases as it approaches 0. Again, these two graphs do not cross at any time.

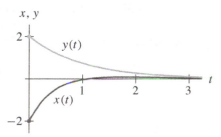

The initial condition $D = (-2, 0)$ lies on the line of eigenvectors associated to the eigenvalue $\lambda_1 = -1$. Therefore, the solution curve remains on the x-axis for all t. Hence, $y(t) = 0$ for all t, and $x(t)$ is the exponential function $-2e^{-2t}$.

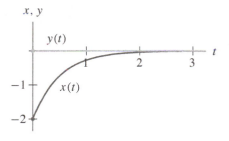

21. (a) The second-order equation is

$$\frac{d^2 y}{dt^2} + 7\frac{dy}{dt} + 6y = 0.$$

Introducing $v = dy/dt$, we obtain the system

$$\frac{dy}{dt} = v$$

$$\frac{dv}{dt} = -6y - 7v.$$

(b) The characteristic polynomial is

$$\lambda^2 + 7\lambda + 6,$$

which factors into $(\lambda + 6)(\lambda + 1)$.

(c) From the characteristic polynomial, we obtain the eigenvalues $\lambda_1 = -6$ and $\lambda_2 = -1$.

(d) To compute the eigenvectors associated to $\lambda_1 = -6$, we solve the simultaneous equations

$$\begin{cases} v = -6y \\ -6y - 7v = -6v. \end{cases}$$

Therefore, any vector on the line $v = -6y$ is an eigenvector associated to the eigenvalue λ_1.

To compute the eigenvectors associated to $\lambda_2 = -1$, we must solve the simultaneous equations

$$\begin{cases} v = -y \\ -6y - 7v = -v. \end{cases}$$

Therefore, any vector on the line $y = -v$ is an eigenvector associated to the eigenvalue λ_2.

Since both eigenvalues are real and negative, we know that origin is a sink, and the solution curve corresponding to the initial condition $(y(0), v(0)) = (2, 0)$ tends toward the origin tangent to the line $y = -v$ in the yv-plane.

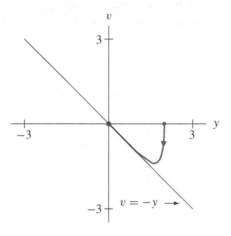

From the phase portrait, we see that the solution curve remains in the fourth quadrant for all $t > 0$. Consequently, it does not cross the line $y = 0$, and the mass cannot cross the equilibrium position. The solution approaches the origin at the rate that is determined by the eigenvalue $\lambda_2 = -1$. In other words, it approaches the origin at the rate of e^{-t}.

23. **(a)** Written in terms of its components, the system is

$$\frac{dx}{dt} = -0.2x - 0.1y$$

$$\frac{dy}{dt} = -0.1y.$$

Since the coefficient of y in dx/dt is negative, the introduction of new fish ($y > 0$) contributes negatively to dx/dt. Hence, the new fish have a negative effect on the native fish population.

Since the equation for dy/dt involves only y, the native fish have no effect on the population of the new fish.

(b) If the new species of fish is not introduced (that is, if $y(t) = 0$ for all t), then the system reduces to $dx/dt = -0.2x$. In this case, we have an exponential decay model as in Section 1.1, and the native fish population tends to its equilibrium level. (Remember: the quantity $x(t)$ is the difference between the native fish population and its equilibrium level, not the actual of fish population.) Thus, the model agrees with the system as described.

(c) There are two lines consisting of straight-line solutions, and the solutions with initial conditions on these lines are asymptotic to the origin as $t \to 0$. To find these lines, we must compute the eigenvalues and eigenvectors.

The characteristic polynomial of the given matrix is

$$(-0.2 - \lambda)(-0.1 - \lambda),$$

and hence the eigenvalues are $\lambda_1 = -0.2$ and $\lambda_2 = -0.1$.

To find an eigenvector for $\lambda_1 = -0.2$, we must solve

$$\begin{pmatrix} -0.2 & -0.1 \\ 0.0 & -0.1 \end{pmatrix} \begin{pmatrix} x_0 \\ y_0 \end{pmatrix} = -0.2 \begin{pmatrix} x_0 \\ y_0 \end{pmatrix}.$$

Rewritten in terms of components, this equation becomes

$$\begin{cases} -0.2x_0 - 0.1y_0 = -0.2x_0 \\ -0.1y_0 = -0.2y_0, \end{cases}$$

which is equivalent to

$$\begin{cases} -0.1y_0 = 0 \\ 0.1y_0 = 0. \end{cases}$$

If we multiply the second equation by -1, we obtain the first equation. Therefore, the equations are redundant and any vector (x_0, y_0) that satisfies the first equation is an eigenvector. Setting $x_0 = 1$ yields the eigenvector $\mathbf{V}_1 = (1, 0)$.

To find an eigenvector for $\lambda_2 = -0.1$, we repeat the process with λ_2 in replace of λ_1, and we obtain the eigenvector $\mathbf{V}_2 = (1, -1)$

Since both eigenvalues are negative, solutions with initial conditions that lie on the lines through the origin determined by the eigenvectors (the x-axis and the line $y = -x$) tend toward the origin.

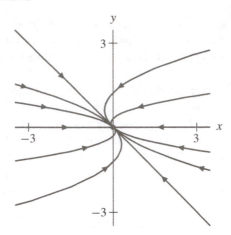

(d) Using the phase portrait shown in part (c), we see that solutions with initial conditions of the form $(0, y)$, $y > 0$, move through the second quadrant and tend toward the equilibrium point at the origin. Thus, our model predicts that, if a small number of new fish are added to the lake, the native population drops below its equilibrium level since x is negative. The new fish will die out and the native fish will return to equilibrium.

25. **(a)** Written in terms of its components, the system is

$$\frac{dx}{dt} = -0.2x + 0.1y$$
$$\frac{dy}{dt} = -0.1y.$$

Since the dy/dt equation does not depend on x, the native fish have no effect on the population of the new fish. Since the coefficient of y in the equation for dx/dt is positive, the introduction of new fish increases the population of the native fish.

(b) If the new species of fish is not introduced (that is, if $y(t) = 0$ for all t), then the above system simply becomes $dx/dt = -0.2x$. In this case, we have an exponential decay model as in Section 1.1, and the native fish population tends to its equilibrium level. (Remember: the quantity $x(t)$ is the difference between the native fish population and its equilibrium level, not the actual of fish population.) Thus, the model agrees with the system as described.

(c) The characteristic equation is

$$(-0.2 - \lambda)(-0.1 - \lambda) = 0,$$

and the eigenvalues are $\lambda_1 = -0.2$ and $\lambda_2 = -0.1$.

To find an eigenvector for $\lambda_1 = -0.2$, we must solve the simultaneous equations

$$\begin{cases} -0.2x + 0.1y = -0.2x \\ -0.1y = -0.2y. \end{cases}$$

Therefore, $y = 0$. One such eigenvector $\mathbf{V}_1$ is $(1, 0)$.

For $\lambda = -0.1$, the simultaneous equations are

$$
\begin{cases}
-0.2x + 0.1y = -0.1x \\
\qquad\quad -0.1y = -0.1y.
\end{cases}
$$

Any vector that satisfies $y = x$ satisfies these equations. One such eigenvector is $\mathbf{V}_2 = (1, 1)$.

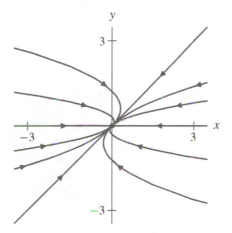

(d) Using the phase portrait shown in part (c), we see that solutions with initial conditions of the form $(0, y), y > 0$, move through the first quadrant and tend toward the equilibrium point at the origin. Thus, our model predicts that, if a small number of new fish are added to the lake, the population of native fish increases above its equilibrium value and the population of new fish decreases. Eventually, the new fish head toward extinction, and the native fish return to their equilibrium population.

27. **(a)** The characteristic equation is

$$(-2 - \lambda)(2 - \lambda) = 0.$$

Therefore, the eigenvalues are $\lambda_1 = -2$ and $\lambda_2 = 2$, and the equilibrium point at the origin is a saddle.

(b) To find the eigenvectors (x_1, y_1) corresponding to $\lambda_1 = -2$, we solve the simultaneous linear equations

$$
\begin{cases}
-2x + y = -2x \\
\qquad\quad 2y = -2y.
\end{cases}
$$

Therefore, the eigenvectors lie on the line $y = 0$, the x-axis. Similarly, the eigenvectors associated to the eigenvalue $\lambda_2 = 2$ lie on the line $y = 4x$.

Using these eigenvalues and eigenvectors, we can give a rough sketch of the phase portrait.

(c)

(d) The eigenvalues are -2 and 2 with eigenvectors $(1, 0)$ and $(1, 4)$ respectively, The initial conditions are on either side of the line of eigenvectors corresponding to eigenvalue -2. Hence, these two solutions will approach the origin along the x-axis, but the y coordinates will grow, at approximately the rate e^{2t}. Since the initial separation is 0.02 and we seek the approximate time t when the separation is 1, we must solve $0.02e^{2t} = 1$, which yields $t = \ln(50)/2 \approx 1.96$.

EXERCISES FOR SECTION 3.4

1. Using Euler's formula, we can write the complex-valued solution $\mathbf{Y}_c(t)$ as

$$
\begin{aligned}
\mathbf{Y}_c(t) &= e^{(1+3i)t} \begin{pmatrix} 2+i \\ 1 \end{pmatrix} \\
&= e^t e^{3it} \begin{pmatrix} 2+i \\ 1 \end{pmatrix} \\
&= e^t (\cos 3t + i \sin 3t) \begin{pmatrix} 2+i \\ 1 \end{pmatrix} \\
&= e^t \begin{pmatrix} 2\cos 3t - \sin 3t \\ \cos 3t \end{pmatrix} + i e^t \begin{pmatrix} 2\sin 3t + \cos 3t \\ \sin 3t \end{pmatrix}.
\end{aligned}
$$

Hence, we have

$$\mathbf{Y}_{\text{re}}(t) = e^t \begin{pmatrix} 2\cos 3t - \sin 3t \\ \cos 3t \end{pmatrix} \quad \text{and} \quad \mathbf{Y}_{\text{im}}(t) = e^t \begin{pmatrix} \cos 3t + 2\sin 3t \\ \sin 3t \end{pmatrix}.$$

The general solution is

$$\mathbf{Y}(t) = k_1 e^t \begin{pmatrix} 2\cos 3t - \sin 3t \\ \cos 3t \end{pmatrix} + k_2 e^t \begin{pmatrix} \cos 3t + 2\sin 3t \\ \sin 3t \end{pmatrix}.$$

3. **(a)** The characteristic equation is

$$(-\lambda)^2 + 4 = \lambda^2 + 4 = 0,$$

and the eigenvalues are $\lambda = \pm 2i$.

(b) Since the real part of the eigenvalues are 0, the origin is a center.

(c) Since $\lambda = \pm 2i$, the natural period is $2\pi/2 = \pi$, and the natural frequency is $1/\pi$.

(d) At $(1, 0)$, the tangent vector is $(-2, 0)$. Therefore, the direction of oscillation is clockwise.

(e) According to the phase plane, $x(t)$ and $y(t)$ are periodic with period π. At the initial condition $(1, 0)$, both $x(t)$ and $y(t)$ are initially decreasing.

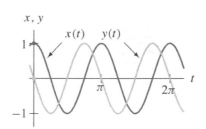

5. **(a)** The characteristic polynomial is

$$(-3 - \lambda)(1 - \lambda) + 15 = \lambda^2 + 2\lambda + 12,$$

so the eigenvalues are $\lambda = -1 \pm i\sqrt{11}$.

(b) The eigenvalues are complex and the real part is negative, so the origin is a spiral sink.

(c) The natural period is $2\pi/\sqrt{11}$. The natural frequency is $\sqrt{11}/(2\pi)$.

(d) At the point $(1, 0)$, the vector field is $(-3, 3)$. Hence, the solution curves must spiral in a counterclockwise fashion.

(e)

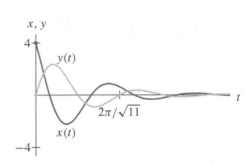

7. **(a)** The characteristic equation is

$$(2 - \lambda)(1 - \lambda) + 12 = \lambda^2 + 3\lambda + 14 = 0,$$

and the eigenvalues are $\lambda = (3 \pm \sqrt{47}\,i)/2$.

(b) Since the real part of the eigenvalues is positive, the origin is a spiral source.

(c) Since $\lambda = (3 \pm \sqrt{47}i)/2$, natural period is $4\pi/\sqrt{47}$, and natural frequency is $\sqrt{47}/(4\pi)$.

(d) At the point $(1, 0)$, the tangent vector is $(2, 2)$. Therefore, the solution curves spiral about the origin in a counterclockwise fashion.

(e) From the phase plane, we see that both $x(t)$ and $y(t)$ oscillate and that the amplitude of these oscillations increases rapidly.

9. **(a)** According to Exercise 3, $\lambda = \pm 2i$. The eigenvectors (x, y) associated to eigenvalue $\lambda = 2i$ must satisfy the equation $2y = 2ix$, which is equivalent to $y = ix$. One such eigenvector is $(1, i)$, and thus we have the complex solution

$$\mathbf{Y}(t) = e^{2it}\begin{pmatrix} 1 \\ i \end{pmatrix} = \begin{pmatrix} \cos 2t \\ -\sin 2t \end{pmatrix} + i \begin{pmatrix} \sin 2t \\ \cos 2t \end{pmatrix}.$$

Taking real and imaginary parts, we obtain the general solution

$$\mathbf{Y}(t) = k_1 \begin{pmatrix} \cos 2t \\ -\sin 2t \end{pmatrix} + k_2 \begin{pmatrix} \sin 2t \\ \cos 2t \end{pmatrix}.$$

(b) From the initial condition, we obtain

$$k_1 \begin{pmatrix} 1 \\ 0 \end{pmatrix} + k_2 \begin{pmatrix} 0 \\ 1 \end{pmatrix} = \begin{pmatrix} 1 \\ 0 \end{pmatrix},$$

and therefore, $k_1 = 1$ and $k_2 = 0$. Consequently, the solution with the initial condition $(1, 0)$ is

$$\mathbf{Y}(t) = \begin{pmatrix} \cos 2t \\ -\sin 2t \end{pmatrix}.$$

(c)

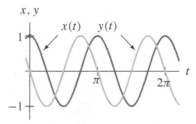

11. **(a)** To find the general solution, we find the eigenvectors from the characteristic polynomial

$$(-3 - \lambda)(1 - \lambda) + 15 = \lambda^2 + 2\lambda + 12.$$

The eigenvalues are $\lambda = -1 \pm i\sqrt{11}$. To find an eigenvector associated to the eigenvector $-1 + i\sqrt{11}$, we must solve the equations

$$\begin{cases} -3x - 5y = \left(-1 + i\sqrt{11}\right)x \\ 3x + y = \left(-1 + i\sqrt{11}\right)y. \end{cases}$$

We see that the eigenvectors must satisfy the equation $3x = (-2 + i\sqrt{11})y$. Using the eigenvector $(-2 + i\sqrt{11}, 3)$, we obtain the complex-valued solution

$$\mathbf{Y}(t) = e^{\left(-1 + i\sqrt{11}\right)t} \begin{pmatrix} -2 + i\sqrt{11} \\ 3 \end{pmatrix}.$$

Using Euler's formula, we write $\mathbf{Y}(t)$ as

$$\mathbf{Y}(t) = e^{-t}\left(\cos\sqrt{11}\,t + i\sin\sqrt{11}\,t\right) \begin{pmatrix} -2 + i\sqrt{11} \\ 3 \end{pmatrix},$$

which can be expressed as

$$\mathbf{Y}(t) = e^{-t} \begin{pmatrix} -2\cos\sqrt{11}\,t - \sqrt{11}\sin\sqrt{11}\,t \\ 3\cos\sqrt{11}\,t \end{pmatrix} + ie^{-t} \begin{pmatrix} \sqrt{11}\cos\sqrt{11}\,t - 2\sin\sqrt{11}\,t \\ 3\sin\sqrt{11}\,t \end{pmatrix}.$$

Taking real and imaginary parts, we can form the general solution

$$k_1 e^{-t} \begin{pmatrix} -2\cos\sqrt{11}\,t - \sqrt{11}\sin\sqrt{11}\,t \\ 3\cos\sqrt{11}\,t \end{pmatrix} + k_2 e^{-t} \begin{pmatrix} \sqrt{11}\cos\sqrt{11}\,t - 2\sin\sqrt{11}\,t \\ 3\sin\sqrt{11}\,t \end{pmatrix}.$$

(b) To find the particular solution with initial condition $(4, 0)$, we solve for k_1 and k_2 and obtain

$$\begin{cases} -2k_1 + \sqrt{11}\,k_2 = 4 \\ \qquad\qquad 3k_1 = 0. \end{cases}$$

We have $k_1 = 0$ and $k_2 = 4/\sqrt{11}$.
The desired solution is

$$\mathbf{Y}(t) = e^{-t} \begin{pmatrix} 4\cos\sqrt{11}\,t - \frac{8}{\sqrt{11}}\sin\sqrt{11}\,t \\ \frac{12}{\sqrt{11}}\sin\sqrt{11}\,t \end{pmatrix}.$$

(c)

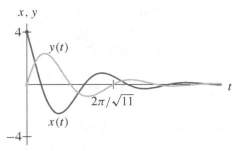

13. **(a)** According to Exercise 7, the eigenvalues are $\lambda = (3 \pm i\sqrt{47})/2$. The eigenvalues (x, y) associated to the eigenvector $(3 + i\sqrt{47})/2$ must satisfy the equation $12y = (1 - i\sqrt{47})x$. Hence, one eigenvector is $(12, 1 - i\sqrt{47})$, and we have the complex-valued solution

$$\mathbf{Y}(t) = e^{(3+i\sqrt{47})t/2} \begin{pmatrix} 12 \\ 1 - i\sqrt{47} \end{pmatrix}$$

$$= e^{3t/2} \begin{pmatrix} 12\cos\left(\frac{\sqrt{47}}{2}t\right) \\ \cos\left(\frac{\sqrt{47}}{2}t\right) + \sqrt{47}\sin\left(\frac{\sqrt{47}}{2}t\right) \end{pmatrix} +$$

$$i e^{3t/2} \begin{pmatrix} 12\sin\left(\frac{\sqrt{47}}{2}t\right) \\ -\sqrt{47}\cos\left(\frac{\sqrt{47}}{2}t\right) + \sin\left(\frac{\sqrt{47}}{2}t\right) \end{pmatrix}.$$

Taking real and imaginary parts

$$\mathbf{Y}_{\mathrm{re}}(t) = e^{3t/2} \begin{pmatrix} 12\cos\left(\frac{\sqrt{47}}{2}t\right) \\ \cos\left(\frac{\sqrt{47}}{2}t\right) + \sqrt{47}\sin\left(\frac{\sqrt{47}}{2}t\right) \end{pmatrix}$$

and

$$\mathbf{Y}_{\mathrm{im}}(t) = e^{3t/2} \begin{pmatrix} 12\sin\left(\frac{\sqrt{47}}{2}t\right) \\ -\sqrt{47}\cos\left(\frac{\sqrt{47}}{2}t\right) + \sin\left(\frac{\sqrt{47}}{2}t\right) \end{pmatrix},$$

we obtain the general solution $k_1\mathbf{Y}_{\mathrm{re}}(t) + k_2\mathbf{Y}_{\mathrm{im}}(t)$.

(b) From the initial condition, we have

$$k_1 \begin{pmatrix} 12 \\ 1 \end{pmatrix} + k_2 \begin{pmatrix} 0 \\ -\sqrt{47} \end{pmatrix} = \begin{pmatrix} 2 \\ 1 \end{pmatrix}.$$

Thus, $k_1 = 1/6$ and $k_2 = -5/(6\sqrt{47})$, and the desired solution is

$$\mathbf{Y}(t) = e^{3t/2} \begin{pmatrix} 2\cos\left(\frac{\sqrt{47}}{2}\,t\right) - \frac{10}{\sqrt{47}}\sin\left(\frac{\sqrt{47}}{2}\,t\right) \\ \cos\left(\frac{\sqrt{47}}{2}\,t\right) + \frac{7}{\sqrt{47}}\sin\left(\frac{\sqrt{47}}{2}\,t\right) \end{pmatrix}.$$

(c)

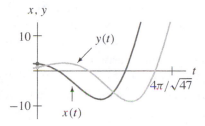

15. **(a)** In the case of complex eigenvalues, the function $x(t)$ oscillates about $x = 0$ with constant period, and the amplitude of successive oscillations is either increasing, decreasing, or constant depending on the sign of the real part of the eigenvalue. The graphs that satisfy these properties are (iv) and (v).

(b) For (iv), the natural period is approximately 1.5, and since the amplitude tends toward zero as t increases, the origin is a sink. For (v), the natural period is approximately 1.25, and since the amplitude increases as t increases, the origin is a source.

(c) (i) The time between successive zeros is not constant.

(ii) Oscillation stops at some t.

(iii) The amplitude is not monotonically decreasing or increasing.

(vi) Oscillation starts at some t. There was no prior oscillation.

17. We know that $\lambda_1 = \alpha + i\beta$ satisfies the equation $\lambda_1^2 + a\lambda_1 + b = 0$. Therefore, if we take the complex conjugate all of the terms in this equation, we obtain

$$(\alpha - i\beta)^2 + a(\alpha - i\beta) + b = 0,$$

since a and b are real. The complex conjugate of λ_1 is $\lambda_2 = \alpha - i\beta$, and we have

$$\lambda_2^2 + a\lambda_2 + b = 0.$$

Therefore, λ_2 is also a root.

19. Suppose $\mathbf{Y}_2 = k\mathbf{Y}_1$ for some constant k. Then, $\mathbf{Y}_0 = (1 + ik)\mathbf{Y}_1$. Since $\mathbf{AY}_0 = \lambda\mathbf{Y}_0$, we have

$$(1 + ik)\mathbf{AY}_1 = \lambda(1 + ik)\mathbf{Y}_1.$$

Thus, $\mathbf{AY}_1 = \lambda\mathbf{Y}_1$. Now note that the left-hand side, $\mathbf{AY}_1$, is a real vector. However, since λ is complex and $\mathbf{Y}_1$ is real, the right-hand side is complex (that is, it has a nonzero imaginary part). Thus, we have a contradiction, and $\mathbf{Y}_1$ and $\mathbf{Y}_2$ must be linearly independent.

21. **(a)** The factor $e^{-\alpha t}$ is positive for all t. Hence, the zeros of $x(t)$ are exactly the zeros of $\sin \beta t$. Suppose t_1 and t_2 are successive zeros (that is, $t_1 < t_2, x(t_1) = x(t_2) = 0$, and $x(t) \neq 0$ for $t_1 < t < t_2$), then $\beta t_2 - \beta t_1 = \pi$. In other words, $t_2 - t_1 = \pi/\beta$.

(b) By the nature of sine function, local maxima and local minima appear alternately. Therefore, we look for t_1 and t_2 such that $x'(t_1) = x'(t_2) = 0$ and $x'(t) \neq 0$ for $t_1 < t < t_2$. From

$$x'(t) = e^{-\alpha t}(-\alpha \sin \beta t + \beta \cos \beta t) = 0,$$

we know that $\tan \beta t = \beta/\alpha$ if t is corresponds to a local extremum. Since the tangent function is periodic with period π, $\beta(t_2 - t_1) = \pi$. Hence, $t_2 - t_1 = \pi/\beta$. Note that the distance between a local minimum and the following local maximum of $x(t)$ is constant over t.

(c) From part (b), we know that the distance between the first local maximum and the first local minimum is π/β and the distance between the first local minimum and the second local maximum is π/β. Therefore, the distance between the first two local maxima of $x(t)$ is $2\pi/\beta$.

(d) From part (b), we know that the first local maximum of $x(t)$ occurs at $t = (\arctan(\beta/\alpha))/\beta$.

23. **(a)** The corresponding first-order system is

$$\frac{dy}{dt} = v$$

$$\frac{dv}{dt} = -qy - pv.$$

(b) The characteristic polynomial is

$$(-\lambda)(-p - \lambda) + q = \lambda^2 + p\lambda + q,$$

so the eigenvalues are $\lambda = (-p \pm \sqrt{p^2 - 4q})/2$. Hence, the eigenvalues are complex if and only if $p^2 < 4q$. Note that q must be positive for this condition to be satisfied.

(c) In order to have a spiral sink, we must have $p^2 < 4q$ (to make the eigenvalues complex) and $p > 0$ (to make the real part of the eigenvalues negative). In other words, the origin is a spiral sink if and only if $q > 0$ and $0 < p < 2\sqrt{q}$. The origin is a center if and only if $q > 0$ and $p = 0$. Finally, the origin is a spiral source if and only if $q > 0$ and $-2\sqrt{q} < p < 0$.

(d) The vector field at $(1, 0)$ is $(0, -q)$. Hence, if $q > 0$, then the vector field points down along the entire y-axis, and the solution curves spiral about the origin in a clockwise fashion. Note that q must be positive for the eigenvalues to be complex, so the solution curves always spiral about the origin in a clockwise fashion as long as the eigenvalues are complex.

25. There is no spiral saddle because a linear saddle is a linear system where some solutions approach the origin and some move away. If one solution spirals toward (or away from) the origin, then we can multiply that solution by any constant, scaling it so that it goes through any point in the plane. This scaled solution is still a solution of the system (recall the Linearity Principle), so every solution spirals in the same way, either toward or away from the origin.

EXERCISES FOR SECTION 3.5

1. **(a)** The characteristic equation is

$$(-3 - \lambda)^2 = 0,$$

and the eigenvalue is $\lambda = -3$.

(b) To find an eigenvector, we solve the simultaneous equations

$$\begin{cases} -3x = -3x \\ x - 3y = -3y. \end{cases}$$

Then, $x = 0$, and one eigenvector is $(0, 1)$.

(c) Note the straight-line solutions along the y-axis.

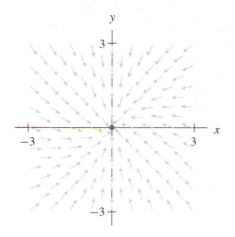

(d) Since the eigenvalue is negative, any solution with an initial condition on the y-axis tends toward the origin as t increases. According to the direction field, every solution tends to the origin as t increases. The solutions with initial conditions in the half-plane $x > 0$ eventually approach the origin along the positive y-axis. Similarly, the solutions with initial conditions in the half-plane $x < 0$ eventually approach the origin along the negative y-axis.

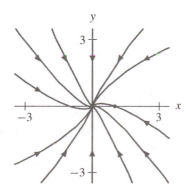

(e) At the point $\mathbf{Y}_0 = (1, 0)$, $d\mathbf{Y}/dt = (-3, 1)$. Therefore, $x(t)$ decreases initially and $y(t)$ increases initially. The solution eventually approaches the origin tangent to the positive y-axis. Therefore, $x(t)$ monotonically decreases to zero and $y(t)$ eventually decreases toward zero. Since the solution with the initial condition $\mathbf{Y}_0$ never crosses y-axis in the phase plane, the function $x(t) > 0$ for all t.

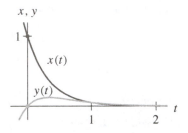

3. (a) The characteristic equation is

$$(-2 - \lambda)(-4 - \lambda) + 1 = (\lambda + 3)^2 = 0,$$

and the eigenvalue is $\lambda = -3$.

(b) To find an eigenvector, we solve the simultaneous equations

$$\begin{cases} -2x - y = -3x \\ x - 4y = -3y. \end{cases}$$

Then, $y = x$, and one eigenvector is $(1, 1)$.

(c) Note the straight-line solutions along the line $y = x$.

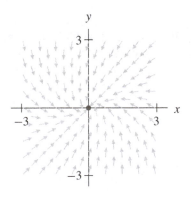

(d) Since the eigenvalue is negative, any solution on the line $y = x$ tends toward the origin along $y = x$ as t increases. According to the direction field, every solution tends to the origin as t increases. The solutions with initial conditions that lie in the half-plane $y > x$ eventually approach the origin tangent to the half-line $y = x$ with $y < 0$. Similarly, the solutions with initial conditions that lie in the half-plane $y < x$ eventually approach the origin tangent to the line $y = x$ with $y > 0$.

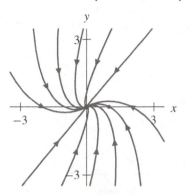

(e) At the point $\mathbf{Y}_0 = (1, 0)$, $d\mathbf{Y}/dt = (-2, 1)$. Therefore, $x(t)$ initially decreases and $y(t)$ initially increases. The solution eventually approaches the origin tangent to the line $y = x$. Since the solution curve never crosses the line $y = x$, the graphs of $x(t)$ and $y(t)$ do not cross.

5. **(a)** According to Exercise 1, there is one eigenvalue, -3, with eigenvectors of the form $(0, y_0)$, where $y_0 \neq 0$.

To find the general solution, we start with an arbitrary initial condition $\mathbf{V}_0 = (x_0, y_0)$. Then

$$\mathbf{V}_1 = \left[\begin{pmatrix} -3 & 0 \\ 1 & -3 \end{pmatrix} + 3 \begin{pmatrix} 1 & 0 \\ 0 & 1 \end{pmatrix} \right] \mathbf{V}_0$$

$$= \begin{pmatrix} 0 & 0 \\ 1 & 0 \end{pmatrix} \begin{pmatrix} x_0 \\ y_0 \end{pmatrix}$$

$$= \begin{pmatrix} 0 \\ x_0 \end{pmatrix}.$$

We obtain the general solution

$$\mathbf{Y}(t) = e^{-3t} \begin{pmatrix} x_0 \\ y_0 \end{pmatrix} + t e^{-3t} \begin{pmatrix} 0 \\ x_0 \end{pmatrix}.$$

(b) The solution that satisfies the initial condition $(x_0, y_0) = (1, 0)$ is

$$\mathbf{Y}(t) = e^{-3t} \begin{pmatrix} 1 \\ 0 \end{pmatrix} + t e^{-3t} \begin{pmatrix} 0 \\ 1 \end{pmatrix}.$$

Hence, $x(t) = e^{-3t}$ and $y(t) = t e^{-3t}$.

(c) Compare the graphs of $x(t) = e^{-3t}$ and $y(t) = te^{-3t}$ with the sketches obtained in part (e) of Exercise 1.

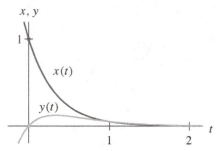

7. **(a)** From Exercise 3, we know that there is only one eigenvalue, $\lambda = -3$, and the eigenvectors (x_0, y_0) satisfy the equation $y_0 = x_0$.

To find the general solution, we start with an arbitrary initial condition $\mathbf{V}_0 = (x_0, y_0)$. Then

$$\mathbf{V}_1 = \left[\begin{pmatrix} -2 & -1 \\ 1 & -4 \end{pmatrix} + 3 \begin{pmatrix} 1 & 0 \\ 0 & 1 \end{pmatrix} \right] \mathbf{V}_0$$

$$= \begin{pmatrix} 1 & -1 \\ 1 & -1 \end{pmatrix} \begin{pmatrix} x_0 \\ y_0 \end{pmatrix}$$

$$= \begin{pmatrix} x_0 - y_0 \\ x_0 - y_0 \end{pmatrix}.$$

We obtain the general solution

$$\mathbf{Y}(t) = e^{-3t} \begin{pmatrix} x_0 \\ y_0 \end{pmatrix} + te^{-3t} \begin{pmatrix} x_0 - y_0 \\ x_0 - y_0 \end{pmatrix}.$$

(b) The solution that satisfies the initial condition $(x_0, y_0) = (1, 0)$ is

$$\mathbf{Y}(t) = e^{-3t} \begin{pmatrix} 1 \\ 0 \end{pmatrix} + te^{-3t} \begin{pmatrix} 1 \\ 1 \end{pmatrix}.$$

Hence, $x(t) = e^{-3t}(t + 1)$ and $y(t) = te^{-3t}$.

(c) Compare the graphs of $x(t) = e^{-3t}(t + 1)$ and $y(t) = te^{-3t}$ with the sketches obtained in part (e) of Exercise 3.

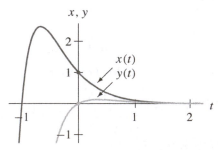

9. **(a)** By solving the quadratic equation, we obtain

$$\lambda = \frac{-\alpha \pm \sqrt{\alpha^2 - 4\beta}}{2}.$$

Therefore, for the quadratic to have a double root, we must have

$$\alpha^2 - 4\beta = 0.$$

(b) If zero is a root, we set $\lambda = 0$ in $\lambda^2 + \alpha\lambda + \beta = 0$, and we obtain $\beta = 0$.

11. The characteristic equation is

$$-\lambda(-p - \lambda) + q = \lambda^2 + p\lambda + q = 0.$$

Solving the quadratic equation, one obtains

$$\lambda = \frac{-p \pm \sqrt{p^2 - 4q}}{2}.$$

(a) Therefore, in order for $\mathbf{A}$ to have two real eigenvalues, p and q must satisfy $p^2 - 4q > 0$.
(b) In order for $\mathbf{A}$ to have complex eigenvalues, p and q must satisfy $p^2 - 4q < 0$.
(c) In order for $\mathbf{A}$ to have only one eigenvalue, p and q must satisfy $p^2 - 4q = 0$.

13. Since every vector is an eigenvector with eigenvalue λ, we substitute $\mathbf{Y} = (1, 0)$ into the equation $\mathbf{AY} = \lambda\mathbf{Y}$ and get

$$\mathbf{A}\begin{pmatrix} 1 \\ 0 \end{pmatrix} = \begin{pmatrix} a \\ c \end{pmatrix} = \lambda\begin{pmatrix} 1 \\ 0 \end{pmatrix}.$$

Hence, $a = \lambda$ and $c = 0$. Similarly, letting $\mathbf{Y} = (0, 1)$, we have

$$\begin{pmatrix} b \\ d \end{pmatrix} = \lambda\begin{pmatrix} 0 \\ 1 \end{pmatrix}.$$

Therefore, $b = 0$ and $d = \lambda$.

15. Since $\mathbf{Y}_1(0) = \mathbf{V}_0$ and $\mathbf{Y}_2(0) = \mathbf{W}_0$, we see that $\mathbf{V}_0 = \mathbf{W}_0$.
Evaluating at $t = 1$ yields

$$\mathbf{Y}_1(1) = e^\lambda(\mathbf{V}_0 + \mathbf{V}_1) \quad \text{and} \quad \mathbf{Y}_2(1) = e^\lambda(\mathbf{W}_0 + \mathbf{W}_1).$$

Since $\mathbf{Y}_1(1) = \mathbf{Y}_2(1)$ and $\mathbf{V}_0 = \mathbf{W}_0$, we see that $\mathbf{V}_1 = \mathbf{W}_1$.

17. **(a)** The characteristic polynomial is

$$(-\lambda)(-1 - \lambda) + 0 = \lambda^2 + \lambda,$$

so the eigenvalues are $\lambda = 0$ and $\lambda = -1$.

(b) To find the eigenvectors $\mathbf{V}_1$ associated to the eigenvalue $\lambda = 0$, we must solve $\mathbf{AV}_1 = 0\mathbf{V}_1 = 0$ where $\mathbf{A}$ is the matrix that defines this linear system. (Note that this is the same calculation we do if we want to locate the equilibrium points.) We get

$$\begin{cases} 2y_1 = 0 \\ -y_1 = 0, \end{cases}$$

where $\mathbf{V}_1 = (x_1, y_1)$. Hence, the eigenvectors associated to $\lambda = 0$ (as well as the equilibrium points) must satisfy the equation $y_1 = 0$.

   To find the eigenvectors $\mathbf{V}_2$ associated to the eigenvalue $\lambda = -1$, we must solve $\mathbf{AV}_2 = -\mathbf{V}_2$. We get

$$\begin{cases} 2y_2 = -x_2 \\ -y_2 = -y_2. \end{cases}$$

where $\mathbf{V}_2 = (x_2, y_2)$. Hence, the eigenvectors associated to $\lambda = -1$ must satisfy $2y_2 = -x_2$.

(c) The equation $y_1 = 0$ specifies a line of equilibrium points. Since the other eigenvalue is negative, solution curves not corresponding to equilibria move toward this line as t increases.

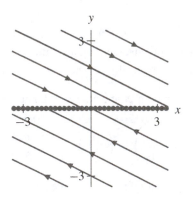

(d) Since $(1, 0)$ is an equilibrium point, it is easy to sketch the corresponding $x(t)$- and $y(t)$-graphs.

(e) To form the general solution, we must pick one eigenvector for each eigenvalue. Using part (b), we pick $\mathbf{V}_1 = (1, 0)$, and $\mathbf{V}_2 = (2, -1)$. We obtain the general solution

$$\mathbf{Y}(t) = k_1 \begin{pmatrix} 1 \\ 0 \end{pmatrix} + k_2 e^{-t} \begin{pmatrix} 2 \\ -1 \end{pmatrix}.$$

(f) To determine the solution whose initial condition is $(1, 0)$, we can substitute $t = 0$ in the general solution and solve for k_1 and k_2. However, since this initial condition is an equilibrium point, we need not make the effort. We simply observe that

$$\mathbf{Y}(t) = \begin{pmatrix} 1 \\ 0 \end{pmatrix}$$

is the desired solution.

19. **(a)** The characteristic polynomial is

$$(4 - \lambda)(1 - \lambda) - 4 = \lambda^2 - 5\lambda,$$

so the eigenvalues are $\lambda = 0$ and $\lambda = 5$.

(b) To find the eigenvectors $\mathbf{V}_1$ associated to the eigenvalue $\lambda = 0$, we must solve $\mathbf{AV}_1 = 0\mathbf{V}_1 = 0$ where $\mathbf{A}$ is the matrix that defines this linear system. (Note that this is the same calculation we do if we want to locate the equilibrium points.) We get

$$\begin{cases} 4x_1 + 2y_1 = 0 \\ 2x_1 + y_1 = 0, \end{cases}$$

where $\mathbf{V}_1 = (x_1, y_1)$. Hence, the eigenvectors associated to $\lambda = 0$ (as well as the equilibrium points) must satisfy the equation $y_1 = -2x_1$.

To find the eigenvectors $\mathbf{V}_2$ associated to the eigenvalue $\lambda = 5$, we must solve $\mathbf{AV}_2 = 5\mathbf{V}_2$. We get

$$\begin{cases} 4x_2 + 2y_2 = 5x_2 \\ 2x_2 + y_2 = 5y_2. \end{cases}$$

where $\mathbf{V}_2 = (x_2, y_2)$. Hence, the eigenvectors associated to $\lambda = 5$ must satisfy $x_2 = 2y_2$.

(c) The equation $y_1 = -2x_1$ specifies a line of equilibrium points. Since the other eigenvalue is positive, solution curves not corresponding to equilibria move away from this line as t increases.

(d) As t increases, both $x(t)$ and $y(t)$ increase exponentially. As t decreases, both x and y approach constants that are determined by the line of equilibrium points.

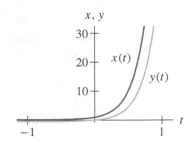

(e) To form the general solution, we must pick one eigenvector for each eigenvalue. Using part (b), we pick $\mathbf{V}_1 = (1, -2)$, and $\mathbf{V}_2 = (2, 1)$. We obtain the general solution

$$\mathbf{Y}(t) = k_1 \begin{pmatrix} 1 \\ -2 \end{pmatrix} + k_2 e^{5t} \begin{pmatrix} 2 \\ 1 \end{pmatrix}.$$

(f) To determine the solution whose initial condition is $(1, 0)$, we let $t = 0$ in the general solution and obtain the equations

$$k_1 \begin{pmatrix} 1 \\ -2 \end{pmatrix} + k_2 \begin{pmatrix} 2 \\ 1 \end{pmatrix} = \begin{pmatrix} 1 \\ 0 \end{pmatrix}.$$

Therefore, $k_1 = 1/5$ and $k_2 = 2/5$, and the particular solution is

$$\mathbf{Y}(t) = \begin{pmatrix} \frac{1}{5} + \frac{4}{5}e^{5t} \\ -\frac{2}{5} + \frac{2}{5}e^{5t} \end{pmatrix}.$$

21. (a) The characteristic polynomial is $\lambda^2 = 0$, so $\lambda = 0$ is the sole eigenvalue. To sketch the phase portrait we note that $dy/dt = 0$, so $y(t)$ is always a constant function. Moreover, $dx/dt = 2y$, so $x(t)$ is increasing if $y > 0$, and it is decreasing if $y < 0$.

(b) This system is exactly the same as the one in part (a) except that the sign of dx/dt has changed. Hence, the phase portrait is the identical except for the fact that the arrows point the other way.

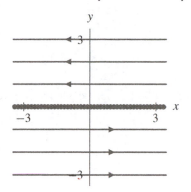

23. (a) The characteristic polynomial is $(a - \lambda)(d - \lambda)$, so the eigenvalues are a and d.

(b) If $a \neq d$, the lines of eigenvectors for a and d are the x- and y-axes respectively.

(c) If $a = d < 0$, every nonzero vector is an eigenvector (see Exercise 14), and all the vectors point toward the origin. Hence, every solution curve is asymptotic to the origin along a straight line.

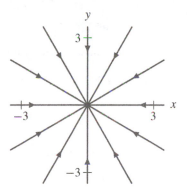

The general solution is $\mathbf{Y}(t) = e^{at}\mathbf{Y}_0$, where $\mathbf{Y}_0$ is the initial condition.

(d) The only difference between this case and part (c) is that the arrows in the vector field are reversed. Every solution tends away from the origin along a straight line.

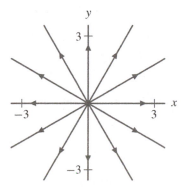

Again the general solution is $\mathbf{Y}(t) = e^{at}\mathbf{Y}_0$, where $\mathbf{Y}_0$ is the initial condition.

EXERCISES FOR SECTION 3.6

1. The characteristic polynomial is

$$s^2 - 6s - 7,$$

so the eigenvalues are $s = -1$ and $s = 7$. Hence, the general solution is

$$y(t) = k_1 e^{-t} + k_2 e^{7t}.$$

3. The characteristic polynomial is

$$s^2 + 6s + 9,$$

so $s = -3$ is a repeated eigenvalue. Hence, the general solution is

$$y(t) = k_1 e^{-3t} + k_2 t e^{-3t}.$$

5. The characteristic polynomial is

$$s^2 + 8s + 25,$$

so the complex eigenvalues are $s = -4 \pm 3i$. Hence, the general solution is

$$y(t) = k_1 e^{-4t} \cos 3t + k_2 e^{-4t} \sin 3t.$$

7. The characteristic polynomial is

$$s^2 + 2s - 3,$$

so the eigenvalues are $s = 1$ and $s = -3$. Hence, the general solution is

$$y(t) = k_1 e^t + k_2 e^{-3t},$$

and we have

$$y'(t) = k_1 e^t - 3k_2 e^{-3t}.$$

From the initial conditions, we obtain the simultaneous equations

$$\begin{cases} k_1 + k_2 = 6 \\ k_1 - 3k_2 = -2. \end{cases}$$

Solving for k_1 and k_2 yields $k_1 = 4$ and $k_2 = 2$. Hence, the solution to our initial-value problem is $y(t) = 4e^t + 2e^{-3t}$.

9. The characteristic polynomial is

$$s^2 - 4s + 13,$$

so the eigenvalues are $s = 2 \pm 3i$. Hence, the general solution is

$$y(t) = k_1 e^{2t} \cos 3t + k_2 e^{2t} \sin 3t.$$

From the initial condition $y(0) = 1$, we see that $k_1 = 1$. Differentiating

$$y(t) = e^{2t} \cos 3t + k_2 e^{2t} \sin 3t$$

and evaluating $y'(t)$ at $t = 0$ yields $y'(0) = 2 + 3k_2$. Since $y'(0) = -4$, we have $k_2 = -2$. Hence, the solution to our initial-value problem is

$$y(t) = e^{2t} \cos 3t - 2e^{2t} \sin 3t.$$

11. The characteristic polynomial is

$$s^2 - 8s + 16,$$

so $s = 4$ is a repeated eigenvalue. Hence, the general solution is

$$y(t) = k_1 e^{4t} + k_2 t e^{4t}.$$

From the initial condition $y(0) = 3$, we see that $k_1 = 3$. Differentiating

$$y(t) = 3e^{4t} + k_2 t e^{4t}$$

and evaluating $y'(t)$ at $t = 0$ yields $y'(0) = 12 + k_2$. Since $y'(0) = 11$, we have $k_2 = -1$. Hence, the solution to our initial-value problem is

$$y(t) = 3e^{4t} - t e^{4t}.$$

13. **(a)** The resulting second-order equation is

$$\frac{d^2 y}{dt^2} + 8\frac{dy}{dt} + 7y = 0,$$

and the corresponding system is

$$\frac{dy}{dt} = v$$

$$\frac{dv}{dt} = -7y - 8v.$$

(b) Recall that we can read off the characteristic equation of the second-order equation straight from the equation without having to revert to the corresponding system. We obtain

$$\lambda^2 + 8\lambda + 7 = 0.$$

Therefore, the eigenvalues are $\lambda_1 = -1$ and $\lambda_2 = -7$.

To find the eigenvectors associated to the eigenvalue λ_1, we solve the simultaneous system of equations

$$\begin{cases} v = -y \\ -7y - 8v = -v. \end{cases}$$

From the first equation, we immediately see that the eigenvectors associated to this eigenvalue must satisfy $v = -y$. Similarly, the eigenvectors associated to the eigenvalue $\lambda_2 = -7$ must satisfy the equation $v = -7y$.

(c) Since the eigenvalues are real and negative, the equilibrium point at the origin is a sink, and the system is overdamped.

(d) We know that all solution curves approach the origin as $t \to \infty$ and, with the exception of those whose initial conditions lie on the line $v = -7y$, these solution curves approach the origin tangent to the line $v = -y$.

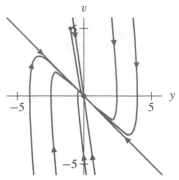

(e) From the phase portrait, we see that $y(t)$ increases monotonically toward 0 as $t \to \infty$. Also, $v(t)$ decreases monotonically toward 0. It is useful to remember that $v = dy/dt$.

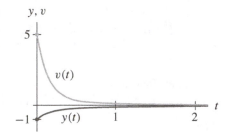

15. (a) The resulting second-order equation is

$$\frac{d^2 y}{dt^2} + 4\frac{dy}{dt} + 5y = 0,$$

and the corresponding system is

$$\frac{dy}{dt} = v$$
$$\frac{dv}{dt} = -5y - 4v.$$

(b) Recall that we can read off the characteristic equation of the second-order equation straight from the equation without having to revert to the corresponding system. We obtain

$$\lambda^2 + 4\lambda + 5 = 0.$$

Therefore, the eigenvalues are $\lambda_1 = -2 + i$ and $\lambda_2 = -2 - i$.
To find the eigenvectors associated to the eigenvalue λ_1, we solve the simultaneous system of equations

$$\begin{cases} v = (-2+i)y \\ -5y - 4v = (-2+i)v. \end{cases}$$

From the first equation, we immediately see that the eigenvectors associated to this eigenvalue must satisfy $v = (-2+i)y$. Similarly, the eigenvectors associated to the eigenvalue $\lambda_2 = -2-i$ must satisfy the equation $v = (-2 - i)y$.

(c) Since the eigenvalues are complex with negative real part, the equilibrium point at the origin is a spiral sink, and the system is underdamped.

(d) All solutions tend to the origin spiralling in the clockwise direction with period 2π. Admittedly, it is difficult to see these oscillations in the picture.

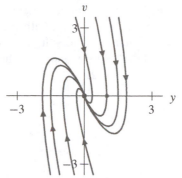

(e) The graph of $y(t)$ initially decreases then oscillates with decreasing amplitude as it tends to 0. Similarly, $v(t)$ initially decreases and becomes negative, then oscillates with decreasing amplitude as it tends to 0.

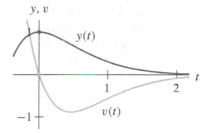

17. (a) The resulting second-order equation is

$$2\frac{d^2y}{dt^2} + 3\frac{dy}{dt} + y = 0,$$

and the corresponding system is

$$\frac{dy}{dt} = v$$
$$\frac{dv}{dt} = -\frac{1}{2}y - \frac{3}{2}v.$$

(b) Recall that we can read off the characteristic equation of the second-order equation straight from the equation without having to revert to the corresponding system. We obtain

$$2\lambda^2 + 3\lambda + 1 = 0.$$

Therefore, the eigenvalues are $\lambda_1 = -1$ and $\lambda_2 = -1/2$.

To find the eigenvectors associated to the eigenvalue λ_1, we solve the simultaneous system of equations

$$\begin{cases} v = -y \\ -\frac{1}{2}y - \frac{3}{2}v = -v. \end{cases}$$

From the first equation, we immediately see that the eigenvectors associated to this eigenvalue must satisfy $v = -y$. Similarly, the eigenvectors associated to the eigenvalue $\lambda_2 = -1/2$ must satisfy the equation $v = -y/2$.

(c) Since the eigenvalues are real and negative, the equilibrium point at the origin is a sink, and the system is overdamped.

(d) We know that all solution curves approach the origin as $t \to \infty$ and, with the exception of those whose initial conditions lie on the line $v = -y$, these solution curves approach the origin tangent to the line $v = -y/2$.

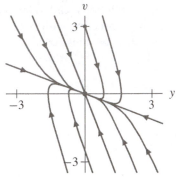

(e) According to the phase plane, $y(t)$ increases initially. Eventually it reaches a maximum value. Then it approaches 0 as $t \to \infty$. Also, $v(t)$ decreases, becomes negative, and then approaches 0 from below. While sketching these graphs, it is useful to remember that $v = dy/dt$.

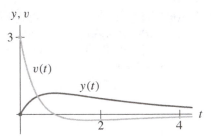

19. (a) The resulting second-order equation is

$$2\frac{d^2y}{dt^2} + 3y = 0,$$

and the corresponding system is

$$\frac{dy}{dt} = v$$
$$\frac{dv}{dt} = -\frac{3}{2}y.$$

(b) Recall that we can read off the characteristic equation of the second-order equation straight from the equation without having to revert to the corresponding system. We obtain

$$2\lambda^2 + 3 = 0.$$

Therefore, we have pure imaginary eigenvalues, $\lambda = \pm i\sqrt{3/2}$.

To find the eigenvectors associated to the eigenvalue $\lambda = i\sqrt{3/2}$, we solve the simultaneous system of equations

$$\begin{cases} v = i\sqrt{\frac{3}{2}}\, y \\ -\frac{3}{2}y = i\sqrt{\frac{3}{2}}\, v. \end{cases}$$

From the first equation, we immediately see that the eigenvectors associated to this eigenvalue must satisfy $v = i\sqrt{3/2}\, y$. Similarly, the eigenvectors associated to the eigenvalue $\lambda = -i\sqrt{3/2}$ must satisfy the equation $v = -i\sqrt{3/2}\, y$.

(c) Since the eigenvalues are pure imaginary, the system is undamped. (Of course, we already knew this because $b = 0$.) The natural period is $2\pi/\sqrt{3/2} = 4\pi/\sqrt{6}$.

(d) Since the eigenvalues are pure imaginary, we know that the solution curves are ellipses.   At the point $(1, 0)$, $d\mathbf{Y}/dt = (0, -3/2)$. Therefore, we know that the oscillation is clockwise.

(e)

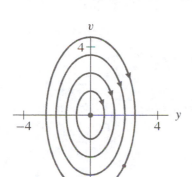

21. (a) The second-order equation is

$$\frac{d^2y}{dt^2} + 8\frac{dy}{dt} + 7y = 0,$$

so the characteristic equation is

$$s^2 + 8s + 7 = 0.$$

The roots are $s = -7$ and $s = -1$. The general solution is

$$y(t) = k_1 e^{-7t} + k_2 e^{-t}.$$

(b) To find the particular solution we compute

$$v(t) = -7k_1 e^{-7t} - k_2 e^{-t}.$$

The particular solution satisfies

$$\begin{cases} -1 = y(0) = k_1 + k_2 \\ 5 = v(0) = -7k_1 - k_2. \end{cases}$$

The first equation yields $k_1 = -k_2 - 1$. Substituting into the second we obtain $5 = 6k_2 + 7$, which implies $k_2 = -1/3$. The first equation then yields $k_1 = -2/3$. The particular solution is

$$y(t) = -\tfrac{2}{3}e^{-7t} - \tfrac{1}{3}e^{-t}.$$

(c) The $y(t)$- and $v(t)$-graphs are displayed in the solution of Exercise 13.

23. **(a)** The second-order equation is

$$\frac{d^2y}{dt^2} + 4\frac{dy}{dt} + 5y = 0,$$

so the characteristic equation is

$$s^2 + 4s + 5 = 0.$$

The roots are $s = -2 + i$ and $s = -2 - i$. A complex-valued solution is

$$y_c(t) = e^{(-2+i)t} = e^{-2t}\cos t + ie^{-2t}\sin t.$$

Therefore the general solution is

$$y(t) = k_1 e^{-2t}\cos t + k_2 e^{-2t}\sin t.$$

(b) To find the particular solution we compute

$$v(t) = (-2k_1 + k_2)e^{-2t}\cos t + (-k_1 - 2k_2)e^{-2t}\sin t.$$

The particular solution satisfies

$$\begin{cases} 1 = y(0) = k_1 \\ 0 = v(0) = -2k_1 + k_2. \end{cases}$$

The first equation yields $k_1 = 1$. Substituting into the second we obtain $k_2 = 2$. The particular solution is

$$y(t) = e^{-2t}\cos t + 2e^{-2t}\sin t.$$

(c) The $y(t)$- and $v(t)$-graphs are displayed in the solution of Exercise 15.

25. **(a)** The second-order equation is

$$2\frac{d^2y}{dt^2} + 3\frac{dy}{dt} + y = 0,$$

so the characteristic equation is

$$2s^2 + 3s + 1 = 0.$$

The roots are $s = -1$ and $s = -1/2$. So the general solution is

$$y(t) = k_1 e^{-t} + k_2 e^{-t/2}.$$

(b) To find the particular solution we compute

$$v(t) = -k_1 e^{-t} - \frac{k_2}{2}e^{-t/2}.$$

The particular solution satisfies

$$\begin{cases} 0 = y(0) = k_1 + k_2 \\ 3 = v(0) = -k_1 - \frac{k_2}{2}. \end{cases}$$

The first equation yields $k_1 = -k_2$. Substituting into the second we obtain $3 = k_2 - k_2/2$, which implies that $k_2 = 6$. The first equation then yields $k_1 = -6$. The particular solution is

$$y(t) = -6e^{-t} + 6e^{-t/2}.$$

(c) The $y(t)$- and $v(t)$-graphs are displayed in the solution of Exercise 17.

27. **(a)** The second-order equation is

$$2\frac{d^2y}{dt^2} + 3y = 0,$$

so the characteristic equation is

$$2s^2 + 3 = 0.$$

The roots are $s = \sqrt{3/2}\,i$ and $s = -\sqrt{3/2}\,i$. A complex-valued solution is

$$y_c(t) = e^{(\sqrt{3/2}\,i)t} = \cos(\sqrt{3/2}\,t) + i\sin(\sqrt{3/2}\,t).$$

Therefore the general solution is

$$y(t) = k_1\cos(\sqrt{3/2}\,t) + k_2\sin(\sqrt{3/2}\,t).$$

(b) To find the particular solution we compute

$$v(t) = -\sqrt{3/2}\,k_1\sin(\sqrt{3/2}\,t) + \sqrt{3/2}\,k_2\cos(\sqrt{3/2}\,t).$$

The particular solution satisfies

$$\begin{cases} 2 = y(0) = k_1 \\ -3 = v(0) = \sqrt{3/2}\,k_2. \end{cases}$$

The first equation yields $k_1 = 2$. Substituting into the second we find $k_2 = -\sqrt{6}$. So the particular solution is

$$y(t) = 2\cos(\sqrt{3/2}\,t) - \sqrt{6}\sin(\sqrt{3/2}\,t).$$

(c) The $y(t)$- and $v(t)$-graphs are displayed in the solution of Exercise 19.

29. Note: We assume that m, k and b are nonnegative—the physically relevant case. All references to graphs and phase portraits are from Sections 3.5 and 3.6.

Table 3.1
Possible harmonic oscillators.

name	eigenvalues	parameters	decay rate	phase portrait and graphs
undamped	pure imaginary	$b = 0$	no decay	Figure 3.41
underdamped	complex with negative real part	$b^2 - 4mk < 0$	$e^{-bt/(2m)}$	Figure 3.42
critically damped	only one eigenvalue	$b^2 - 4mk = 0$	$e^{-bt/(2m)}$	Figure 3.34
overdamped	two negative real	$b^2 - 4mk > 0$	$e^{\lambda t}$ where $\lambda = \dfrac{-b + \sqrt{b^2 - 4mk}}{2m}$	Figures 3.43–3.45 and Exercise 13

31. Note that

$$\frac{dy}{dt} = \frac{d}{dt}(y_{re} + iy_{im}) = \frac{dy_{re}}{dt} + i\frac{dy_{im}}{dt}$$

and

$$\frac{d^2y}{dt^2} = \frac{d^2}{dt^2}(y_{re} + iy_{im}) = \frac{d^2y_{re}}{dt^2} + i\frac{d^2y_{im}}{dt^2}.$$

Then note that

$$\frac{d^2y}{dt^2} + p\frac{dy}{dt} + qy = \left(\frac{d^2y_{re}}{dt^2} + p\frac{dy_{re}}{dt} + qy_{re}\right) + i\left(\frac{d^2y_{im}}{dt^2} + p\frac{dy_{im}}{dt} + qy_{im}\right).$$

Both

$$\frac{d^2y_{re}}{dt^2} + p\frac{dy_{re}}{dt} + qy_{re} = 0 \quad \text{and} \quad \frac{d^2y_{im}}{dt^2} + p\frac{dy_{im}}{dt} + qy_{im} = 0$$

because a complex number is zero only if both its real and imaginary parts vanish. In other words, $y_{re}(t)$ and $y_{im}(t)$ are solutions of the original equation.

33. **(a)** If we let $v = dy/dt$, then the corresponding first-order system is

$$\frac{dy}{dt} = v$$

$$\frac{dv}{dt} = -qy - pv,$$

and the corresponding matrix is

$$\mathbf{A} = \begin{pmatrix} 0 & 1 \\ -q & -p \end{pmatrix}.$$

If λ_0 is a repeated eigenvalue, then the characteristic polynomial is

$$\lambda^2 + p\lambda + q = (\lambda - \lambda_0^2) = \lambda^2 - 2\lambda_0\lambda + \lambda_0^2.$$

Consequently, $p = -2\lambda_0, q = \lambda_0^2$, and

$$\mathbf{A} = \begin{pmatrix} 0 & 1 \\ -\lambda_0^2 & 2\lambda_0 \end{pmatrix}.$$

(b) To compute the general solution of the corresponding first-order system, we consider an arbitrary initial condition $\mathbf{V}_0 = (y_0, v_0)$ and calculate

$$(\mathbf{A} - \lambda_0\mathbf{I})\mathbf{V}_0 = \begin{pmatrix} -\lambda_0 & 1 \\ -\lambda_0^2 & \lambda_0 \end{pmatrix}\begin{pmatrix} y_0 \\ v_0 \end{pmatrix}$$

$$= \begin{pmatrix} -\lambda_0 y_0 + v_0 \\ -\lambda_0^2 y_0 + \lambda_0 v_0 \end{pmatrix}.$$

The general solution of the first-order system is

$$\mathbf{Y}(t) = e^{\lambda_0 t}\begin{pmatrix} y_0 \\ v_0 \end{pmatrix} + te^{\lambda_0 t}\begin{pmatrix} -\lambda_0 y_0 + v_0 \\ -\lambda_0^2 y_0 + \lambda_0 v_0 \end{pmatrix}.$$

(c) From the first component of the result in part (b), we obtain the general solution of the original second-order equation in the form

$$y(t) = y_0 e^{\lambda_0 t} + (-\lambda_0 y_0 + v_0) t e^{\lambda_0 t}.$$

(d) Let $k_1 = y_0$ and $k_2 = -\lambda_0 y_0 + v_0$. Clearly, all k_1 are possible. Moreover, once the value of k_1 is determined, k_2 can be determined from v_0 using $k_2 = -\lambda_0 k_1 + v_0$, and v_0 can be determined by k_2 using $v_0 = k_2 + \lambda_0 k_1$. Hence, k_1 and k_2 are arbitrary constants because y_0 and v_0 are arbitrary.

35. The characteristic equation for this harmonic oscillator is

$$s^2 + bs + 3 = 0,$$

and the roots are

$$s_1 = \frac{-b - \sqrt{b^2 - 12}}{2} \quad \text{and} \quad s_2 = \frac{-b + \sqrt{b^2 - 12}}{2}.$$

If $b^2 < 12$, these roots are complex. In this case, all solutions include a factor of the form $e^{(-b/2)t}$, and they tend to the equilibrium at this rate.

If $b^2 > 12$, the roots are real, and the general solution is

$$y(t) = k_1 e^{s_1 t} + k_2 e^{s_2 t}.$$

The first exponential in this expression tends to 0 most quickly, so if $k_2 = 0$, we have solutions that tend to 0 at the rate of $e^{s_1 t}$. This rate is the quickest approach to 0.

The roots are repeated if $b^2 - 12 = 0$, that is, if $b = 2\sqrt{3}$. The fastest approach is then given by a term of the form $e^{-\sqrt{3}\,t}$.

37. (a) Since the fluid causes the object to accelerate as it moves and the force causing this acceleration is proportional to the velocity, the force equation for this "mass-spring" system is

$$m\frac{d^2 y}{dt^2} = -ky + b_{mf}\frac{dy}{dt},$$

which can be written as

$$m\frac{d^2 y}{dt^2} - b_{mf}\frac{dy}{dt} + ky = 0.$$

(b) The equivalent first-order system is

$$\frac{dy}{dt} = v$$

$$\frac{dv}{dt} = -\frac{k}{m}y + \frac{b_{mf}}{m}v.$$

(c) The characteristic equation is

$$m\lambda^2 - b_{mf}\lambda + k = 0,$$

and the eigenvalues are

$$\frac{b_{mf} \pm \sqrt{b_{mf}^2 - 4mk}}{2m}.$$

Since m, b_{mf}, and k are all positive parameters, the eigenvalues are either positive real numbers or complex numbers with a positive real part. If both eigenvalues are real, then the origin is called an "overstimulated" source. The magnitudes of $y(t)$ and $v(t)$ tend to infinity without oscillation. If the eigenvalues are complex, then the origin is a spiral source and the oscillator is called understimulated. The solutions spiral away from the origin with natural period $4m\pi/\sqrt{b_{mf}^2 - 4mk}$.

39. The differential equation is

$$m\frac{d^2y}{dt^2} + 2y = 0,$$

and the characteristic equation is $m\lambda^2 + 2 = 0$. Hence, the eigenvalues are $\lambda = \pm i\sqrt{2/m}$. The natural period is $2\pi/\sqrt{2/m} = \pi\sqrt{2m}$. For natural period to be 1, we must have $m = 1/(2\pi^2)$.

EXERCISES FOR SECTION 3.7

1.

Table 3.2

Possibilities for linear systems

type	condition on λ	examples
sink	$\lambda_1 < \lambda_2 < 0$	Sec. 3.7, Fig. 3.52
saddle	$\lambda_1 < 0 < \lambda_2$	Sec. 3.3, Fig. 3.12–3.14
source	$0 < \lambda_1 < \lambda_2$	Sec. 3.3, Fig. 3.19
spiral sink	$\lambda = \alpha \pm i\beta, \alpha < 0, \beta \neq 0$	Sec. 3.1, Fig. 3.2 and 3.4
spiral source	$\lambda = \alpha \pm i\beta, \alpha > 0, \beta \neq 0$	Sec. 3.4, Fig. 3.29–3.30
center	$\lambda_1 = \pm i\beta, \beta \neq 0$	Sec. 3.1, Fig. 3.1 and 3.3
		Sec. 3.4, Fig. 3.28
sink	$\lambda_1 = \lambda_2 < 0$	Sec. 3.5, Fig. 3.35–3.36
(special case)	One line of eigenvectors	
source	$0 < \lambda_1 = \lambda_2$	Sec. 3.5, Ex. 2
(special case)	One line of eigenvectors	
sink	$\lambda_1 = \lambda_2 < 0$	Sec. 3.5, Ex. 23
(special case)	Every vector is eigenvector	
source	$0 < \lambda_1 = \lambda_2$	Sec. 3.5, Ex. 23
(special case)	Every vector is eigenvector	
no name	$\lambda_1 < \lambda_2 = 0$	Sec. 3.5, Fig. 3.39–3.40
no name	$0 = \lambda_1 < \lambda_2$	Sec. 3.5, Ex. 19
no name	$\lambda_1 = \lambda_2 = 0$	Sec. 3.5, Ex. 21
	One line of eigenvectors	
no name	$\lambda_1 = \lambda_2 = 0$	entire plane of equilibrium points
	Every vector is an eigenvector	

3. (a)

(b) The trace T is $2a$, and the determinant D is $-a$. Therefore, the curve in the trace-determinant plane is $D = -T/2$. This line crosses the parabola $T^2 - 4D = 0$ at two points—at $(T, D) = (0, 0)$ if $a = 0$ and at $(T, D) = (-2, 1)$ if $a = -1$.

The portion of the line for which $a < -1$ corresponds to a positive determinant and a negative trace such that $T^2 - 4D > 0$. The corresponding phase portraits are real sinks. If $a = -1$, we have a sink with repeated eigenvalues. If $-1 < a < 0$, we have complex eigenvalues with negative real parts. Therefore, the phase portraits are spiral sinks. If $a = 0$, we have a degenerate case with an entire line of equilibrium points. Finally, if $a > 0$, the corresponding portion of the line is below the T-axis, and the phase portraits are saddles.

(c) Bifurcations occur at $a = -1$, where we have a sink with repeated eigenvalues, and at $a = 0$, where we have zero as a repeated eigenvalue. For $a = 0$, the y-axis is entirely composed of equilibrium points.

5. (a)

(b) The curve in the trace-determinant plane is the portion of the unit circle centered at 0 that lies in the half-plane $y \le 0$.

A glance at the trace-determinant plane shows that for $-1 < a < 1$, we have a saddle. If $a = 1$, the eigenvalues are 0 and 1. If $a = -1$, the eigenvalues are 0 and -1.

(c) Bifurcations occur only at $a = \pm 1$. For these two special values of a, we have a line of equilibrium points. The nonzero equilibrium points disappear if $-1 < a < 1$.

7. (a)

(b) The trace T is $2a$, and the determinant D is $a^2 - a$. Therefore, the curve in the trace-determinant plane is

$$D = a^2 - a$$

$$= \left(\frac{T}{2}\right)^2 - \frac{T}{2}$$

$$= \frac{T^2}{4} - \frac{T}{2}.$$

This curve is a parabola. It meets the repeated-eigenvalue parabola (the parabola $D = T^2/4$) if

$$\frac{T^2}{4} - \frac{T}{2} = \frac{T^2}{4}.$$

Solving this equation yields $T = 0$, which corresponds to $a = 0$.
This curve also meets the T-axis (the line $D = 0$) if

$$\frac{T^2}{4} - \frac{T}{2} = 0,$$

so if $T = 0$ or $T = 2$, then $D = 0$.
From the location of the parabola $D = T^2/4 - T/2$ in the trace-determinant plane, we see that the phase portrait is a spiral sink if $a < 0$ since $T < 0$, a saddle if $0 < a < 1$ since $0 < T < 2$, and a source with distinct real eigenvalues if $a > 1$ since $T > 2$.

(c) Bifurcations occur at $a = 0$, where we have repeated zero eigenvalues, and at $a = 1$, where we have a single zero eigenvalue.

9. The eigenvalues are roots of the equation $\lambda^2 - 2a\lambda + a^2 - b^2 = 0$. These roots are

$$a \pm \sqrt{b^2} = a \pm |b|.$$

So we have a repeated zero eigenvalue if $a = b = 0$.
If $a = \pm b$, then one of the eigenvalues is 0, and as long as $a \neq 0$ (so $b \neq 0$), the other eigenvalue is nonzero.
The eigenvalues are repeated (both equal to a) if $b = 0$. The eigenvalues are never complex since $\sqrt{b^2} \geq 0$.
If $a > |b|$, then $a \pm |b| > 0$, so we have a source with real eigenvalues. If $a < 0$ and $-a > |b|$, then $a \pm |b| < 0$, so we have a sink with real eigenvalues. In all other cases we have a saddle.

11. **(a)** This second-order equation is equivalent to the system

$$\frac{dy}{dt} = v$$

$$\frac{dv}{dt} = -3y - bv.$$

Therefore, $T = -b$ and $D = 3$. So the corresponding curve in the trace-determinant plane is $D = 3$.

(b)

(c) The line $D = 3$ in the trace-determinant plane crosses the repeated-eigenvalue parabola $D = T^2/4$ if $b^2 = 12$, which implies that $b = 2\sqrt{3}$ since b is a nonnegative parameter. If $b = 0$, we have pure imaginary eigenvalues—the undamped case. If $0 < b < 2\sqrt{3}$, the eigenvalues are complex with a negative real part—the underdamped case. If $b = 2\sqrt{3}$, the eigenvalues are repeated and negative—the critically damped case. Finally, if $b > 2\sqrt{3}$, the eigenvalues are real and negative—the overdamped case.

13. (a) The second-order equation reduces to the first-order system

$$\frac{dy}{dt} = v$$

$$\frac{dv}{dt} = -\frac{2}{m}y + \frac{1}{m}v.$$

Hence $T = -1/m$, $D = 2/m$, and as the parameter m varies, the systems move along the line $D = -2T$ in the second quadrant of the trace-determinant plane.

(b)

(c) The line $D = -2T$ intersects the repeated-eigenvalue parabola $D = T^2/4$ at the point (T, D) that satisfies $-2T = T^2/4$. We have

$$\frac{T^2}{4} + 2T = T\left(\frac{T}{4} + 2\right) = 0,$$

which yields $T = 0$ or $T = -8$.

For $-8 < T < 0$, the system is underdamped; for $T = -8$, the system is critically damped; and for $T < -8$, the system is overdamped. Since $T = -1/m$, the system is overdamped if $0 < m < 1/8$; it is critically damped if $m = 1/8$; and it is underdamped if $m > 1/8$.

EXERCISES FOR SECTION 3.8

1. To check that the given vector-valued functions are solutions, we differentiate each coordinate and check that the system is satisfied. More precisely, we must check that $dx/dt = 0.1y$, $dy/dt = 0.2z$, and $dz/dt = 0.4x$.

To check that these equations are satisfied, we simply differentiate and simplify. For example, to check that $\mathbf{Y}_2(t)$ is a solution, we first differentiate the x-coordinate, and we obtain

$$\frac{dx}{dt} = -0.1\,e^{-0.1t}\left(-\cos\left(\sqrt{0.03}\,t\right) - \sqrt{3}\,\sin\left(\sqrt{0.03}\,t\right)\right) +$$
$$e^{-0.1t}\left(\sqrt{0.03}\,\sin\left(\sqrt{0.03}\,t\right) - \sqrt{3}\,\sqrt{0.03}\,\cos\left(\sqrt{0.03}\,t\right)\right)$$
$$= e^{-0.1t}\left(\left(0.1 - \sqrt{0.09}\right)\cos\left(\sqrt{0.03}\,t\right) + \left(0.1\sqrt{3} + \sqrt{0.03}\right)\sin\left(\sqrt{0.03}\,t\right)\right)$$
$$= e^{-0.1t}\left(-0.2\cos\left(\sqrt{0.03}\,t\right) + 0.2\sqrt{3}\,\sin\left(\sqrt{0.03}\,t\right)\right)$$
$$= 0.1\,e^{-0.1t}\left(-2\cos\left(\sqrt{0.03}\,t\right) + 2\sqrt{3}\,\sin\left(\sqrt{0.03}\,t\right)\right).$$

Note that this last expression is just $0.1y(t)$. Hence, the first component of the differential equation is satisfied.

In order to complete the verification that $\mathbf{Y}_2(t)$ is a solution, we must also verify the equations for dy/dt and dz/dt. The calculations are similar.

3. (a) Suppose

$$k_1\begin{pmatrix}1\\2\\1\end{pmatrix} + k_2\begin{pmatrix}1\\3\\1\end{pmatrix} + k_3\begin{pmatrix}1\\4\\1\end{pmatrix} = \begin{pmatrix}0\\0\\0\end{pmatrix}.$$

We obtain the simultaneous equations

$$\begin{cases} k_1 + k_2 + k_3 = 0 \\ 2k_1 + 3k_2 + 4k_3 = 0. \end{cases}$$

We have two equations with three unknowns. Therefore, we cannot uniquely determine the values of k_1, k_2, and k_3. In other words, we can find infinitely many triples (k_1, k_2, k_3) such that $k_1\mathbf{Y}_1 + k_2\mathbf{Y}_2 + k_3\mathbf{Y}_3 = 0$. For example, $k_1 = -1$, $k_2 = 2$, and $k_3 = -1$ is one such triple.

(b) Suppose

$$k_1\begin{pmatrix}2\\0\\1\end{pmatrix} + k_2\begin{pmatrix}3\\2\\2\end{pmatrix} + k_3\begin{pmatrix}1\\-2\\-3\end{pmatrix} = \begin{pmatrix}0\\0\\0\end{pmatrix}.$$

In matrix notation, we can write this as

$$\mathbf{A}\begin{pmatrix}k_1\\k_2\\k_3\end{pmatrix} = \begin{pmatrix}-2 & 3 & 0\\ 3 & -2 & 0\\ 0 & 0 & -1\end{pmatrix}\begin{pmatrix}k_1\\k_2\\k_3\end{pmatrix} = \begin{pmatrix}0\\0\\0\end{pmatrix}.$$

Since $\det \mathbf{A} = -12 \neq 0$, $\mathbf{A}$ has an inverse matrix, and by multiplying the inverse matrix, we obtain $(k_1, k_2, k_3) = (0, 0, 0)$. Hence, the vectors are linearly independent.

(c) Suppose

$$k_1 \begin{pmatrix} 1 \\ 2 \\ 0 \end{pmatrix} + k_2 \begin{pmatrix} 0 \\ 1 \\ 2 \end{pmatrix} + k_3 \begin{pmatrix} 2 \\ 0 \\ 1 \end{pmatrix} = \begin{pmatrix} 0 \\ 0 \\ 0 \end{pmatrix}.$$

In scalar form, this vector equation is equivalent to the scalar equations

$$k_1 + k_3 = 0$$
$$2k_1 + k_2 = 0$$
$$2k_2 + k_3 = 0.$$

From the third equation, $-2k_3 = k_1$, and substitution of this equation into the first equation yields $k_1 = 0$. Then, $k_3 = 0$, and using the second equation, we conclude that $k_2 = 0$. Therefore, the three vectors are independent.

(d) Suppose

$$k_1 \begin{pmatrix} -3 \\ \pi \\ 1 \end{pmatrix} + k_2 \begin{pmatrix} 0 \\ 1 \\ 0 \end{pmatrix} + k_3 \begin{pmatrix} -2 \\ -2 \\ -2 \end{pmatrix} = \begin{pmatrix} 0 \\ 0 \\ 0 \end{pmatrix}.$$

This vector equation can be written as the simultaneous equations

$$-3k_1 - 2k_3 = 0$$
$$\pi k_1 + k_2 - 2k_3 = 0$$
$$k_1 - 2k_3 = 0.$$

From the third equation, $2k_3 = k_1$ and substitution of this equation into the first equation yields $k_1 = 0$. Then, $k_3 = 0$, and using the second equation, we obtain $k_2 = 0$. Therefore, the three vectors are independent.

5. **(a)** The characteristic equation is

$$\det(\mathbf{A} - \lambda \mathbf{I}) = (-2 - \lambda)(-2 - \lambda)(-1 - \lambda) - (3)(3)(-1 - \lambda) = 0,$$

which reduces to

$$-(\lambda + 1)(\lambda + 5)(\lambda - 1) = 0.$$

Therefore, the eigenvalues are $\lambda = \pm 1$ and $\lambda = -5$.

(b) Writing the differential equation in coordinates, one obtains

$$\frac{dx}{dt} = -2x + 3y$$
$$\frac{dy}{dt} = 3x - 2y$$
$$\frac{dz}{dt} = -z.$$

Since dx/dt and dy/dt do not depend on z and dz/dt does not depend on x or y, the system decouples into a two-dimensional system in the xy-plane and a one-dimensional system on the z-axis.

(c) In the xy-plane, the characteristic equation is $(-2 - \lambda)^2 - 9 = \lambda^2 + 4\lambda - 5 = 0$, and the eigenvalues are $\lambda = -5$ and $\lambda = 1$. Therefore, the system is a saddle in the xy-plane. The eigenvectors (x, y, z) for $\lambda = -5$ satisfy the equations $y = -x$ and $z = 0$, and the eigenvectors for $\lambda = 1$ satisfy the equations $x = y$ and $z = 0$.

The z-axis is the phase line for the equation $dz/dt = -z$. Therefore, there is a single equilibrium point at the origin, and every solution curve lying on the z-axis is asymptotic to 0 as $t \to \infty$.

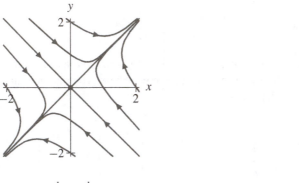

xy-phase plane z-phase line

(d) The xy-phase plane and the z-phase line can be combined to obtain the xyz-phase space.

7. (a) The characteristic equation is

$$\det(\mathbf{A} - \lambda\mathbf{I}) = (1 - \lambda)((2 - \lambda)(2 - \lambda) - 1) = 0,$$

which simplifies to

$$-(\lambda - 3)(\lambda - 1)(\lambda - 1) = 0.$$

Therefore, the eigenvalues are $\lambda = 1$ and $\lambda = 3$.

(b) Writing the differential equation in coordinates, one obtains

$$\frac{dx}{dt} = x$$
$$\frac{dy}{dt} = 2y - z$$
$$\frac{dz}{dt} = -y + 2z.$$

Since dy/dt and dz/dt do not depend on x, and dx/dt does not depend on y or z, the system decouples into a two-dimensional system in the yz-plane and a one-dimensional system on the x-axis.

(c) In yz-plane, the characteristic equation is $(2-\lambda)^2 - 1 = (\lambda-3)(\lambda-1) = 0$, and the eigenvalues are $\lambda = 1$ and $\lambda = 3$. Therefore, in the yz-phase plane, the system is a source. The eigenvectors (x, y, z) for $\lambda = 1$ satisfy the equations $x = 0$ and $y = z$, and the eigenvectors for $\lambda = 3$ satisfy the equations $x = 0$ and $y = -z$.

    The x-axis is the phase line for the equation $dx/dt = x$. Therefore, there is a single equilibrium point at the origin, and every solution curve lying on the x-axis tends to infinity as $t \to \infty$.

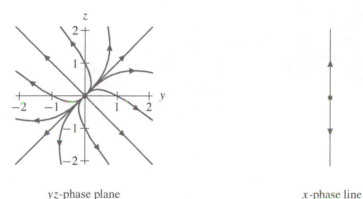

yz-phase plane x-phase line

(d) The x-phase line and the yz-phase plane can be combined to obtain the xyz-phase space.

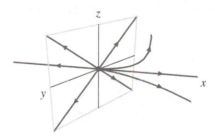

9. (Note: This exercise can also be done by taking the complex conjugate of both sides of the equation $p(a + ib) = 0$.)

    Using the definition of $p(x)$, we compute

$$p(a + ib) = (\alpha a^3 - 3\alpha ab^2 + \beta a^2 - \beta b^2 + \gamma a + \delta) + i(3\alpha a^2 b - \alpha b^3 + 2\beta ab + \gamma b)$$

and

$$p(a - bi) = (\alpha a^3 - 3\alpha ab^2 + \beta a^2 - \beta b^2 + \gamma a + \delta) - i(3\alpha a^2 b - \alpha b^3 + 2\beta ab + \gamma b).$$

Since $p(a + bi) = 0$,

$$\alpha a^3 - 3\alpha ab^2 + \beta a^2 - \beta b^2 + \gamma a + \delta = 3\alpha a^2 b - \alpha b^3 + 2\beta ab + \gamma b = 0.$$

Therefore, $p(a - ib) = 0$, and $a - ib$ is also a root.

11. **(a)** The characteristic equation is

$$(-2 - \lambda)(-2 - \lambda)(1 - \lambda) = 0,$$

and the eigenvalues are $\lambda = -2$ and $\lambda = 1$.

(b) In the xy-plane, the system has a repeated eigenvalue, $\lambda = -2$, and the eigenvectors (x, y, z) associated to that eigenvalue satisfy the equations $y = z = 0$. The origin is a sink in the xy-phase plane.

   The z-axis is the phase line for the equation $dz/dt = z$. Therefore, there is a single equilibrium point at the origin, and every solution curve lying on the z-axis tends to infinity as $t \to \infty$.

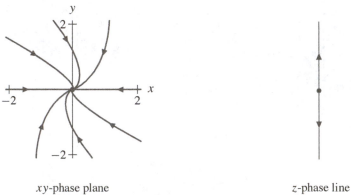

xy-phase plane z-phase line

(c) Combining the xy-phase plane and z-phase line, we obtain a picture of the xyz-phase space.

13. **(a)** The characteristic equation is

$$(-1 - \lambda)(-4 - \lambda)(-\lambda) - 4(-\lambda) = -\lambda^2(\lambda + 5) = 0.$$

One eigenvalue is $\lambda = -5$, and the other eigenvalue, $\lambda = 0$, is a repeated eigenvalue.

(b) In xy-plane, the eigenvalues are $\lambda = 0$ and $\lambda = -5$. The eigenvectors (x, y, z) associated to the eigenvalue $\lambda = 0$ satisfy the equations $z = 0$ and $-x + 2y = 0$. The eigenvectors associated to the eigenvalue $\lambda = -5$ satisfy the equations $z = 0$ and $4x + 2y = 0$. Hence, the line $y = x/2$ is a line of equilibrium points, and every solution curve lying in the xy-plane tends toward one of the equilibrium points on $y = x/2$ as $t \to \infty$.

The z-axis is the phase line for $dz/dt = 0$. Note that this line consists entirely of equilibrium points.

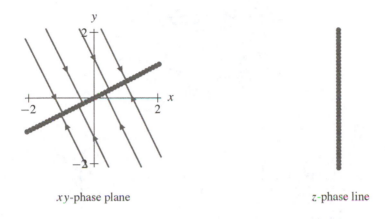

xy-phase plane   z-phase line

(c) Combining the xy-phase plane and z-phase line, we obtain a picture of the xyz-phase space.

15. (a) The characteristic equation is $-\lambda^3 = 0$. Consequently, there is only one eigenvalue, $\lambda = 0$.

(b) For $\lambda = 0$, the eigenvectors (x, y, z) must satisfy both $y = 0$ and $z = 0$. Therefore, the x-axis is both a line of eigenvectors and a line of equilibrium points.

(c) Since $dz/dt = 0$, $z(t)$ is a constant function. That is, if $z(0) = z_0$, then $z(t) = z_0$ for all t. Since $dy/dt = z$ and z is constant, we have $y(t) = z_0 t + y_0$, where $y(0) = y_0$ is the initial condition for $y(t)$. Finally, since $dx/dt = y = z_0 t + y_0$, $x(t) = z_0 t^2/2 + y_0 t + x_0$, where $x(0) = x_0$ is the initial condition for $x(t)$.

For $z_0 = 0$, $y(t)$ is constant, and the solution curves lie on straight lines parallel to the x-axis. For $y_0 > 0$, $x(t)$ is increasing, and for $y_0 < 0$, $x(t)$ is decreasing. For $z_0 \neq 0$, $x(t)$ is quadratic in y. Therefore, solution curves that satisfy the initial condition $z(0) = z_0$ stay on the plane $z = z_0$ and lie on a parabola.

17. (a) We compute

$$\mathbf{AV}_1 = \mathbf{A} \begin{pmatrix} 1 \\ 1 \\ 0 \end{pmatrix} = \begin{pmatrix} -1 \\ -1 \\ 0 \end{pmatrix} = -1 \begin{pmatrix} 1 \\ 1 \\ 0 \end{pmatrix}.$$

Hence, $\mathbf{V}_1$ is an eigenvector associated to the eigenvalue -1.

(b) The characteristic equation is

$$(-4 - \lambda)((-1 - \lambda)(-\lambda) + 5) + 15 = (\lambda + 1)(\lambda^2 + 4\lambda + 5) = 0.$$

The eigenvalues are $\lambda = -1$ and $\lambda = -2 \pm i$.

(c) Since one eigenvalue is a negative real number and the other two are complex with a negative real part, the system is a spiral sink.

(d) To determine all eigenvectors $\mathbf{V}_2$ associated to the eigenvalue $\lambda = -2 + i$, we must solve the vector equation $\mathbf{AV}_2 = (-2+i)\mathbf{V}_2$. This vector equation is equivalent to the three simultaneous equations

$$\begin{cases} -4x + 3y = (-2 + i)x \\ -y + z = (-2 + i)y \\ -x + 3y - z = (-2 + i)z. \end{cases}$$

Then, all complex eigenvectors $\mathbf{V}_2 = (x, y, z)$ must satisfy the equations $y = (2 + i)x/3$ and $z = (-3 + i)x/3$. One such eigenvector is $(3, 2 + i, -3 + i) = (3, 2, -3) + i(0, 1, 1)$. Taking real and imaginary parts, we obtain two vectors, $(3, 2, -3)$ and $(0, 1, 1)$, that determine a plane on which the solutions spiral toward the origin with natural period 2π.

19. (a) If Glen makes a profit, z is positive. Since the coefficients of z in the equations for dx/dt and dy/dt are positive, these terms contribute positively to dx/dt and dy/dt. In other words, Glen's profitability helps Paul and Bob be profitable (and they need all the help they can get).

(b) Since dz/dt does not have either an x or a y term, the values of x and y do not contribute to dz/dt. Hence, the profitability of either Paul or Bob makes no difference to Glen's future profits.

21. (a) For $z = 0$, $dz/dt = 0$ and, therefore, $z(t) = 0$ for all t. Consequently, we can analyze this system as if it has only two dependent variables, x and y. We have

$$\frac{dx}{dt} = -y$$
$$\frac{dy}{dt} = -x.$$

This system is a saddle in the xy-plane.

The eigenvalues for the xy-system are ± 1, and the eigenvectors for the eigenvalue -1 satisfy the equation $x = y$. Since we are assuming $x(0) = y(0)$, the given solution tends to the origin along the line $x = y$ in the xy-plane.

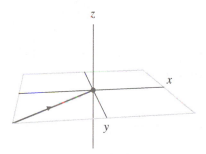

(b) Since $x(0) = y(0)$ and $z(0) = 0$ is an initial condition that is an eigenvector associated to the eigenvalue -1, we know that $x(t) = y(t)$ and $z(t) = 0$ for all t. We also know that both $x(t)$ and $y(t)$ decay to 0 like the function e^{-t}.

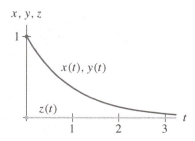

The graphs of $x(t)$ and $y(t)$ are identical, and $z(t) = 0$ for all t.

(c) Glen continues to break even. Both Paul's and Bob's profits tend to the break-even point as $t \to \infty$.

REVIEW EXERCISES FOR CHAPTER 3

1. The characteristic polynomial is $(1 - \lambda)(2 - \lambda)$, so the eigenvalues are $\lambda = 1$ and $\lambda = 2$.

3. The system has eigenvalues -2 and 3. One eigenvector associated with $\lambda = 3$ is $(1, 0)$, and one eigenvector associated with $\lambda = -2$ is $(0, 1)$. The general solution is

$$\mathbf{Y}(t) = k_1 e^{3t} \begin{pmatrix} 1 \\ 0 \end{pmatrix} + k_2 e^{-2t} \begin{pmatrix} 0 \\ 1 \end{pmatrix}.$$

5. Note that $b \geq 0$ by assumption. The characteristic polynomial is

$$s^2 + bs + 5,$$

so the eigenvalues are

$$s = \frac{-b \pm \sqrt{b^2 - 20}}{2}.$$

If $b > \sqrt{20}$, the harmonic oscillator is overdamped. If $b = \sqrt{20}$, the harmonic oscillator is critically damped. If $0 < b < \sqrt{20}$, the harmonic oscillator is underdamped, and if $b = 0$, the harmonic oscillator is undamped.

7. Every linear system has the origin as an equilibrium point, so the solution to the initial-value problem is the equilibrium solution $\mathbf{Y}(t) = (0, 0)$ for all t.

9. Letting $x(t) = 3 \cos 2t$ and $y(t) = \sin 2t$, we have

$$\frac{dx}{dt} = \frac{d(3 \cos 2t)}{dt} = -6 \sin 2t = -6y$$

and

$$\frac{dy}{dt} = \frac{d(\sin 2t)}{dt} = 2 \cos t = \frac{2}{3}x.$$

Hence, $\mathbf{Y}(t)$ satisfies the linear system

$$\frac{d\mathbf{Y}}{dt} = \begin{pmatrix} 0 & -6 \\ 2/3 & 0 \end{pmatrix} \mathbf{Y}.$$

11. False. For example, the linear system

$$\frac{d\mathbf{Y}}{dt} = \begin{pmatrix} 3 & 0 \\ 0 & 0 \end{pmatrix} \mathbf{Y}$$

has a line of equilibria (the y-axis). Another example is the linear system

$$\frac{d\mathbf{Y}}{dt} = \begin{pmatrix} 0 & 0 \\ 0 & 0 \end{pmatrix} \mathbf{Y}.$$

Every point is an equilibrium point for this system.

13. True. Linear systems have solutions that consist of just sine and cosine functions only when the eigenvalues are purely imaginary (that is, of the form $\pm i\omega$). In this case, the sine and cosine terms are of the form $\sin \omega t$ and $\cos \omega t$. For the first coordinate of $\mathbf{Y}(t)$ to be part of a solution, we would have to have $\omega = 2$, but the second coordinate would force $\omega = 1$. So this function cannot be the solution of a linear system.

15. False. In the graph, the amount of time between consecutive crossings of the t-axis decreases as t increases. Even though solutions of underdamped harmonic oscillators oscillate, the amount of time between consecutive crossings of the t-axis is constant.

17. True. If the matrix has a real eigenvalue, then there the corresponding system has at least a line of eigenvectors. Any initial condition that is an eigenvector corresponds to a solution that stays on the line of eigenvectors for all time. Hence, a system with a real eigenvalue has infinitely many straight-line solutions.

19. First, we compute the characteristic polynomials and eigenvalues for each matrix.

(i) The characteristic polynomial is $\lambda^2 + 1$, and the eigenvalues are $\lambda = \pm i$. Center.

(ii) The characteristic polynomial is $\lambda^2 + 2\lambda - 2$, and the eigenvalues are $\lambda = 1 \pm \sqrt{3}$. Saddle.

(iii) The characteristic polynomial is $\lambda^2 + 3\lambda + 1$, and the eigenvalues are $\lambda = (-3 \pm \sqrt{5})/2$. Sink.

(iv) The characteristic polynomial is $\lambda^2 + 1$, and the eigenvalues are $\lambda = \pm i$. Center.

(v) The characteristic polynomial is $\lambda^2 - \lambda - 2$, and the eigenvalues are $\lambda = -1$ and $\lambda = 2$. Saddle.

(vi) The characteristic polynomial is $\lambda^2 - 3\lambda + 1$, and the eigenvalues are $\lambda = (3 \pm \sqrt{5})/2$. Source.

(vii) The characteristic polynomial is $\lambda^2 + 4\lambda + 4$. The eigenvalue $\lambda = -2$ is a repeated eigenvalue. Sink.

(viii) The characteristic polynomial is $\lambda^2 + 2\lambda + 3$, and the eigenvalues are $\lambda = -1 \pm i\sqrt{2}$. Spiral sink.

Given this information, we can match the matrices with the phase portraits.

(a) This portrait is a center. There are two possibilities, (i) and (iv). At $(1, 0)$, the vector for (i) is $(1, -2)$, and the vector for (iv) is $(-1, -2)$. This phase portrait corresponds to matrix (iv).

(b) This portrait is a sink with two lines of eigenvectors. The only possibility is matrix (iii).

(c) This portrait is a saddle. The only possibilities are (ii) and (v). However, in (v), all vectors on the y-axis are eigenvectors corresponding to the eigenvalue $\lambda = -1$. Therefore, the phase portrait cannot correspond to (v).

(d) This portrait is a sink with a single line of eigenvectors. The only possibility is matrix (vii).

21. First, we compute the characteristic polynomials and eigenvalues for each matrix.

(i) The characteristic polynomial is $\lambda^2 - 3\lambda - 4$, and the eigenvalues are $\lambda = -1$ and $\lambda = 4$. Saddle.

(ii) The characteristic polynomial is $\lambda^2 - 7\lambda + 10$, and the eigenvalues are $\lambda = 2$ and $\lambda = 5$. Source.

(iii) The characteristic polynomial is $\lambda^2 + 4\lambda + 3$, and the eigenvalues are $\lambda = -3$ and $\lambda = -1$. Sink.

(iv) The characteristic polynomial is $\lambda^2 + 4$, and the eigenvalues are $\lambda = \pm 2i$. Center.

(v) The characteristic polynomial is $\lambda^2 + 9$, and the eigenvalues are $\lambda = \pm 3i$. Center.

(vi) The characteristic polynomial is $\lambda^2 - 2\lambda + \frac{15}{16}$, and the eigenvalues are $\lambda = 3/4$ and $\lambda = 5/4$.

Source.

(vii) The characteristic polynomial is $\lambda^2 + 2.2\lambda + 5.21$. The eigenvalues are $\lambda = -1.1 \pm 2i$. Spiral sink.

(viii) The characteristic polynomial is $\lambda^2 + 0.2\lambda + 4.01$, and the eigenvalues are $\lambda = -0.1 \pm 2i$. Spiral sink.

Given this information, we can match the matrices with the $x(t)$- and $y(t)$-graphs.

(a) This solution approaches equilibrium without oscillating. Therefore, the system has at least one negative real eigenvalue. Matrices (i) and (iii) are the only matrices with a negative eigenvalue. Since matrix (i) corresponds to a saddle, its only solutions that approach equilibrium are straight line-solutions. However, this solution is not a straight-line solution because $y(t)/x(t)$ is not constant. This solution must correspond to matrix (iii).

(b) Note that $y(t) = -x(t)$ for all t. Therefore, this solution corresponds to a straight-line solution for a source or a saddle with eigenvector $(1, -1)$. Direct computation shows that $(1, -1)$ is not an eigenvector for matrices (i) and (ii), and it is an eigenvector corresponding to eigenvalue $\lambda = 3/4$ for matrix (vi).

(c) This solution is periodic. Therefore, the corresponding matrix has purely imaginary eigenvalues. Matrices (iv) and (v) are the only matrices with purely imaginary eigenvalues. The solution oscillates three times over any interval of length 2π. Hence, the period of the solution is $2\pi/3$. Therefore, the eigenvalues must be $\pm 3i$, and this solution corresponds to matrix (v).

(d) This solution oscillates as it approaches equilibrium. Therefore, the corresponding matrix has complex eigenvalues with a negative real part. Matrices (vii) and (viii) are the only possibilities. Since the real part of the eigenvalues for matrix (vii) is -1.1, its solutions decay at a rate of $e^{-1.1t}$. Similarly, the real part of the eigenvalues for matrix (viii) is -0.1, so its solutions decay at a rate of $e^{-0.1t}$. The rate of decay of the solution graphed is $e^{-0.1t}$. Consequently, these graphs correspond to matrix (viii).

23. The characteristic polynomial is

$$s^2 + 5s + 6,$$

so the eigenvalues are $s = -2$ and $s = -3$. Hence, the general solution is

$$y(t) = k_1 e^{-2t} + k_2 e^{-3t},$$

and we have

$$y'(t) = -2k_1 e^{-2t} - 3k_2 e^{-3t}.$$

From the initial conditions, we obtain the simultaneous equations

$$\begin{cases} k_1 + k_2 = 0 \\ -2k_1 - 3k_2 = 2. \end{cases}$$

Solving for k_1 and k_2 yields $k_1 = 2$ and $k_2 = -2$. Hence, the solution to our initial-value problem is $y(t) = 2e^{-2t} - 2e^{-3t}$.

25. The characteristic polynomial is

$$s^2 + 2s + 1,$$

so $s = -1$ is a repeated eigenvalue. Hence, the general solution is

$$y(t) = k_1 e^{-t} + k_2 t e^{-t}.$$

From the initial condition $y(0) = 1$, we see that $k_1 = 1$. Differentiating

$$y(t) = e^{-t} + k_2 t e^{-t}$$

and evaluating $y'(t)$ at $t = 0$ yields $y'(0) = -1 + k_2$. Since $y'(0) = 1$, we have $k_2 = 2$. Hence, the solution to our initial-value problem is

$$y(t) = e^{-t} + 2t e^{-t}.$$

27. (a) The characteristic polynomial is

$$(1 - \lambda)(-1 - \lambda) - 3 = \lambda^2 - 4$$

so the eigenvalues are $\lambda_1 = 2$ and $\lambda_2 = -2$. The equilibrium point at the origin is a saddle. The eigenvectors for $\lambda_1 = 2$ satisfy the equations

$$\begin{cases} x + 3y = 2x \\ x - y = 2y. \end{cases}$$

Consequently, the eigenvectors (x, y) for this eigenvalue satisfy $x = 3y$. The eigenvector $(3, 1)$ is one such point.

The eigenvectors for $\lambda_2 = -2$ satisfy the equations

$$\begin{cases} x + 3y = -2x \\ x - y = -2y. \end{cases}$$

Consequently, the eigenvectors (x, y) for this eigenvalue satisfy $y = -x$. The eigenvector $(1, -1)$ is one such point.

Hence, the general solution of the system is

$$\mathbf{Y}(t) = k_1 e^{2t} \begin{pmatrix} 3 \\ 1 \end{pmatrix} + k_2 e^{-2t} \begin{pmatrix} 1 \\ -1 \end{pmatrix}.$$

(b)

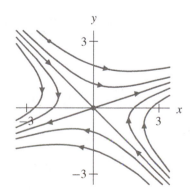

(c) To solve the initial-value problem, we solve for k_1 and k_2 in the equation

$$\begin{pmatrix} -2 \\ 3 \end{pmatrix} = \mathbf{Y}(0) = k_1 \begin{pmatrix} 3 \\ 1 \end{pmatrix} + k_2 \begin{pmatrix} 1 \\ -1 \end{pmatrix}.$$

This vector equation is equivalent to the two scalar equations

$$\begin{cases} 3k_1 + k_2 = -2 \\ k_1 - k_2 = 3, \end{cases}$$

so $k_1 = 1/4$ and $k_2 = -11/4$. The solution to the initial-value problem is

$$\mathbf{Y}(t) = \tfrac{1}{4} e^{2t} \begin{pmatrix} 3 \\ 1 \end{pmatrix} - \tfrac{11}{4} e^{-2t} \begin{pmatrix} 1 \\ -1 \end{pmatrix}$$

(d)

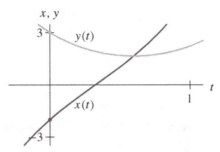

29. (a) The characteristic polynomial is

$$(-2 - \lambda)(2 - \lambda) + 6 = \lambda^2 + 2,$$

so the eigenvalues are $\lambda = \pm i\sqrt{2}$. The equilibrium point at the origin is a center.
The eigenvectors (x, y) corresponding to $\lambda = i\sqrt{2}$ are solutions of the equations

$$\begin{cases} -2x + 3y = i\sqrt{2}\,x \\ -2x + 2y = i\sqrt{2}\,y. \end{cases}$$

These equations are equivalent to the equation $(2 + i\sqrt{2})x = 3y$. Consequently, $(3, 2 + i\sqrt{2})$
is one eigenvector. Linearly independent solutions are given by the real and imaginary parts of

$$e^{i\sqrt{2}\,t} \begin{pmatrix} 3 \\ 2 + i\sqrt{2} \end{pmatrix},$$

which by Euler's formula is

$$(\cos \sqrt{2}\,t + i \sin \sqrt{2}\,t) \begin{pmatrix} 3 \\ 2 + i\sqrt{2} \end{pmatrix}.$$

Hence, the general solution is

$$\mathbf{Y}(t) = k_1 \begin{pmatrix} 3 \cos \sqrt{2}\,t \\ 2 \cos \sqrt{2}\,t - \sqrt{2} \sin \sqrt{2}\,t \end{pmatrix} + k_2 \begin{pmatrix} 3 \sin \sqrt{2}\,t \\ \sqrt{2} \cos \sqrt{2}\,t + 2 \sin \sqrt{2}\,t \end{pmatrix}.$$

(b)

(c) To satisfy the initial condition, we solve

$$\begin{pmatrix} -2 \\ 2 \end{pmatrix} = \mathbf{Y}(0) = k_1 \begin{pmatrix} 3 \\ 2 \end{pmatrix} + k_2 \begin{pmatrix} 0 \\ \sqrt{2} \end{pmatrix},$$

which is equivalent to the two scalar equations

$$\begin{cases} 3k_1 = -2 \\ 2k_1 + \sqrt{2}\,k_2 = 2. \end{cases}$$

We get $k_1 = -2/3$ and $k_2 = 5\sqrt{2}/3$, and the solution to the initial-value problem is

$$\mathbf{Y}(t) = \begin{pmatrix} -2\cos\sqrt{2}\,t + 5\sqrt{2}\,\sin\sqrt{2}\,t \\ 2\cos\sqrt{2}\,t + 4\sqrt{2}\,\sin\sqrt{2}\,t \end{pmatrix}$$

(d)

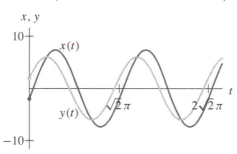

31. **(a)** The characteristic polynomial is

$$(-3 - \lambda)(-1 - \lambda) + 1 = (\lambda + 2)^2,$$

so $\lambda = -2$ is a repeated eigenvalue. The equilibrium point at the origin is a sink.

   To find the general solution, we start with an arbitrary initial condition (x_0, y_0), and we calculate

$$\begin{pmatrix} -1 & 1 \\ -1 & 1 \end{pmatrix} \begin{pmatrix} x_0 \\ y_0 \end{pmatrix} = \begin{pmatrix} y_0 - x_0 \\ y_0 - x_0 \end{pmatrix}.$$

We obtain the general solution

$$\mathbf{Y}(t) = e^{-2t} \begin{pmatrix} x_0 \\ y_0 \end{pmatrix} + te^{-2t} \begin{pmatrix} y_0 - x_0 \\ y_0 - x_0 \end{pmatrix}.$$

(b)

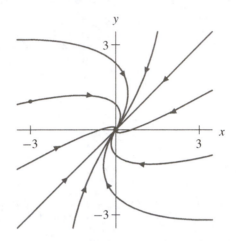

(c) The solution that satisfies the initial condition $(x_0, y_0) = (-3, 1)$ is

$$\mathbf{Y}(t) = e^{-2t} \begin{pmatrix} -3 \\ 1 \end{pmatrix} + te^{-2t} \begin{pmatrix} 4 \\ 4 \end{pmatrix} = e^{-2t} \begin{pmatrix} 4t - 3 \\ 4t + 1 \end{pmatrix}.$$

(d)

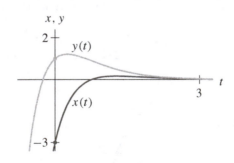

CHAPTER 4

Forcing and Resonance

EXERCISES FOR SECTION 4.1

1. To compute the general solution of the unforced equation, we use the method of Section 3.6. The characteristic polynomial is

$$s^2 - s - 6,$$

so the eigenvalues are $s = -2$ and $s = 3$. Hence, the general solution of the homogeneous equation is

$$k_1 e^{-2t} + k_2 e^{3t}.$$

To find a particular solution of the forced equation, we guess $y_p(t) = ke^{4t}$. Substituting into the left-hand side of the differential equation gives

$$\frac{d^2 y_p}{dt^2} - \frac{dy_p}{dt} - 6y_p = 16ke^{4t} - 4ke^{4t} - 6ke^{4t}$$

$$= 6ke^{4t}.$$

In order for $y_p(t)$ to be a solution of the forced equation, we must take $k = 1/6$. The general solution of the forced equation is

$$y(t) = k_1 e^{-2t} + k_2 e^{3t} + \tfrac{1}{6} e^{4t}.$$

3. To compute the general solution of the unforced equation, we use the method of Section 3.6. The characteristic polynomial is

$$s^2 - s - 2,$$

so the eigenvalues are $s = -1$ and $s = 2$. Hence, the general solution of the homogeneous equation is

$$k_1 e^{-t} + k_2 e^{2t}.$$

To find a particular solution of the forced equation, we guess $y_p(t) = ke^{3t}$. Substituting into the left-hand side of the differential equation gives

$$\frac{d^2 y_p}{dt^2} - \frac{dy_p}{dt} - 2y_p = 9ke^{3t} - 3ke^{3t} - 2ke^{3t}$$

$$= 4ke^{3t}.$$

In order for $y_p(t)$ to be a solution of the forced equation, we must take $k = 5/4$. The general solution of the forced equation is

$$y(t) = k_1 e^{-t} + k_2 e^{2t} + \tfrac{5}{4} e^{3t}.$$

5. To compute the general solution of the unforced equation, we use the method of Section 3.6. The characteristic polynomial is

$$s^2 + 4s + 13,$$

so the eigenvalues are $s = -2 \pm 3i$. Hence, the general solution of the homogeneous equation is

$$k_1 e^{-2t} \cos 3t + k_2 e^{-2t} \sin 3t.$$

To find a particular solution of the forced equation, we guess $y_p(t) = ke^{-2t}$. Substituting into the left-hand side of the differential equation gives

$$\frac{d^2y_p}{dt^2} + 4\frac{dy_p}{dt} + 13y_p = 4ke^{-2t} - 8ke^{-2t} + 13ke^{-2t}$$
$$= 9ke^{-2t}.$$

In order for $y_p(t)$ to be a solution of the forced equation, we must take $k = -1/3$. The general solution of the forced equation is

$$y(t) = k_1e^{-2t}\cos 3t + k_2e^{-2t}\sin 3t - \tfrac{1}{3}e^{-2t}.$$

7. To compute the general solution of the unforced equation, we use the method of Section 3.6. The characteristic polynomial is

$$s^2 - 5s + 4,$$

so the eigenvalues are $s = 1$ and $s = 4$. Hence, the general solution of the homogeneous equation is

$$k_1e^t + k_2e^{4t}.$$

To find a particular solution of the forced equation, a reasonable looking guess is $y_p(t) = ke^{4t}$. However, this guess is a solution of the homogeneous equation, so it is doomed to fail. We make the standard second guess of $y_p(t) = kte^{4t}$. Substituting into the left-hand side of the differential equation gives

$$\frac{d^2y_p}{dt^2} - 5\frac{dy_p}{dt} + 4y_p = (8ke^{4t} + 16kte^{4t}) - 5(ke^{4t} + 4kte^{4t}) + 4kte^{4t}$$
$$= 3ke^{4t}.$$

In order for $y_p(t)$ to be a solution of the forced equation, we must take $k = 1/3$. The general solution of the forced equation is
$$y(t) = k_1e^t + k_2e^{4t} + \tfrac{1}{3}te^{4t}.$$

9. First we derive the general solution. The characteristic polynomial is

$$s^2 + 6s + 8,$$

so the eigenvalues are $s = -2$ and $s = -4$. To find the general solution of the forced equation, we also need a particular solution. We guess $y_p(t) = ke^{-t}$ and find that $y_p(t)$ is a solution only if $k = 1/3$. Therefore, the general solution is

$$y(t) = k_1e^{-2t} + k_2e^{-4t} + \tfrac{1}{3}e^{-t}.$$

To find the solution with the initial conditions $y(0) = y'(0) = 0$, we compute

$$y'(t) = -2k_1e^{-2t} - 4k_2e^{-4t} - \tfrac{1}{3}e^{-t}.$$

Then we evaluate at $t = 0$ and obtain the simultaneous equations

$$\begin{cases} k_1 + k_2 + \frac{1}{3} = 0 \\ -2k_1 - 4k_2 - \frac{1}{3} = 0. \end{cases}$$

Solving, we have $k_1 = -1/2$ and $k_2 = 1/6$, so the solution of the initial-value problem is

$$y(t) = -\tfrac{1}{2}e^{-2t} + \tfrac{1}{6}e^{-4t} + \tfrac{1}{3}e^{-t}.$$

11. This is the same equation as Exercise 5. The general solution is

$$y(t) = k_1 e^{-2t} \cos 3t + k_2 e^{-2t} \sin 3t - \tfrac{1}{3}e^{-2t}.$$

To find the solution with the initial conditions $y(0) = y'(0) = 0$, we compute

$$y'(t) = -2k_1 e^{-2t} \cos 3t - 3k_1 e^{-2t} \sin 3t - 2k_2 e^{-2t} \sin 3t + 3k_2 e^{-2t} \cos 3t + \tfrac{2}{3}e^{-2t}.$$

Then we evaluate at $t = 0$ and obtain the simultaneous equations

$$\begin{cases} k_1 - \frac{1}{3} = 0 \\ -2k_1 + 3k_2 + \frac{2}{3} = 0. \end{cases}$$

Solving, we have $k_1 = 1/3$ and $k_2 = 0$, so the solution of the initial-value problem is

$$y(t) = \tfrac{1}{3}e^{-2t} \cos 3t - \tfrac{1}{3}e^{-2t}.$$

13. **(a)** The characteristic polynomial of the unforced equation is

$$s^2 + 4s + 3.$$

So the eigenvalues are $s = -1$ and $s = -3$, and the general solution of the unforced equation is

$$k_1 e^{-t} + k_2 e^{-3t}.$$

To find a particular solution of the forced equation, we guess $y_p(t) = ke^{-t/2}$. Substituting $y_p(t)$ into the left-hand side of the differential equation gives

$$\frac{d^2 y_p}{dt^2} + 4\frac{dy_p}{dt} + 3y_p = \tfrac{1}{4}ke^{-t/2} - 2ke^{-t/2} + 3ke^{-t/2}$$
$$= \tfrac{5}{4}ke^{-t/2}.$$

So $k = 4/5$ yields a solution of the forced equation.
The general solution of the forced equation is therefore

$$y(t) = k_1 e^{-t} + k_2 e^{-3t} + \tfrac{4}{5}e^{-t/2}.$$

(b) The derivative of the general solution is

$$y'(t) = -k_1 e^{-t} - 3k_2 e^{-3t} - \tfrac{2}{5} e^{-t/2}.$$

To find the solution with $y(0) = y'(0) = 0$, we evaluate at $t = 0$ and obtain the simultaneous equations

$$\begin{cases} k_1 + k_2 + \tfrac{4}{5} = 0 \\ -k_1 - 3k_2 - \tfrac{2}{5} = 0. \end{cases}$$

Solving, we find that $k_1 = -1$ and $k_2 = 1/5$, so the solution of the initial-value problem is

$$y(t) = -e^{-t} + \tfrac{1}{5} e^{-3t} + \tfrac{4}{5} e^{-t/2}.$$

(c) Every solution tends to zero as t increases. Of the three terms that sum to the general solution, $\tfrac{4}{5} e^{-t/2}$ dominates when t is large, so all solutions are approximately $\tfrac{4}{5} e^{-t/2}$ for t large.

15. (a) The characteristic polynomial of the unforced equation is

$$s^2 + 4s + 3.$$

So the eigenvalues are $s = -1$ and $s = -3$, and the general solution of the unforced equation is

$$k_1 e^{-t} + k_2 e^{-3t}.$$

To find a particular solution of the forced equation, we guess $y_p(t) = ke^{-4t}$. Substituting $y_p(t)$ into the left-hand side of the differential equation gives

$$\frac{d^2 y_p}{dt^2} + 4\frac{dy_p}{dt} + 3y_p = 16ke^{-4t} - 16ke^{-4t} + 3ke^{-4t}$$
$$= 3ke^{-4t}.$$

So $k = 1/3$ yields a solution of the forced equation.

The general solution of the forced equation is therefore

$$y(t) = k_1 e^{-t} + k_2 e^{-3t} + \tfrac{1}{3} e^{-4t}.$$

(b) The derivative of the general solution is

$$y'(t) = -k_1 e^{-t} - 3k_2 e^{-3t} - \tfrac{4}{3} e^{-4t}.$$

To find the solution with $y(0) = y'(0) = 0$, we evaluate at $t = 0$ and obtain the simultaneous equations

$$\begin{cases} k_1 + k_2 + \tfrac{1}{3} = 0 \\ -k_1 - 3k_2 - \tfrac{4}{3} = 0. \end{cases}$$

Solving, we find that $k_1 = 1/6$ and $k_2 = -1/2$, so the solution of the initial-value problem is

$$y(t) = \tfrac{1}{6} e^{-t} - \tfrac{1}{2} e^{-3t} + \tfrac{1}{3} e^{-4t}.$$

(c) In the general solution, all three terms tend to zero, so the solution tends to zero. We can say a little more by noting that the term $k_1 e^{-t}$ is much larger (provided $k_1 \neq 0$). Hence, most solutions tend to zero at the rate of e^{-t}. If $k_1 = 0$, then solutions tend to zero at the rate of e^{-3t} provided $k_2 \neq 0$.

17.   **(a)** The characteristic polynomial of the unforced equation is

$$s^2 + 4s + 20.$$

So the eigenvalues are $s = -2 \pm 4i$, and the general solution of the unforced equation is

$$k_1 e^{-2t} \cos 4t + k_2 e^{-2t} \sin 4t.$$

To find a particular solution of the forced equation, we guess $y_p(t) = ke^{-2t}$. Substituting $y_p(t)$ into the left-hand side of the differential equation gives

$$\frac{d^2 y_p}{dt^2} + 4\frac{dy_p}{dt} + 20y_p = 4ke^{-2t} - 8ke^{-2t} + 20ke^{-2t}$$

$$= 16ke^{-2t}.$$

So $k = 1/16$ yields a solution of the forced equation.
The general solution of the forced equation is therefore

$$y(t) = k_1 e^{-2t} \cos 4t + k_2 e^{-2t} \sin 4t + \tfrac{1}{16}e^{-2t}.$$

(b) The derivative of the general solution is

$$y'(t) = -2k_1 e^{-2t} \cos 4t - 4k_1 e^{-2t} \sin 4t - 2k_2 e^{-2t} \sin 4t + 4k_2 e^{-2t} \cos 4t - \tfrac{1}{8}e^{-2t}.$$

To find the solution with $y(0) = y'(0) = 0$, we evaluate at $t = 0$ and obtain the simultaneous equations

$$\begin{cases} k_1 + \tfrac{1}{16} = 0 \\ -2k_1 + 4k_2 - \tfrac{1}{8} = 0. \end{cases}$$

Solving, we find that $k_1 = -1/16$ and $k_2 = 0$, so the solution of the initial-value problem is

$$y(t) = -\tfrac{1}{16}e^{-2t} \cos 4t + \tfrac{1}{16}e^{-2t}.$$

(c) Every solution tends to zero like e^{-2t} and all but one exponential solution oscillates with frequency $2/\pi$.

19. The natural guesses of $y_p(t) = ke^{-t}$ and $y_p(t) = kte^{-t}$ fail to be solutions of the forced equation because they are both solutions of the unforced equation. (The characteristic polynomial of the unforced equation is

$$s^2 + 2s + 1,$$

which has -1 as a double root.)
So we guess $y_p(t) = kt^2 e^{-t}$. Substituting this guess into the left-hand side of the differential equation gives

$$\frac{d^2 y_p}{dt^2} + 2\frac{dy_p}{dt} + y_p = (2ke^{-t} - 4kte^{-t} + kt^2 e^{-t}) + 2(2kte^{-t} - kt^2 e^{-t}) + kt^2 e^{-t}$$

$$= 2ke^{-t}.$$

So $k = 1/2$ yields the solution

$$y_p(t) = \tfrac{1}{2}t^2 e^{-t}.$$

From the characteristic polynomial, we know that the general solution of the unforced equation is

$$k_1 e^{-t} + k_2 t e^{-t}.$$

Consequently, the general solution of the forced equation is

$$y(t) = k_1 e^{-t} + k_2 t e^{-t} + \tfrac{1}{2} t^2 e^{-t}.$$

21. **(a)** The characteristic polynomial of the unforced equation is

$$s^2 - 5s + 4.$$

So the eigenvalues are $s = 1$ and $s = 4$, and the general solution of the unforced equation is

$$k_1 e^t + k_2 e^{4t}.$$

To find one solution of the forced equation, we guess the constant function $y_p(t) = k$. Substituting $y_p(t)$ into the left-hand side of the differential equation, we obtain

$$\frac{d^2 y_p}{dt^2} - 5 \frac{d y_p}{dt} + 4 y_p = 0 - 5 \cdot 0 + 4k = 4k.$$

Hence, $k = 5/4$ yields a solution of the forced equation. The general solution of the forced equation is

$$y(t) = k_1 e^t + k_2 e^{4t} + \tfrac{5}{4}.$$

(b) To find the solution satisfying the initial conditions $y(0) = y'(0) = 0$, we compute the derivative of the general solution

$$y'(t) = k_1 e^t + 4 k_2 e^{4t}.$$

Using the initial conditions and evaluating $y(t)$ and $y'(t)$ at $t = 0$, we obtain the simultaneous equations

$$\begin{cases} k_1 + k_2 + \tfrac{5}{4} = 0 \\[6pt] k_1 + 4 k_2 = 0. \end{cases}$$

Solving for k_1 and k_2 gives $k_1 = -5/3$ and $k_2 = 5/12$. The solution of the initial-value problem is

$$y(t) = \tfrac{5}{4} - \tfrac{5}{3} e^t + \tfrac{5}{12} e^{4t}.$$

23. **(a)** The characteristic polynomial of the unforced equation is

$$s^2 + 2s + 10.$$

So the eigenvalues are $s = -1 \pm 3i$, and the general solution of the unforced equation is

$$k_1 e^{-t} \cos 3t + k_2 e^{-t} \sin 3t.$$

To find one solution of the forced equation, we guess the constant function $y_p(t) = k$. Substituting $y_p(t)$ into the left-hand side of the differential equation, we obtain

$$\frac{d^2 y_p}{dt^2} + 2 \frac{d y_p}{dt} + 10 y_p = 0 + 2 \cdot 0 + 10k = 10k.$$

Hence, $k = 1$ yields a solution of the forced equation. The general solution of the forced equation is

$$y(t) = k_1 e^{-t} \cos 3t + k_2 e^{-t} \sin 3t + 1.$$

(b) To find the solution satisfying the initial conditions $y(0) = y'(0) = 0$, we compute the derivative of the general solution

$$y'(t) = -k_1 e^{-t} \cos 3t - 3k_1 e^{-t} \sin 3t - k_2 e^{-t} \sin 3t + 3k_2 e^{-t} \cos 3t.$$

Using the initial conditions and evaluating $y(t)$ and $y'(t)$ at $t = 0$, we obtain the simultaneous equations

$$\begin{cases} k_1 + 1 = 0 \\ -k_1 + 3k_2 = 0. \end{cases}$$

Solving for k_1 and k_2 gives $k_1 = -1$ and $k_2 = -1/3$. The solution of the initial-value problem is

$$y(t) = -e^{-t} \cos 3t - \tfrac{1}{3} e^{-t} \sin 3t + 1.$$

25. (a) The characteristic polynomial of the unforced equation is

$$s^2 + 9.$$

So the eigenvalues are $s = \pm 3i$, and the general solution of the unforced equation is

$$k_1 \cos 3t + k_2 \sin 3t.$$

To find one solution of the forced equation, we guess $y_p(t) = ke^{-t}$. Substituting $y_p(t)$ into the left-hand side of the differential equation, we obtain

$$\frac{d^2 y_p}{dt^2} + 9y_p = ke^{-t} + 9ke^{-t}$$
$$= 10ke^{-t}.$$

Hence, $k = 1/10$ yields a solution of the forced equation. The general solution of the forced equation is

$$y(t) = k_1 \cos 3t + k_2 \sin 3t + \tfrac{1}{10} e^{-t}.$$

(b) To find the solution satisfying the initial conditions $y(0) = y'(0) = 0$, we compute the derivative of the general solution

$$y'(t) = -3k_1 \sin 3t + 3k_2 \cos 3t - \tfrac{1}{10} e^{-t}.$$

Using the initial conditions and evaluating $y(t)$ and $y'(t)$ at $t = 0$, we obtain the simultaneous equations

$$\begin{cases} k_1 + \tfrac{1}{10} = 0 \\ 3k_2 - \tfrac{1}{10} = 0. \end{cases}$$

Solving for k_1 and k_2 gives $k_1 = -1/10$ and $k_2 = 1/30$. The solution of the initial-value problem is

$$y(t) = -\tfrac{1}{10} \cos 3t + \tfrac{1}{30} \sin 3t + \tfrac{1}{10} e^{-t}.$$

(c) Since the function $e^{-t}/10 \to 0$ quickly, the solution quickly approaches a solution of the unforced oscillator.

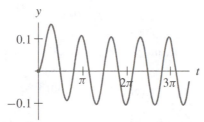

27. (a) The characteristic polynomial of the unforced equation is

$$s^2 + 2.$$

So the eigenvalues are $s = \pm i\sqrt{2}$, and the general solution of the unforced equation is

$$k_1 \cos \sqrt{2}\, t + k_2 \sin \sqrt{2}\, t.$$

To find one solution of the forced equation, we guess $y_p(t) = k$. Substituting into the left-hand side of the differential equation, we obtain

$$\frac{d^2 y_p}{dt^2} + 2y_p = 0 + 2k$$
$$= 2k.$$

Hence, $k = -3/2$ yields a solution of the forced equation. The general solution of the forced equation is

$$y(t) = k_1 \cos \sqrt{2}\, t + k_2 \sin \sqrt{2}\, t - \tfrac{3}{2}.$$

(b) To find the solution satisfying the initial conditions $y(0) = y'(0) = 0$, we compute the derivative of the general solution

$$y'(t) = -\sqrt{2}\, k_1 \sin \sqrt{2}\, t + \sqrt{2}\, k_2 \cos \sqrt{2}\, t.$$

Using the initial conditions and evaluating $y(t)$ and $y'(t)$ at $t = 0$, we obtain the simultaneous equations

$$\begin{cases} k_1 - \tfrac{3}{2} = 0 \\ \sqrt{2}\, k_2 = 0. \end{cases}$$

Solving for k_1 and k_2 gives $k_1 = 3/2$ and $k_2 = 0$. The solution of the initial-value problem is

$$y(t) = \tfrac{3}{2} \cos \sqrt{2}\, t - \tfrac{3}{2}.$$

(c) The solution oscillates about the constant $y = -3/2$ with oscillations of amplitude $3/2$.

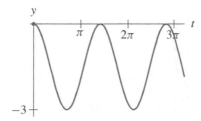

29. **(a)** The characteristic polynomial of the unforced equation is

$$s^2 + 9.$$

So the eigenvalues are $s = \pm 3i$, and the general solution of the unforced equation is

$$k_1 \cos 3t + k_2 \sin 3t.$$

To find one solution of the forced equation, we guess $y_p(t) = k$, where k is a constant. Substituting this guess into the left-hand side of the differential equation, we obtain

$$\frac{d^2 y_p}{dt^2} + 9y_p = 9k.$$

Hence, $k = 2/3$ yields a solution of the forced equation. The general solution of the forced equation is

$$y(t) = k_1 \cos 3t + k_2 \sin 3t + \tfrac{2}{3}.$$

(b) To find the solution satisfying the initial conditions $y(0) = y'(0) = 0$, we compute the derivative of the general solution

$$y'(t) = -3k_1 \sin 3t + 3k_2 \cos 3t.$$

Using the initial conditions and evaluating $y(t)$ and $y'(t)$ at $t = 0$, we obtain the simultaneous equations

$$\begin{cases} k_1 + \tfrac{2}{3} = 0 \\ 3k_2 = 0. \end{cases}$$

Solving for k_1 and k_2 gives $k_1 = -2/3$ and $k_2 = 0$. The solution of the initial-value problem is

$$y(t) = -\tfrac{2}{3} \cos 3t + \tfrac{2}{3}.$$

(c) The solution oscillates about the constant function $y = 2/3$ with amplitude $2/3$.

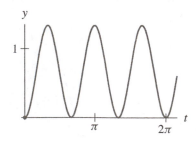

31. **(a)** The general solution for the homogeneous equation is

$$k_1 \cos 2t + k_2 \sin 2t.$$

Suppose $y_p(t) = at^2 + bt + c$. Substituting $y_p(t)$ into the differential equation, we get

$$\frac{d^2 y_p}{dt^2} + 4y_p = -3t^2 + 2t + 3$$

$$2a + 4(at^2 + bt + c) = -3t^2 + 2t + 3$$

$$4at^2 + 4bt + (2a + 4c) = -3t^2 + 2t + 3.$$

Therefore, $y_p(t)$ is a solution if and only if

$$\begin{cases} 4a = -3 \\ 4b = 2 \\ 2a + 4c = 3. \end{cases}$$

Therefore, $a = -3/4, b = 1/2$, and $c = 9/8$. The general solution is

$$y(t) = k_1 \cos 2t + k_2 \sin 2t - \tfrac{3}{4}t^2 + \tfrac{1}{2}t + \tfrac{9}{8}.$$

(b) To solve the initial-value problem, we use the initial conditions $y(0) = 2$ and $y'(0) = 0$ along with the general solution to form the simultaneous equations

$$\begin{cases} k_1 + \tfrac{9}{8} = 2 \\ 2k_2 + \tfrac{1}{2} = 0. \end{cases}$$

Therefore, $k_1 = 7/8$ and $k_2 = -1/4$. The solution is

$$y(t) = \tfrac{7}{8} \cos 2t - \tfrac{1}{4} \sin 2t - \tfrac{3}{4}t^2 + \tfrac{1}{2}t + \tfrac{9}{8}.$$

33. **(a)** For the unforced equation, the general solution is

$$k_1 \cos 2t + k_2 \sin 2t.$$

To find a particular solution of the forced equation, we guess $y_p(t) = at + b$. Substituting this guess into the differential equation, we get

$$\frac{d^2 y_p}{dt} + 4y_p = 3t + 2$$

$$0 + 4(at + b) = 3t + 2$$

$$4at + 4b = 3t + 2.$$

Therefore, $a = 3/4$ and $b = 1/2$ yield a solution. The general solution for the forced equation is

$$y(t) = k_1 \cos 2t + k_2 \sin 2t + \tfrac{3}{4}t + \tfrac{1}{2}.$$

(b) To solve the initial-value problem, we compute

$$y'(t) = -2k_1 \sin 2t + 2k_2 \cos 2t + \tfrac{3}{4}.$$

Evaluating $y(t)$ and $y'(t)$ at $t = 0$ and using the initial conditions, we obtain the simultaneous equations

$$\begin{cases} k_1 + \tfrac{1}{2} = 0 \\ 2k_2 + \tfrac{3}{4} = 0. \end{cases}$$

Hence, $k_1 = -1/2$ and $k_2 = -3/8$ provide the desired solution

$$y(t) = -\tfrac{1}{2} \cos 2t - \tfrac{3}{8} \cos 2t + \tfrac{3}{4}t + \tfrac{1}{2}.$$

(c) The solution tends to ∞ as it oscillates about the line $y = \tfrac{3}{4}t + \tfrac{1}{2}$.

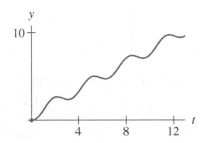

35. **(a)** The general solution of the homogeneous equation is

$$k_1 \cos 2t + k_2 \sin 2t.$$

To find a particular solution to the nonhomogeneous equation, we guess

$$y_p(t) = at^2 + bt + c.$$

Substituting $y_p(t)$ into the differential equation yields

$$\frac{d^2 y_p}{dt^2} + 4y_p = t - \frac{t}{20}$$

$$2a + 4(at^2 + bt + c) = t - \frac{t}{20}$$

$$(4a)t^2 + (4b)t + (2a + 4c) = t - \frac{t}{20}.$$

Equating coefficients, we obtain the simultaneous equations

$$\begin{cases} 4a = -\tfrac{1}{20} \\ 4b = 1 \\ 2a + 4c = 0. \end{cases}$$

Therefore, $a = -1/80$, $b = 1/4$, and $c = 1/160$ yield a solution to the nonhomogeneous equation, and the general solution of the nonhomogeneous equation is

$$y(t) = k_1 \cos 2t + k_2 \sin 2t - \tfrac{1}{80}t^2 + \tfrac{1}{4}t + \tfrac{1}{160}.$$

(b) To solve the initial-value problem with $y(0) = 0$ and $y'(0) = 0$, we have

$$\begin{cases} k_1 + \tfrac{1}{160} = 0 \\[2mm] 2k_2 + \tfrac{1}{4} = 0. \end{cases}$$

Therefore, $k_1 = -1/160$ and $k_2 = -1/8$, and the solution is

$$y(t) = -\tfrac{1}{160}\cos 2t - \tfrac{1}{8}\sin 2t - \tfrac{1}{80}t^2 + \tfrac{1}{4}t + \tfrac{1}{160}.$$

(c) Since the solution to the homogeneous equation is periodic with a small amplitude and since the solution to the nonhomogeneous equation goes to $-\infty$ at a rate determined by $-t^2/80$, the solution tends to $-\infty$.

37. **(a)** We must find a particular solution. Using the result of Exercise 36, we guess $y_p(t) = ae^{-t} + b$, where a and b are constants to be determined. (We could solve two separate problems and add the answers, but this approach is more efficient.) Hence we have $dy_p/dt = -ae^{-t}$ and $d^2y_p/dt^2 = ae^{-t}$. Substituting these derivatives into the differential equation, we obtain

$$(a - 5a + 6a)e^{-t} + 6b = e^{-t} + 4,$$

which is satisfied if $2a = 1$ and $6b = 4$. Hence, $a = 1/2$ and $b = 2/3$ yield the particular solution $y_p(t) = e^{-t}/2 + 2/3$.

The general solution of the homogeneous equation is obtained from the characteristic polynomial

$$s^2 + 5s + 6,$$

whose roots are $s = -2$ and $s = -3$.

Hence the general solution is

$$y(t) = k_1 e^{-2t} + k_2 e^{-3t} + \tfrac{1}{2}e^{-t} + \tfrac{2}{3}.$$

(b) To obtain the solution to the initial-value problem specified, we note that

$$y(0) = k_1 + k_2 + 1/2 + 2/3 \quad \text{and} \quad y'(0) = -2k_1 - 3k_2 - 1/2.$$

Using the initial conditions $y(0) = 0$ and $y'(0) = 0$, we have $k_1 = -3$ and $k_2 = 11/6$. The solution is

$$y(t) = -3e^{-2t} + \tfrac{11}{6}e^{-3t} + \tfrac{1}{2}e^{-t} + \tfrac{2}{3}.$$

(c) All of the exponential terms in the solution to the initial-value problem tend to 0. Hence, the solution tends to the constant $y = 2/3$. The rate that this solution tends to the constant is determined by $e^{-t}/2$, which is the largest of the terms that tend to zero when t is large.

39. **(a)** First, to find a particular solution of the forced equation, we guess

$$y_p(t) = at + b + ce^{-t}.$$

For y_p, $dy_p/dt = a - ce^{-t}$ and $d^2y_p/dt^2 = ce^{-t}$. Substituting these derivatives into the differential equation and collecting terms gives

$$(c - 6c + 8c)e^{-t} + (8a)t + (6a + 8b) = 2t + e^{-t},$$

which holds if $3c = 1$, $8a = 2$, and $6a + 8b = 0$. Hence, $c = 1/3$, $a = 1/4$, and $b = -3/16$ yield the solution

$$y_p(t) = -\tfrac{3}{16} + \tfrac{1}{4}t + \tfrac{1}{3}e^{-t}.$$

The characteristic polynomial of the homogeneous equation is

$$s^2 + 6s + 8,$$

which has roots $s = -4$ and $s = -2$, so the general solution of the forced equation is

$$y(t) = k_1 e^{-4t} + k_2 e^{-2t} - \tfrac{3}{16} + \tfrac{1}{4}t + \tfrac{1}{3}e^{-t}.$$

(b) To find the solution for the initial conditions $y(0) = 0$ and $y'(0) = 0$, we solve

$$\begin{cases} k_1 + k_2 - \tfrac{3}{16} + \tfrac{1}{3} = 0 \\ -4k_1 - 2k_2 + \tfrac{1}{4} - \tfrac{1}{3} = 0. \end{cases}$$

Thus, $k_1 = 5/48$ and $k_2 = -1/4$ yield the solution

$$y(t) = \tfrac{5}{48}e^{-4t} - \tfrac{1}{4}e^{-2t} - \tfrac{3}{16} + \tfrac{1}{4}t + \tfrac{1}{3}e^{-t}$$

of the initial-value problem.

(c) All exponential terms in the solution tend to zero. Hence, the solution tends to infinity linearly in t and is close to $t/4$ for large t.

41. **(a)** To find the general solution, we first guess

$$y_p(t) = ae^{-t} + bt + c,$$

where a, b and c are constants to be determined. For y_p,

$$\frac{dy_p}{dt} = -ae^{-t} + b \quad \text{and} \quad \frac{d^2y_p}{dt^2} = ae^{-t}.$$

Substituting these derivatives into the differential equation and collecting terms gives

$$(a + 4a)e^{-t} + (4b)t + (4c) = t + e^{-t},$$

which is satisfied if $5a = 1$, $4b = 1$, and $4c = 0$. Hence, a solution is

$$y_p(t) = \tfrac{1}{5}e^{-t} + \tfrac{1}{4}t.$$

To find the general solution of the homogeneous equation, we note that the characteristic polynomial $s^2 + 4$ has roots $s = \pm 2i$. Hence, the general solution of the forced equation is

$$y(t) = k_1 \cos 2t + k_2 \sin 2t + \tfrac{1}{5}e^{-t} + \tfrac{1}{4}t.$$

(b) To find the solution with the desired initial conditions, we note that $y(0) = k_1 + 0 + 1/5$ and $y'(0) = 2k_2 - 1/5 + 1/4$. We must solve the simultaneous equations

$$\begin{cases} k_1 + \frac{1}{5} = 0 \\ 2k_2 + \frac{1}{20} = 0. \end{cases}$$

Thus, $k_1 = -1/5$ and $k_2 = -1/40$ yield the solution

$$y(t) = -\tfrac{1}{5}\cos 2t - \tfrac{1}{40}\sin 2t + \tfrac{1}{5}e^{-t} + \tfrac{1}{4}t.$$

(c) Since all of the terms in the solution except $t/4$ are bounded for $t > 0$, the solution tends to infinity at a rate that is determined by $t/4$.

EXERCISES FOR SECTION 4.2

1. Recalling that the real part of e^{it} is $\cos t$, we see that the complex version of this equation is

$$\frac{d^2y}{dt^2} + 3\frac{dy}{dt} + 2y = e^{it}.$$

To find a particular solution, we guess $y_c(t) = ae^{it}$. Then $dy_c/dt = iae^{it}$ and $d^2y_c/dt^2 = -ae^{it}$. Substituting these derivatives into the equation and collecting terms yields

$$(-a + 3ia + 2a)e^{-it} = e^{it},$$

which is satisfied if

$$(1 + 3i)a = 1.$$

Hence, we must have

$$a = \frac{1}{1+3i} = \frac{1}{10} - \frac{3}{10}i.$$

So

$$y_c(t) = \frac{1-3i}{10}e^{it} = \frac{1-3i}{10}(\cos t + i\sin t)$$

is a particular solution of the complex version of the equation. Taking the real part, we obtain the solution

$$y(t) = \tfrac{1}{10}\cos t + \tfrac{3}{10}\sin t.$$

To produce the general solution of the homogeneous equation, we note that the characteristic polynomial $s^2 + 3s + 2$ has roots $s = -2$ and $s = -1$. So the general solution is

$$y(t) = k_1 e^{-2t} + k_2 e^{-t} + \tfrac{1}{10}\cos t + \tfrac{3}{10}\sin t.$$

3. Recalling that the imaginary part of e^{it} is $\sin t$, the complex version of the equation is

$$\frac{d^2y}{dt^2} + 3\frac{dy}{dt} + 2y = e^{it}.$$

This equation is precisely the same complex equation as in Exercise 1. Hence, we have already computed the solution

$$y_c(t) = \frac{1 - 3i}{10}(\cos t + i \sin t).$$

In this case we take the imaginary part

$$y(t) = -\tfrac{3}{10}\cos t + \tfrac{1}{10}\sin t$$

to obtain a solution of the original differential equation.

The general solution of the homogeneous equation is the same as in Exercise 1, so the general solution is

$$y(t) = k_1 e^{-2t} + k_2 e^{-t} - \tfrac{3}{10}\cos t + \tfrac{1}{10}\sin t.$$

5. The complex version of this equation is

$$\frac{d^2y}{dt^2} + 6\frac{dy}{dt} + 8y = e^{it}.$$

We guess a particular solution of the form $y_c(t) = ae^{it}$. Then $dy_c/dt = iae^{it}$ and $d^2y/dt^2 = -ae^{it}$. Substituting these derivatives into the complex differential equation yields

$$(-a + 6ia + 8a)e^{it} = e^{it},$$

which is satisfied if $(7 + 6i)a = 1$. Then $a = 1/(7 + 6i)$, and

$$y_c(t) = \frac{7 - 6i}{85}e^{it} = \frac{7 - 6i}{85}(\cos t + i \sin t).$$

The real part

$$y(t) = \tfrac{7}{85}\cos t + \tfrac{6}{85}\sin t$$

is a solution of the original differential equation.

To find the general solution of the homogeneous equation, we note that the characteristic polynomial $s^2 + 6s + 8$ has roots $s = -4$ and $s = -2$. Consequently, the general solution of the original nonhomogeneous equation is

$$y(t) = k_1 e^{-4t} + k_2 e^{-2t} + \tfrac{7}{85}\cos t + \tfrac{6}{85}\sin t.$$

7. The complex version of this equation is

$$\frac{d^2y}{dt^2} + 4\frac{dy}{dt} + 13y = 3e^{2it},$$

so we guess $y_c(t) = ae^{2it}$ to find a particular solution. Substituting $y_c(t)$ into the differential equation gives

$$(-4a + 8ai + 13a)e^{2it} = 3e^{2it},$$

which is satisfied if $(9 + 8i)a = 3$. Thus, $y_c(t)$ is a solution if

$$y_c(t) = \frac{3}{9 + 8i}e^{2it} = \frac{27 - 24i}{145}(\cos 2t + i \sin 2t).$$

A particular solution of the original equation is the real part

$$y(t) = \tfrac{27}{145}\cos 2t + \tfrac{24}{145}\sin 2t.$$

To find the general solution of the homogeneous equation, we note that the characteristic polynomial $s^2 + 4s + 13$ has roots $s = -2 \pm 3i$. Hence, the general solution of the original forced equation is

$$y(t) = k_1 e^{-2t}\cos 3t + k_2 e^{-2t}\sin 3t + \tfrac{27}{145}\cos 2t + \tfrac{24}{145}\sin 2t.$$

9. The complex version of this equation is

$$\frac{d^2y}{dt^2} + 4\frac{dy}{dt} + 20y = -3e^{2it},$$

and we guess that there is a solution of the form $y_c(t) = ae^{2it}$. Substituting this guess into the differential equation yields

$$(-4a + 8ia + 20a)e^{2it} = -3e^{2it},$$

which can be simplified to

$$(16 + 8i)ae^{2it} = -3e^{2it}.$$

Thus, $y_c(t)$ is a solution if $a = -3/(16 + 8i)$. We have

$$y_c(t) = \frac{-3}{16 + 8i}e^{2it} = \left(-\frac{3}{20} + \frac{3}{40}i\right)\left(\cos 2t + i \sin 2t\right).$$

We take the imaginary part to obtain a solution

$$y(t) = \tfrac{3}{40}\cos 2t - \tfrac{3}{20}\sin 2t$$

of the original equation.

To find the general solution of the homogeneous equation, we note that the characteristic polynomial is $s^2 + 4s + 20$, whose roots are $s = -2 \pm 4i$. Hence, the general solution of the original equation is

$$y(t) = k_1 e^{-2t}\cos 4t + k_2 e^{-2t}\sin 4t - \tfrac{3}{20}\sin 2t + \tfrac{3}{40}\cos 2t.$$

11. From Exercise 5, we know that the general solution of this equation is

$$y(t) = k_1 e^{-4t} + k_2 e^{-2t} + \tfrac{7}{85} \cos t + \tfrac{6}{85} \sin t.$$

To find the desired solution, we must solve for k_1 and k_2 using the initial conditions. We have

$$\begin{cases} k_1 + k_2 + \tfrac{7}{85} = 0 \\ -4k_1 - 2k_2 + \tfrac{6}{85} = 0. \end{cases}$$

We obtain $k_1 = 2/17$ and $k_2 = -1/5$. The desired solution is

$$y(t) = \tfrac{2}{17} e^{-4t} - \tfrac{1}{5} e^{-2t} + \tfrac{7}{85} \cos t + \tfrac{6}{85} \sin t.$$

13. From Exercise 9, we know that the general solution of this equation is

$$y(t) = k_1 e^{-2t} \cos 4t + k_2 e^{-2t} \sin 4t - \tfrac{3}{20} \sin 2t + \tfrac{3}{40} \cos 2t.$$

To find the desired solution, we must solve for k_1 and k_2 using the initial conditions. We have

$$\begin{cases} k_1 + \tfrac{3}{40} = 0 \\ -2k_1 + 4k_2 - \tfrac{6}{20} = 0. \end{cases}$$

We obtain $k_1 = -3/40$ and $k_2 = 3/80$. The desired solution is

$$y(t) = -\tfrac{3}{40} e^{-2t} \cos 4t + \tfrac{3}{80} e^{-2t} \sin 4t - \tfrac{3}{20} \sin 2t + \tfrac{3}{40} \cos 2t.$$

15. **(a)** If we guess

$$y_p(t) = a \cos 3t + b \sin 3t,$$

then

$$y_p'(t) = -3a \sin 3t + 3b \cos 3t$$

and

$$y_p''(t) = -9a \cos 3t - 9b \sin 3t.$$

Substituting this guess and its derivatives into the differential equation gives

$$(-8a + 9b) \cos 3t + (-9a - 8b) \sin 3t = \cos 3t.$$

Thus $y_p(t)$ is a solution if a and b satisfy the simultaneous equations

$$\begin{cases} -8a + 9b = 1 \\ -9a - 8b = 0. \end{cases}$$

Solving these equations for a and b, we obtain $a = -8/145$ and $b = 9/145$, so

$$y_p(t) = -\tfrac{8}{145} \cos 3t + \tfrac{9}{145} \sin 3t$$

is a solution.

(b) If we guess

$$y_p(t) = A\cos(3t + \phi),$$

then

$$y_p'(t) = -3A\sin(3t + \phi)$$

and

$$y_p''(t) = -9A\cos(3t + \phi).$$

Substituting this guess and its derivatives into the differential equation yields

$$-8A\cos(3t + \phi) - 9A\sin(3t + \phi) = \cos 3t.$$

Using the trigonometric identities for the sine and cosine of the sum of two angles, we have

$$-8A\,(\cos 3t\cos\phi - \sin 3t\sin\phi) - 9A\,(\sin 3t\cos\phi + \cos 3t\sin\phi) = \cos 3t.$$

This equation can be rewritten as

$$(-8A\cos\phi - 9A\sin\phi)\cos 3t + (8A\sin\phi - 9A\cos\phi)\sin 3t = \cos 3t.$$

It holds if

$$\begin{cases} -8A\cos\phi - 9A\sin\phi = 1 \\ 9A\cos\phi - 8A\sin\phi = 0. \end{cases}$$

Multiplying the first equation by 9 and the second by 8 and adding yields

$$145A\sin\phi = -9.$$

Similarly, multiplying the first equation by -8 and the second by 9 and adding yields

$$145A\cos\phi = -8.$$

Taking the ratio gives

$$\frac{\sin\phi}{\cos\phi} = \tan\phi = \frac{9}{8}.$$

Also, squaring both $145A\sin\phi = -9$ and $145A\cos\phi = -8$ yields

$$145^2 A^2\cos^2\phi + 145^2 A^2\sin^2\phi = 145,$$

so $A^2 = 1/145$.

We can use either $A = 1/\sqrt{145}$ or $A = -1/\sqrt{145}$, but this choice of sign for A effects the value of ϕ. If we pick $A = -1/\sqrt{145}$, then $\sqrt{145}\,\sin\phi = 9$, $\sqrt{145}\,\cos\phi = 8$, and $\tan\phi = 9/8$. In this case, $\phi = \arctan(9/8)$. Hence, a particular solution of the original equation is

$$y_p(t) = \frac{1}{\sqrt{145}}\cos\left(3t + \arctan\frac{9}{8}\right).$$

17. Since p and q are both positive, the solution of the homogeneous equation (the unforced response) tends to zero. Hence, we can match solutions to equations by considering the period (or frequency) and the amplitude of the steady-state solution (forced response). We also need to consider the rate at which solutions tend to the steady-state solution.

(a) The steady-state solution has period $2\pi/3$, and since the period of the steady-state solution is the same as the period of the forcing function, these solutions correspond to equations (v) or (vi). Moreover, this observation applies to the solutions in part (d) as well. Therefore, we need to match equations (v) and (vi) with the solutions in parts (a) and (d).

Solutions approach the steady-state faster in (d) than in (a). To distinguish (v) from (vi), we consider their characteristic polynomials. The characteristic polynomial for (v) is

$$s^2 + 5s + 1,$$

which has eigenvalues $(-5 \pm \sqrt{21})/2$. The characteristic polynomial for (vi) is

$$s^2 + s + 1,$$

which has eigenvalues $(-1 \pm i\sqrt{3})/2$. The rate of approach to the steady-state for (v) is determined by the slow eigenvalue $(-5 + \sqrt{21})/2 \approx -0.21$. The rate of approach to the steady-state for (vi) is determined by the real part of the eigenvalue, -0.5. Therefore, the graphs in part (a) come from equation (v), and the graphs in part (d) come from equation (vi).

(b) The steady-state solution has period 2π, and since the period of the steady-state solution is the same as the period of the forcing function, these solutions correspond to equations (i) or (ii). Moreover, this observation applies to the solutions in part (c) as well. Therefore, we need to match equations (i) and (ii) with the solutions in parts (b) and (c).

The amplitude of the steady-state solution is larger in (b) than in (c). To distinguish (i) from (ii), we calculate the amplitudes of the steady-state solutions for (i) and (ii). If we complexify these equations, we get

$$\frac{d^2y}{dt^2} + p\frac{dy}{dt} + qy = e^{it}.$$

Guessing a solution of the form $y_c(t) = ae^{it}$, we see that

$$a = \frac{1}{(q-1) + pi}.$$

The amplitude of the steady-state solution is $|a|$. For equation (i), $|a| = 1/\sqrt{29} \approx 0.19$, and for equation (ii), it is $1/\sqrt{5} \approx 0.44$. Therefore, the graphs in part (b) correspond to equation (ii), and the graphs in part (c) correspond to equation (i).

(c) See the answer to part (b).

(d) See the answer to part (a).

19. (a) Using the fact that the real part of $e^{(-1+i)t}$ is $e^{-t}\cos t$, the complex version of this equation is

$$\frac{d^2y}{dt^2} + 4\frac{dy}{dt} + 20y = e^{(-1+i)t}.$$

Guessing $y_c(t) = ae^{(-1+i)t}$ yields

$$a(-1+i)^2 e^{(-1+i)t} + 4a(-1+i)e^{(-1+i)t} + 20ae^{(-1+i)t} = e^{(-1+i)t}.$$

Simplifying we have

$$a(16 + 2i)e^{(-1+i)t} = e^{(-1+i)t}.$$

Thus, $y_c(t)$ is a solution of the complex differential equation if $a = 1/(16 + 2i)$, and we have

$$y_c(t) = \left(\tfrac{4}{65} - \tfrac{1}{130}i\right)e^{-t}(\cos t + i\sin t).$$

So one solution of the original equation is

$$y_p(t) = \tfrac{4}{65}e^{-t}\cos t + \tfrac{1}{130}e^{-t}\sin t.$$

To find the general solution of the homogeneous equation, we note that the characteristic polynomial $s^2 + 4s + 20$ has roots $s = -2 \pm 4i$.

Hence, the general solution of the original equation is

$$y(t) = k_1 e^{-2t}\cos 4t + k_2 e^{-2t}\sin 4t + \tfrac{4}{65}e^{-t}\cos t + \tfrac{1}{130}e^{-t}\sin t.$$

(b) All four terms in the general solution tend to zero as $t \to \infty$. Hence, all solutions tend to zero as $t \to \infty$. The terms with factors of e^{-2t} tend to zero very quickly, which leaves the terms of the particular solution $y_p(t)$ as the largest terms, so all solutions are asymptotic to $y_p(t)$. Since the solution $y_p(t)$ oscillates with period 2π and the amplitude of its oscillations decreases at the rate of e^{-t}, all solutions oscillate with this period and decaying amplitude.

21. Note that the real part of

$$(a - bi)(\cos \omega t + i\sin \omega t)$$

is $g(t)$. Hence, we must find k and ϕ such that

$$ke^{i\phi} = a - bi.$$

Using the polar form of the complex number $z = a - bi$, we see that

$$ke^{i\phi} = a - bi = z = |z|e^{i\theta},$$

where θ is the polar angle for z (see Appendix C). Therefore, we can choose

$$k = |z| = \sqrt{a^2 + b^2} \quad \text{and} \quad \phi = \theta.$$

23. Note that the real part of

$$(k_1 - ik_2)e^{i\beta t} = (k_1 - ik_2)(\cos \beta t + i\sin \beta t)$$

is

$$y(t) = k_1 \cos \beta t + k_2 \sin \beta t.$$

Let $Ke^{i\phi}$ be the polar form of the complex number $k_1 + ik_2$. Then the polar form of $k_1 - ik_2$ is $Ke^{-i\phi}$. Using the Laws of Exponents and Euler's formula, we have

$$(k_1 - ik_2)e^{i\beta t} = Ke^{-i\phi}e^{i\beta t}$$
$$= Ke^{i(\beta t - \phi)}$$
$$= K(\cos(\beta t - \phi) + i\sin(\beta t - \phi),$$

and the real part is $K\cos(\beta t - \phi)$. Hence, we see that

$$y(t) = k_1 \cos \beta t + k_2 \sin \beta t$$

can be rewritten as

$$y(t) = K\cos(\beta t - \phi).$$

EXERCISES FOR SECTION 4.3

1. The complex version of this equation is

$$\frac{d^2y}{dt^2} + 9y = e^{it}.$$

Guessing $y_c(t) = ae^{it}$ as a particular solution and substituting this guess into the left-hand side of the differential equation yields

$$8ae^{it} = e^{it}.$$

Thus, $y_c(t)$ is a solution if $8a = 1$. The real part of

$$y_c(t) = \tfrac{1}{8}e^{it} = \tfrac{1}{8}(\cos t + i\sin t)$$

is $y(t) = \tfrac{1}{8}\cos t$. This $y(t)$ is a solution to the original differential equation. [Because there is no dy/dt-term (no damping), we could have guessed a solution of the form $y(t) = a\cos t$ instead of using the complex version of the equation.]

To find the general solution of the homogeneous equation, we note that the characteristic polynomial is $s^2 + 9$, which has roots $s = \pm 3i$. So the general solution of the original equation is

$$y(t) = k_1 \cos 3t + k_2 \sin 3t + \tfrac{1}{8}\cos t.$$

3. The complex version of this equation is

$$\frac{d^2y}{dt^2} + 4y = -e^{it/2}.$$

Guessing $y_c(t) = ae^{it/2}$ as a particular solution and substituting into the equation yields

$$\tfrac{15}{4}ae^{it/2} = -e^{it/2}.$$

Thus, $y_c(t)$ is a solution if $\tfrac{15}{4}a = -1$. The real part of

$$y_c(t) = -\frac{4}{15}e^{it/2} = -\frac{4}{15}\left(\cos\frac{t}{2} + i\sin\frac{t}{2}\right)$$

is

$$y(t) = -\frac{4}{15} \cos \frac{t}{2}.$$

This $y(t)$ is a solution to the original differential equation. [Because there is no dy/dt-term (no damping), we could have guessed a solution of the form $y(t) = a \cos t/2$ instead of using the complex version of the equation.]

To find the general solution of the homogeneous equation, we note that the characteristic polynomial is $s^2 + 4$, which has roots $s = \pm 2i$. So the general solution of the original equation is

$$y(t) = k_1 \cos 2t + k_2 \sin 2t - \frac{4}{15} \cos \frac{t}{2}.$$

5. The complex version of the equation is

$$\frac{d^2y}{dt^2} + 9y = 2e^{3it}.$$

Guessing $y_c(t) = ae^{3it}$ as a particular solution and substituting this guess into the left-hand side of the differential equation, we see that a must satisfy

$$-9a + 9a = 2,$$

which is impossible. Hence, the forcing is in resonance with the associated homogeneous equation.

We must make a second guess of $y_c(t) = ate^{3it}$. This guess gives

$$y_c'(t) = a(1 + 3it)e^{3it}$$

and

$$y_c''(t) = a(6i - 9t)e^{3it}.$$

Substituting $y_c(t)$ and its second derivative into the differential equation, we obtain

$$a(6i - 9t)e^{3it} + 9ate^{3it} = 2e^{3it},$$

which simplifies to

$$6aie^{3it} = 2e^{3it}.$$

Thus, $y_c(t)$ is a solution if $a = 2/(6i) = -i/3$. Taking the real part of

$$y_c(t) = -\tfrac{1}{3}it(\cos 3t + i \sin 3t),$$

we obtain the solution

$$y(t) = \tfrac{1}{3}t \sin 3t$$

of the original equation.

To find the general solution of the homogeneous equation, we note that the characteristic polynomial is $s^2 + 9$, which has roots $s = \pm 3i$.

Hence, the general solution of the original equation is

$$y(t) = k_1 \cos 3t + k_2 \sin 3t + \tfrac{1}{3}t \sin 3t.$$

7. The complex version of this equation is

$$\frac{d^2 y}{dt^2} + 3y = e^{3it}.$$

Guessing $y_c(t) = ae^{3it}$ as a particular solution and substituting this guess into the left-hand side of the differential equation yields

$$-6ae^{3it} = e^{3it}.$$

Thus, $y_c(t)$ is a solution if $-6a = 1$. The real part of

$$y_c(t) = -\tfrac{1}{6}e^{3it} = -\tfrac{1}{6}(\cos 3t + i \sin 3t)$$

is $y(t) = -\tfrac{1}{6}\cos 3t$. This $y(t)$ is a solution to the original differential equation. [Because there is no dy/dt-term (no damping), we could have guessed a solution of the form $y(t) = a \cos 3t$ instead of using the complex version of the equation.]

To find the general solution of the homogeneous equation, we note that the characteristic polynomial is $s^2 + 3$, which has roots $s = \pm\sqrt{3}\,i$.

Hence, the general solution of the original equation is

$$y(t) = k_1 \cos \sqrt{3}\,t + k_2 \sin \sqrt{3}\,t - \tfrac{1}{6}\cos 3t.$$

9. From Exercise 1, we know that the general solution is

$$y(t) = k_1 \cos 3t + k_2 \sin 3t + \tfrac{1}{8}\cos t.$$

So

$$y'(t) = -3k_1 \sin 3t + 3k_2 \cos 3t - \tfrac{1}{8}\sin t.$$

Using the initial conditions $y(0) = 0$ and $y'(0) = 0$, we obtain the simultaneous equations

$$\begin{cases} k_1 + \tfrac{1}{8} = 0 \\ 3k_2 = 0, \end{cases}$$

which imply that $k_1 = -1/8$ and $k_2 = 0$. The solution to the initial-value problem is

$$y(t) = -\tfrac{1}{8}\cos 3t + \tfrac{1}{8}\cos t.$$

11. First we find the general solution by considering the complex version of the equation

$$\frac{d^2 y}{dt^2} + 5y = 3e^{2it}.$$

Guessing a particular solution of the form $y_c(t) = ae^{2it}$ and substituting this guess into the left-hand side of the equation yields

$$ae^{2it} = 3e^{2it}.$$

Thus, $y_c(t)$ is a solution if $a = 3$. The real part of

$$y_c(t) = 3(\cos 2t + i \sin 2t)$$

is $y(t) = 3 \cos 2t$. This $y(t)$ is a solution to the original differential equation. [Because there is no dy/dt-term (no damping), we could have guessed a solution of the form $y(t) = a \cos 2t$ instead of using the complex version of the equation.]

To find the general solution of the homogeneous equation, we note that the characteristic polynomial is $s^2 + 5$, which has roots $s = \pm\sqrt{5}\,i$. Hence, the general solution is

$$y(t) = k_1 \cos \sqrt{5}\,t + k_2 \sin \sqrt{5}\,t + 3 \cos 2t.$$

Note that

$$y'(t) = -\sqrt{5}\,k_1 \sin \sqrt{5}\,t + \sqrt{5}\,k_2 \cos \sqrt{5}\,t - 6 \sin 2t.$$

Using the initial conditions $y(0) = 0$ and $y'(0) = 0$, we obtain the simultaneous equations

$$\begin{cases} k_1 + 3 = 0 \\ \sqrt{5}\,k_2 = 0, \end{cases}$$

which imply that $k_1 = -3$ and $k_2 = 0$. The solution to the initial-value problem is

$$y(t) = -3 \cos \sqrt{5}\,t + 3 \cos 2t.$$

13. From Exercise 5, we know that the general solution is

$$y(t) = k_1 \cos 3t + k_2 \sin 3t + \tfrac{1}{3}t \sin 3t.$$

So

$$y'(t) = -3k_1 \sin 3t + 3k_2 \cos 3t + \tfrac{1}{3} \sin 3t + t \cos 3t.$$

From the initial condition $y(0) = 2$, we see that $k_1 = 2$. Using the initial condition $y'(0) = -9$, we have $3k_2 = -9$. Hence, $k_2 = -3$. The solution to the initial-value problem is

$$y(t) = 2 \cos 3t - 3 \sin 3t + \tfrac{1}{3}t \sin 3t.$$

15. The characteristic polynomial of the unforced equation is $s^2 + 4$, which has roots $s = \pm 2i$. So the natural frequency is $2/(2\pi)$, and the forcing frequency is $9/(8\pi)$.

(a) The frequency of the beats is

$$\frac{\frac{9}{4} - 2}{4\pi} = \frac{1}{16\pi},$$

and therefore, the period of one beat is $16\pi \approx 50$.

(b) The frequency of the rapid oscillations is

$$\frac{\frac{9}{4} + 2}{4\pi} = \frac{17}{16\pi}.$$

Therefore, there are 17 rapid oscillations in each beat.

(c)

17. The characteristic polynomial of the unforced equation is $s^2 + 5$, which has roots $s = \pm i\sqrt{5}$. So the natural frequency is $\sqrt{5}/(2\pi)$, and the forcing frequency is $2/(2\pi)$.

(a) The frequency of the beats is

$$\frac{\sqrt{5} - 2}{4\pi},$$

and therefore, the period of one beat is approximately 53.

(b) The frequency of the rapid oscillations is

$$\frac{\sqrt{5} + 2}{4\pi}.$$

Therefore, there are approximately 18 rapid oscillations in each beat.

(c)

19. **(a)** To find the general solution, we deal with each of the forcing terms separately. In other words, we find solutions to

$$\frac{d^2y}{dt^2} + 12y = 3\cos 4t \quad \text{and} \quad \frac{d^2y}{dt^2} + 12y = 2\sin t$$

separately and add them to get a solution to the original equation (see Exercise 36 in Section 4.1).

First consider the equation whose forcing term is $3\cos 4t$. The complex version of the equation is

$$\frac{d^2y}{dt^2} + 12y = 3e^{4it}.$$

We guess a solution of the form $y_c(t) = ae^{4it}$ and substitute it into the differential equation to obtain

$$a(-16 + 12)e^{4it} = 3e^{4it},$$

which is satisfied if $a = -3/4$. Hence, by taking the real part of

$$y_c(t) = -\tfrac{3}{4}(\cos 4t + i \sin 4t),$$

we obtain the solution $y_1(t) = -\tfrac{3}{4}\cos 4t$.

Similarly, to find a solution of the equation whose forcing term is $2 \sin t$, we consider

$$\frac{d^2 y}{dt^2} + 12y = 2e^{it}$$

and guess $y_c(t) = be^{it}$. Substituting this guess into the equation yields

$$b(-1 + 12)e^{it} = 2e^{it},$$

which is satisfied if $b = 2/11$. So, by taking the imaginary part of

$$y_c(t) = \tfrac{2}{11}(\cos t + i \sin t),$$

we obtain the solution $y_2(t) = \tfrac{2}{11}\sin t$.

To obtain the general solution of the unforced equation, we note that the characteristic polynomial is $s^2 + 12$, which has roots $s = \pm i\sqrt{12}$. So the general solution of the original differential equation is

$$y(t) = k_1 \cos \sqrt{12}\,t + k_2 \sin \sqrt{12}\,t - \tfrac{3}{4}\cos 4t + \tfrac{2}{11}\sin t.$$

(b) To obtain the initial conditions $y(0) = 0$ and $y'(0) = 0$, we note that

$$y'(t) = -\sqrt{12}\,k_1 \sin \sqrt{12}\,t + \sqrt{12}\,k_2 \cos \sqrt{12}\,t + 3 \sin 4t + \tfrac{2}{11}\cos t.$$

Hence, we must solve the simultaneous equations

$$\begin{cases} k_1 - \tfrac{3}{4} = 0 \\[2mm] \sqrt{12}\,k_2 + \tfrac{2}{11} = 0 \end{cases}$$

for k_1 and k_2. We obtain $k_1 = 3/4$ and $k_2 = -2/(11\sqrt{12})$. Hence, the solution to the initial-value problem is

$$y(t) = \tfrac{3}{4}\cos \sqrt{12}\,t - \tfrac{2}{11\sqrt{12}}\sin \sqrt{12}\,t - \tfrac{3}{4}\cos 4t + \tfrac{2}{11}\sin t.$$

(c)

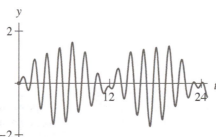

(d) The solution is the sum of the general solution of the unforced equation and particular solutions of the forced equations with each forcing term considered separately. Since the forcing function $\cos 4t$ is close to resonance, we expect to see large amplitude beats with frequency $(4 - \sqrt{12})/(4\pi)$ and rapid oscillations with frequency $(4 + \sqrt{12})/(4\pi)$. The particular solution for the $2 \sin t$ forcing term has a relatively small amplitude, and therefore it does not have a significant effect on the final oscillations most of the time.

21. **(a)** The graph shows either the solution of a resonant equation or one with beats whose period is very large. The period of the beats in equation (iii) is 4π, and the period of the beats in equation (iv) is $4\pi/(4 - \sqrt{14}) \approx 48.6$. Hence this graph must correspond to a solution of the resonant equation—equation (v).

(b) The graph has beats with period 4π. Therefore, this graph corresponds to equation (iii).

(c) This solution has no beats and no change in amplitude. Therefore, it corresponds to either (i), (ii), or (vi). Note that the general solution of equation (i) is

$$k_1 \cos 4t + k_2 \sin 4t + \tfrac{5}{8},$$

and the general solution of equation (ii) is

$$k_1 \cos 4t + k_2 \sin 4t - \tfrac{5}{8}.$$

Equation (iv) has a steady-state solution whose oscillations are centered about $y = 0$. Since the oscillations shown are centered around a positive constant, this function is a solution of equation (i).

(d) The graph has beats with a period that is approximately 50. Therefore, this graph corresponds to equation (iv) (see part (a)).

23. The equation of motion for the unforced mass-spring system is

$$\frac{d^2 y}{dt^2} + 16y = 0,$$

so the natural period is $2\pi/4 = \pi/2 \approx 1.57$.

Tapping with the hammer as shown increases the velocity if the mass is moving to the right at the time of the tap and decreases the velocity if the mass is moving to the left at the time of the tap. Faster motion results in higher amplitude oscillations. Since none of the tapping periods is exactly $\pi/2$, the taps sometimes increase the amplitude and sometimes decrease the amplitude of the oscillations (that is, resonance does not occur).

The period $T = 3/2$ is closest to the natural period and hence for taps with this period we expect the largest amplitude oscillations.

EXERCISES FOR SECTION 4.4

1. Rubbing a finger around the edge of the glass starts the glass vibrating. The finger then skips along the glass, giving a forcing term which has the same frequency as the natural frequency of vibration of the glass. Pressing harder changes the size of the forcing. Since the frequency is determined by the motion of the glass itself, pushing harder does not alter the frequency.

3. Given that

$$y_p''(t) + p y_p'(t) + q y_p(t) = g(t)$$

for all t, we have

$$y_p''(t + \theta) + p y_p'(t + \theta) + q y_p(t + \theta) = g(t + \theta).$$

Let $z(t) = y_p(t + \theta)$. Then

$$z'(t) = y_p'(t + \theta) \quad \text{and} \quad z''(t) = y_p''(t + \theta)$$

by the Chain Rule. Thus,

$$z''(t) + pz'(t) + qz(t) = g(t + \theta),$$

and $z(t) = y_p(t + \theta)$ is a solution of

$$\frac{d^2y}{dt^2} + p\frac{dy}{dt} + qy = g(t + \theta).$$

5. As in Exercise 4, the forced response is

$$y(t) = A_1 \cos \omega_1 t + A_2 \cos \omega_2 t.$$

where

$$A_i = \frac{1}{\sqrt{(q - \omega_i^2)^2 + p^2\omega_i^2}}$$

for $i = 1$ and 2.

   (a) Setting $q = \omega_1^2$, we have

$$\frac{A_1}{A_2} = \frac{\sqrt{(\omega_1^2 - \omega_2^2)^2 + p^2\omega_2^2}}{p\omega_1}.$$

   (b) Let $R(p) = A_1/A_2$. We see that $R(p) \to \infty$ as $p \to 0$ and $R(p) \to \omega_2/\omega_1$ as $p \to \infty$. Moreover,

$$\frac{dR}{dp} = \frac{-(\omega_1^2 - \omega_2^2)^2}{p^2\omega_1^2\sqrt{(\omega_1^2 - \omega_2^2)^2 + p^2\omega_2^2}},$$

   which is negative for all $p > 0$. Hence, $R(p)$ is monotonically decreasing.

7. **(a)**

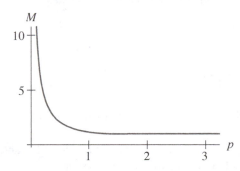

   (b) If q is fixed and $p \to 0$, we need only consider the formula

$$M(p, q) = \frac{1}{\sqrt{p^2q - p^4/4}}.$$

For small $p > 0$,

$$M(p, q) \approx \frac{1}{\sqrt{p^2q}} = \frac{1}{\sqrt{q}}\frac{1}{p}.$$

9. We guess a solution of the form $y(t) = \alpha \sin \omega t + \beta \cos \omega t$. Substituting $y(t)$ into the left-hand side of the equation, we obtain

$$-\omega^2 \alpha \sin \omega t - \omega^2 \beta \cos \omega t + b(\omega\alpha \cos \omega t - \omega\beta \sin \omega t) + k(\alpha \sin \omega t + \beta \cos \omega t).$$

Our guess is a solution if this expression agrees with the right-hand side of the equation, namely

$$k A \cos \omega t - b\omega A \sin \omega t.$$

Equating the coefficients of sine and cosine from both sides of the equation, we obtain

$$-\omega^2 \beta + b\omega\alpha + k\beta = k A$$
$$-\omega^2 \alpha - b\omega\beta + k\alpha = -b\omega A,$$

which can be rewritten as

$$b\omega\alpha + (k - \omega^2)\beta = k A$$
$$(k - \omega^2)\alpha - b\omega\beta = -b\omega A.$$

Solving this system of two equations for α and β yields

$$\alpha = A\frac{b\omega^3}{(k - \omega^2)^2 + b^2\omega^2} \quad \text{and} \quad \beta = A\frac{b\omega^3}{(k - \omega^2)^2 + b^2\omega^2}.$$

So the particular solution is

$$y(t) = A\frac{b\omega^3}{(k - \omega^2)^2 + b^2\omega^2} \sin \omega t + A\frac{b\omega^3}{(k - \omega^2)^2 + b^2\omega^2} \cos \omega t.$$

11. If ω is large, the coefficients of both terms in the particular solution are small because they both have ω^4 in the denominator and at most ω^3 in the numerator. Hence, the oscillations are small. The weak spring and dash pot do not transmit the forcing to the mass fast enough and the pushes and pulls "average" out before they can effect the motion of the mass.

EXERCISES FOR SECTION 4.5

1. **(a)** The stiffness of the roadbed is measured by the coefficient of y. Increasing the stiffness corresponds to increasing β.

  (b) As β increases, the term corresponding to the stretch in cable becomes less important. Therefore, the system behaves more like a linear system.

3. **(a)** As the strength of the cable is increased, the spring constant γ for the cables (which appears in the definition of $c(y)$) increases.

  (b) Increasing γ in the definition of $c(y)$ has several effects. First, it makes the jump between the behavior of the system for $y < 0$ and $y > 0$ (that is, between taut and loose cables) more pronounced. This effect increases the system's "nonlinearity." Also, the equilibrium point moves closer to $y = 0$. On the other hand, larger γ makes it harder to displace the bridge from its rest position, so in most situations, we only see small amplitude oscillations.

5. The buoyancy force is proportional to the volume of the water displaced. Suppose p is that proportionality constant and a is the length of one side of a cube. Let y be the distance between the bottom of the cube and the surface of the water. Since the cube always stays in contact with the water and is never completely submerged, the buoyancy force is $-pa^2y$, and the force equation $F = ma$ is

$$mg - pa^2y - \epsilon\frac{dy}{dt} = m\frac{d^2y}{dt^2},$$

where the term $\epsilon(dy/dt)$ measures the damping. This equation can be rewritten as

$$m\frac{d^2y}{dt^2} + \epsilon\frac{dy}{dt} + pa^2y = mg.$$

7. If the cube rises completely out of the water, the buoyancy force vanishes. If we represent the buoyancy force by the function $b(y)$, we have

$$b(y) = \begin{cases} -pa^2y & \text{if } y > 0; \\ 0 & \text{if } y < 0. \end{cases}$$

Therefore, the differential equation is

$$m\frac{d^2y}{dt^2} + \epsilon\frac{dy}{dt} = mg + b(y).$$

REVIEW EXERCISES FOR CHAPTER 4

1. The natural guess for a particular solution is a constant function $y_p(t) = a$. For this function, $dy_p/dt = d^2y_p/dt^2 = 0$, so it is a solution if and only if $ka = 1$. Hence, the constant function $y_p(t) = 1/k$ for all t is a solution.

3. The frequency of the steady-state solution is the same as the frequency of the forcing function. The frequency of $4\cos 2t$ is $2/(2\pi) = 1/\pi$.

5. Yes, there is a steady-state response for this equation. Note that the general solution to the associated homogeneous equation is $ke^{-\lambda t}$, which tends to zero as $t \to \infty$.

   We can compute one particular solution to forced equation using the techniques of Section 1.8, or we can complexify the equation and guess a solution to

$$\frac{dy}{dt} + \lambda y = e^{i\omega t}.$$

To find one solution to the complexified equation, we guess $y_c(t) = ae^{i\omega t}$. Then

$$\frac{dy_c}{dt} + \lambda y_c = a(i\omega)e^{i\omega t} + \lambda(ae^{i\omega t})$$

$$= a(\lambda + i\omega)e^{i\omega t}.$$

We have a solution if $a = 1/(\lambda + i\omega)$.

To obtain the steady-state solution of the original equation, we take the real part of

$$y_c(t) = \frac{1}{\lambda + i\omega} e^{i\omega t} = \frac{\lambda - i\omega}{\lambda^2 + \omega^2} e^{i\omega t}.$$

We obtain

$$\frac{\lambda}{\lambda^2 + \omega^2} \cos \omega t + \frac{\omega}{\lambda^2 + \omega^2} \sin \omega t.$$

See the discussion and example on page 118 of Section 1.8.

7. The eigenvalues of the associated homogeneous equation are $-1 \pm i\sqrt{3}$. Hence, the difference $|y_1(t) - y_2(t)|$ between solutions decays at a rate of e^{-t}. To get a rough estimate of T, we solve

$$e^{-T} = \frac{1}{100}$$

and obtain $T \approx \ln 100 \approx 4.6$

9. By observing the amplitude of the oscillations for different forcing frequencies, you can determine the resonant frequency. Since the damping is very small, the model is approximately

$$m\frac{d^2y}{dt^2} + ky = \cos \omega t,$$

and resonance occurs if $\omega = \sqrt{k/m}$. Hence, you can determine the ratio of the spring constant and the mass.

11. True. The eigenvalues of the associated homogeneous equation are

$$s = \frac{-1 \pm i\sqrt{23}}{2}.$$

Hence, the general solution has the form

$$y(t) = k_1 e^{-t/2} \cos \sqrt{23}\,t + k_2 e^{-t/2} \sin \sqrt{23}\,t + \alpha \cos t + \beta \sin t,$$

where α and β are determined by the equation and k_1 and k_2 are determined by the initial condition. Unless both k_1 and k_2 are zero, $y(t)$ is unbounded as $t \to -\infty$. If $k_1 = k_2 = 0$, then $y(t)$ is the steady-state solution. Using the standard way to compute the amplitude of the steady-state solution, we see that it is $1/\sqrt{26}$, which is definitely less than 1.

13. False. The frequency of the forcing function and the natural frequency of the oscillator agree. Consequently, the system is in resonance, and the amplitude of the forced response increases without bound as $t \to \infty$. However, this amplitude grows linearly. Repeatedly doubling over equal intervals of time results in exponential growth rather than linear growth.

15. To compute the general solution of the unforced equation, we use the method of Section 3.6. The characteristic polynomial is

$$s^2 + 6s + 8,$$

so the eigenvalues are $s = -2$ and $s = -4$. Hence, the general solution of the homogeneous equation is

$$k_1 e^{-2t} + k_2 e^{-4t}.$$

To find a particular solution of the forced equation, we guess $y_p(t) = ke^{-t}$. Substituting into the left-hand side of the differential equation gives

$$\frac{d^2 y_p}{dt^2} + 6\frac{dy_p}{dt} + 8y_p = ke^{-t} - 6ke^{-t} + 8ke^{-t}$$
$$= 3ke^{-t}.$$

In order for $y_p(t)$ to be a solution of the forced equation, we must take $k = 1/3$. The general solution of the forced equation is

$$y(t) = k_1 e^{-2t} + k_2 e^{-4t} + \tfrac{1}{3}e^{-t}.$$

17. To compute the general solution of the unforced equation, we use the method of Section 3.6. The characteristic polynomial is

$$s^2 - 2s - 3,$$

so the eigenvalues are $s = -1$ and $s = 3$. Hence, the general solution of the homogeneous equation is

$$k_1 e^{-t} + k_2 e^{3t}.$$

To find a particular solution of the forced equation, a reasonable looking guess is $y_p(t) = ke^{3t}$. However, this guess is a solution of the homogeneous equation, so it is doomed to fail. We make the standard second guess of $y_p(t) = kte^{3t}$. Substituting into the left-hand side of the differential equation gives

$$\frac{d^2 y_p}{dt^2} - 2\frac{dy_p}{dt} - 3y_p = (6ke^{3t} + 9kte^{3t}) - 2(ke^{3t} + 3kte^{3t}) - 3kte^{3t}$$
$$= 4ke^{3t}.$$

In order for $y_p(t)$ to be a solution of the forced equation, we must take $k = 1/4$. The general solution of the forced equation is

$$y(t) = k_1 e^{-t} + k_2 e^{3t} + \tfrac{1}{4}te^{3t}.$$

19. To compute the general solution of the unforced equation, we use the method of Section 3.6. The characteristic polynomial is

$$s^2 + 6s + 8.$$

So the eigenvalues are $s = -2$ and $s = -4$, and the general solution of the homogeneous equation is

$$k_1 e^{-2t} + k_2 e^{-4t}.$$

To find one solution of the forced equation, we guess the constant function $y_p(t) = k$. Substituting $y_p(t)$ into the left-hand side of the differential equation, we obtain

$$\frac{d^2 y_p}{dt^2} + 6\frac{dy_p}{dt} + 8y_p = 0 + 6 \cdot 0 + 8k = 8k.$$

Hence, $k = 5/8$ yields a solution of the forced equation. The general solution of the forced equation is

$$y(t) = k_1 e^{-2t} + k_2 e^{-4t} + \tfrac{5}{8}.$$

21. To compute the general solution of the unforced equation, we use the method of Section 3.6. The characteristic polynomial is

$$s^2 - 4s + 13,$$

so the eigenvalues are $s = 2 \pm 3i$. Hence, the general solution of the homogeneous equation is

$$k_1 e^{2t} \cos 3t + k_2 e^{2t} \sin 3t.$$

To find one particular solution of the forced equation, we complexify the equation and obtain

$$\frac{d^2 y}{dt^2} - 4\frac{dy}{dt} + 13y = 5e^{4it},$$

and we guess that there is a solution of the form $y_c(t) = ae^{4it}$. Substituting this guess into the differential equation yields

$$a(-16 - 16i + 13)e^{4it} = 5e^{4it},$$

which can be simplified to $a(-3 - 16i)e^{4it} = 5e^{4it}$. Thus, $y_c(t)$ is a solution if $a = 5/(-3 - 16i)$. We have

$$y_c(t) = \frac{5}{-3 - 16i}e^{4it} = \left(-\frac{3}{53} + \frac{16}{53}i\right)\left(\cos 4t + i \sin 4t\right).$$

We take the real part to obtain a solution

$$y(t) = -\tfrac{3}{53} \cos 4t - \tfrac{16}{53} \sin 4t$$

of the original nonhomogeneous equation.

The general solution of the forced equation is

$$y(t) = k_1 e^{2t} \cos 3t + k_2 e^{2t} \sin 3t - \tfrac{3}{53} \cos 4t - \tfrac{16}{53} \sin 4t.$$

23. We begin with some observations about the various equations:

(i) Solutions are periodic with period $\pi/2$.

(ii) The period of the steady-state solution is π.

(iii) The period of the steady-state solution is $\pi/2$.

(iv) The equilibrium point at the origin is a spiral source.

(v) The equilibrium point at the origin is a spiral sink.

(vi) The period of the beats is approximately 1.37.

(vii) Solutions are periodic with period $2\pi/3$.

(viii) The period of the beats is approximately 39.27.

Now we match the graphs with the equations.

(a) This graph is either a solution of an undamped homogeneous equation, or it is the steady-state solution of a damped homogeneous equation. So we consider equations (i), (ii), (iii), and (vii). The period of the solution is $2\pi/3$. Consequently, this solution comes from equation (vii).

(b) This graph corresponds to a forced equation with damping. So we consider equations (ii) and (iii). The period of the steady-state solution is π. Consequently, this solution comes from equation (ii).

(c) This graph corresponds to a homogeneous equation with damping. So we consider equations (iv) and (v). Equation (iv) has a spiral source at the origin, and equation (v) has a spiral sink at the origin. Consequently, this solution comes from equation (v).

(d) This graph is the solution of an undamped sinusoidally forced equation. So we consider equations (vi) and (viii). The period of the beat is approximately 40. Consequently, this solution comes from equation (viii).

25. **(a)** To compute the general solution of the unforced equation, we use the method of Section 3.6. The characteristic polynomial is

$$s^2 + 2s + 3,$$

so the eigenvalues are $s = -1 \pm i\sqrt{2}$. Hence, the general solution of the homogeneous equation is

$$k_1 e^{-t} \cos \sqrt{2}\,t + k_2 e^{-t} \sin \sqrt{2}\,t.$$

To find one particular solution of the nonhomogeneous equation, we complexify the equation and obtain

$$\frac{d^2 y}{dt^2} + 2\frac{dy}{dt} + 3y = e^{2it},$$

and we guess that there is a solution of the form $y_c(t) = ae^{2it}$. Substituting this guess into the differential equation yields

$$a(-4 + 4i + 3)e^{2it} = e^{2it},$$

which can be simplified to $a(-1 + 4i)e^{2it} = e^{2it}$. Thus, $y_c(t)$ is a solution if $a = 1/(-1 + 4i)$. We have

$$y_c(t) = \frac{1}{-1 + 4i}\, e^{2it} = \left(-\frac{1}{17} - \frac{4}{17}i\right)\left(\cos 2t + i\sin 2t\right).$$

We take the real part to obtain a solution

$$y(t) = -\tfrac{1}{17}\cos 2t + \tfrac{4}{17}\sin 2t$$

of the original nonhomogeneous equation.

The general solution of the forced equation is

$$y(t) = k_1 e^{-t}\cos\sqrt{2}\,t + k_2 e^{-t}\sin\sqrt{2}\,t - \tfrac{1}{17}\cos 2t + \tfrac{4}{17}\sin 2t.$$

(b) The amplitude of the steady-state solution is

$$|a| = \frac{1}{|-1 + 4i|} = \frac{1}{\sqrt{17}} \approx 0.24.$$

The phase angle ϕ is the polar angle of a. In this case, it is the angle in the third quadrant such that $\tan\phi = (-4)/(-1)$. We get $\phi \approx -104°$. (Note that this angle is expressed in terms of degrees.)

27. **(a)** To compute the general solution of the unforced equation, we use the method of Section 3.6. The characteristic polynomial is

$$s^2 + 4s + 3,$$

so the eigenvalues are $s = -3$ and $s = -1$. Hence, the general solution of the homogeneous equation is

$$k_1 e^{-3t} + k_2 e^{-t}.$$

To find one particular solution of the nonhomogeneous equation, we complexify the equation and obtain

$$\frac{d^2y}{dt^2} + 4\frac{dy}{dt} + 3y = 5e^{2it},$$

and we guess that there is a solution of the form $y_c(t) = ae^{2it}$. Substituting this guess into the differential equation yields

$$a(-4 + 8i + 3)e^{2it} = 5e^{2it},$$

which can be simplified to $a(-1+8i)e^{2it} = 5e^{2it}$. Thus, $y_c(t)$ is a solution if $a = 5/(-1+8i)$. We have

$$y_c(t) = \frac{5}{-1+8i} e^{2it} = \left(-\frac{1}{13} - \frac{8}{13}i\right)\left(\cos 2t + i\sin 2t\right).$$

We take the imaginary part to obtain a solution

$$y(t) = -\tfrac{8}{13}\cos 2t - \tfrac{1}{13}\sin 2t$$

of the original nonhomogeneous equation.

The general solution of the forced equation is

$$y(t) = k_1 e^{-3t} + k_2 e^{-t} - \tfrac{8}{13}\cos 2t - \tfrac{1}{13}\sin 2t.$$

(b) The amplitude of the steady-state solution is

$$|a| = \frac{5}{|-1+8i|} = \frac{5}{\sqrt{65}} = \frac{\sqrt{65}}{13} \approx 0.62.$$

Since we got the steady-state solution by taking the imaginary part of $y_c(t)$, the phase angle is the polar angle of the complex number

$$-\tfrac{8}{13} + \tfrac{1}{13}i.$$

In this case, it is the angle in the second quadrant such that $\tan\phi = 1/(-8)$. We get $\phi \approx -187°$. (Note that this angle is expressed in terms of degrees. It is not between $-180°$ and $0°$ due to the fact that the forcing is in terms of sine rather than cosine.)

CHAPTER 5

Nonlinear Systems

EXERCISES FOR SECTION 5.1

1. The linearizations of systems (i) and (iii) are both

$$\frac{dx}{dt} = 2x + y$$
$$\frac{dy}{dt} = -y,$$

so these two systems have the same "local picture" near $(0, 0)$. This system has eigenvalues 2 and -1; hence, $(0, 0)$ is a saddle for these systems. System (ii) has linearization

$$\frac{dx}{dt} = 2x + y$$
$$\frac{dy}{dt} = y,$$

which has eigenvalues 2 and 1, hence, $(0, 0)$ is a source for this system.

3.   (a) The linearized system is

$$\frac{dx}{dt} = -2x + y$$
$$\frac{dy}{dt} = -y.$$

We can see this either by "dropping higher-order terms" or by computing the Jacobian matrix

$$\begin{pmatrix} -2 & 1 \\ 2x & -1 \end{pmatrix}$$

and evaluating it at $(0, 0)$.

   (b) The eigenvalues of the linearized system are -2 and -1, so $(0, 0)$ is a sink.

   (c) The vector $(1, 0)$ is an eigenvector for eigenvalue -2 and $(1, 1)$ is an eigenvector for the eigenvalue -1.

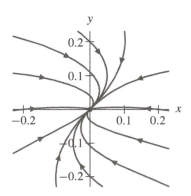

(d) By computing the Jacobian matrix

$$\begin{pmatrix} -2 & 1 \\ 2x & -1 \end{pmatrix}$$

and evaluating at $(2, 4)$, we see that linearized system at $(2, 4)$ is

$$\frac{dx}{dt} = -2x + y$$
$$\frac{dy}{dt} = 4x - y.$$

Its eigenvalues are $(-3 \pm \sqrt{17})/2$, so $(2, 4)$ is a saddle.

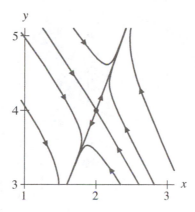

5. **(a)** Using separation of variables (or simple guessing), we have $x(t) = x_0 e^{-t}$.
   (b) Using the result in part (a), we can rewrite the equation for dy/dt as

$$\frac{dy}{dt} = y - 4x_0^3 e^{-3t}.$$

This first-order equation is a nonhomogeneous linear equation.

The general solution of its associated homogeneous equation is ke^t. To find a particular solution to the nonhomogeneous equation, we rewrite it as

$$\frac{dy}{dt} - y = -4x_0^3 e^{-3t},$$

and we guess a solution of the form $y_p = \alpha e^{-3t}$. Substituting this guess into the left-hand side of the equation yields

$$\frac{dy_p}{dt} - y_p = -4\alpha e^{-3t}.$$

Therefore, y_p is a solution if $\alpha = x_0^3$. The general solution of the original equation is

$$y(t) = x_0^3 e^{-3t} + ke^t.$$

To express this result in terms of the initial condition $y(0) = y_0$, we evaluate at $t = 0$ and note that $k = y_0 - x_0^3$. We conclude that

$$y(t) = x_0^3 e^{-3t} + (y_0 - x_0^3)e^t.$$

(c) The general solution of the system is

$$x(t) = x_0 e^{-t}$$
$$y(t) = x_0^3 e^{-3t} + (y_0 - x_0^3)e^t.$$

(d) For all solutions, $x(t) \to 0$ as $t \to \infty$. For a solution to tend to the origin as $t \to \infty$, we must have $y(t) \to 0$, and this can happen only if $y_0 - x_0^3 = 0$.

(e) Since $x = x_0 e^{-t}$, we see that a solution will tend toward the origin as $t \to -\infty$ only if $x_0 = 0$. In that case, $y(t) = y_0 e^t$, and $y(t) \to 0$ as $t \to -\infty$.

(f)

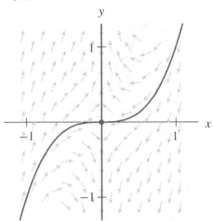

(g) Solutions tend away from the origin along the y-axis in both systems. In the nonlinear system, solutions approach the origin along the curve $y = x^3$ which is tangent to the x-axis. For the linearized system, solutions tend to the origin along the x-axis. Near the origin, the phase portraits are almost the same.

7. (a) The equilibrium points are $(0, 0)$, $(0, 100)$, $(150, 0)$, and $(30, 40)$. To determine the type of an equilibrium point, we compute the Jacobian matrix, which is

$$\begin{pmatrix} -2x - 3y + 150 & -3x \\ -2y & -2x - 2y + 100 \end{pmatrix},$$

and evaluate at the point. At $(0, 0)$, the Jacobian is

$$\begin{pmatrix} 150 & 0 \\ 0 & 100 \end{pmatrix},$$

and the eigenvalues are 150 and 100. Hence, the origin is a source. At $(0, 100)$, the Jacobian matrix is

$$\begin{pmatrix} -150 & 0 \\ -200 & -100 \end{pmatrix},$$

and the eigenvalues are -150 and -100. So $(0, 100)$ is a sink. The Jacobian at $(150, 0)$ is

$$\begin{pmatrix} -150 & -450 \\ 0 & -200 \end{pmatrix},$$

and the eigenvalues are -150 and -200. Therefore, $(150, 0)$ is a sink. Finally, the Jacobian matrix at $(30, 40)$ is

$$\begin{pmatrix} -30 & -90 \\ -80 & -40 \end{pmatrix},$$

and the eigenvalues are approximately -120 and 50. So $(30, 40)$ is a saddle.

(b)

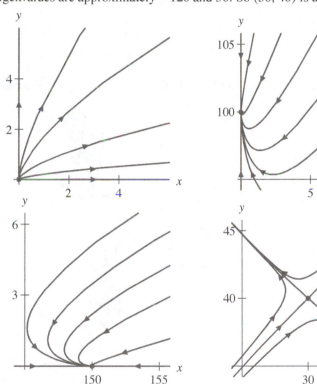

9. **(a)** The equilibrium points are $(0, 0)$, $(0, 25)$, $(100, 0)$ and $(75, 12.5)$. We classify these equilibrium points by computing the Jacobian matrix, which is

$$\begin{pmatrix} 100 - 2x - 2y & -2x \\ -y & 150 - x - 12y \end{pmatrix},$$

and evaluating it at each of the equilibrium points. At $(0, 0)$, the Jacobian matrix is

$$\begin{pmatrix} 100 & 0 \\ 0 & 150 \end{pmatrix},$$

and the eigenvalues are 100 and 150. So this point is a source. At $(0, 25)$, the Jacobian matrix is

$$\begin{pmatrix} 50 & 0 \\ -25 & -150 \end{pmatrix},$$

and the eigenvalues are 50 and -150. Hence, this point is a saddle. At $(100, 0)$, the Jacobian matrix is

$$\begin{pmatrix} -100 & -200 \\ 0 & 50 \end{pmatrix},$$

and the eigenvalues are -100 and 50. Therefore, this point is a saddle. Finally, at $(75, 12.5)$, the Jacobian matrix is

$$\begin{pmatrix} -75 & -150 \\ -12.5 & -75 \end{pmatrix},$$

and the eigenvalues are approximately -32 and -118. So this point is a sink.

(b)

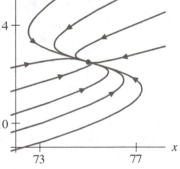

11. **(a)** The equilibrium points in the first quadrant are $(0, 0)$, $(0, 50)$ and $(40, 0)$. To classify these equilibrium points, we compute the Jacobian matrix, which is

$$\begin{pmatrix} -2x - y + 40 & -x \\ -2xy & -x^2 - 3y^2 + 2500 \end{pmatrix},$$

and we evaluate it at each of the points. At $(0, 0)$, the Jacobian matrix is

$$\begin{pmatrix} 40 & 0 \\ 0 & 2500 \end{pmatrix},$$

which has eigenvalues 40 and 2500. Therefore, $(0, 0)$ is a source. At $(0, 50)$, the Jacobian matrix is

$$\begin{pmatrix} -10 & 0 \\ 0 & -5000 \end{pmatrix},$$

which has eigenvalues -10 and -5000. So $(0, 50)$ is a sink. At $(40, 0)$, the Jacobian matrix is

$$\begin{pmatrix} -40 & -40 \\ 0 & 900 \end{pmatrix},$$

which has eigenvalues -40 and 900. Hence, $(40, 0)$ is a saddle.

(b)

13. (a) The equilibrium points (in the first quadrant) are $(0, 0)$, $(0, 50)$, $(60, 0)$, $(30, 40)$ and $(234/5, 88/5)$. To classify these points, we compute the Jacobian matrix, which is,

$$\begin{pmatrix} -16x - 6y + 480 & -6x \\ -2xy & -x^2 - 3y^2 + 2500 \end{pmatrix},$$

and evaluate it at each point. At $(0, 0)$, the Jacobian is

$$\begin{pmatrix} 480 & 0 \\ 0 & 2500 \end{pmatrix},$$

which has eigenvalues 480 and 2500. Thus, $(0, 0)$ is a source. At $(0, 50)$, the Jacobian matrix is

$$\begin{pmatrix} 180 & 0 \\ 0 & -5000 \end{pmatrix},$$

which has eigenvalues 180 and -5000. So $(0, 50)$ is a saddle. At $(60, 0)$, the Jacobian matrix is

$$\begin{pmatrix} -480 & -360 \\ 0 & -1100 \end{pmatrix},$$

which has eigenvalues -480 and -1100. Hence, $(60, 0)$ is a sink. At $(30, 40)$, the Jacobian matrix is

$$\begin{pmatrix} -240 & -180 \\ -2400 & -3200 \end{pmatrix},$$

which has eigenvalues approximately equal to -3339 and -101. So, $(30, 40)$ is a sink. At $(234/5, 88/5)$, the Jacobian matrix is

$$\begin{pmatrix} -1872/5 & -1404/5 \\ -41184/25 & -15488/25 \end{pmatrix},$$

which has eigenvalues approximately equal to -1188 and 194. Therefore, $(234/5, 88/5)$ is a saddle.

(b)

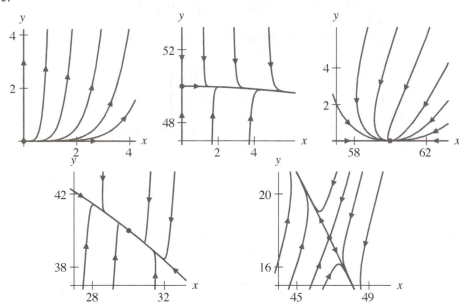

15. **(a)** The equilibrium points are $(0, 0)$, $(1, 1)$ and $(2, 0)$. We determine the type of each of these points by computing the Jacobian, which is

$$\begin{pmatrix} 2 - 2x - y & -x \\ -y & 2y - x \end{pmatrix},$$

and evaluating it at the points. At $(0, 0)$, the Jacobian is

$$\begin{pmatrix} 2 & 0 \\ 0 & 0 \end{pmatrix},$$

which has eigenvalues 2 and 0. An eigenvector for the eigenvalue 2 is $(1, 0)$, so solutions move away from the origin parallel to the x-axis. On the line $x = 0$, we have $dy/dt = y^2$ so solutions move upwards when $y \neq 0$. Hence, $(0, 0)$ is a node. However, solutions near the origin in the first quadrant move away from the origin as t increases. At $(1, 1)$, the Jacobian is

$$\begin{pmatrix} -1 & -1 \\ -1 & 1 \end{pmatrix},$$

which has eigenvalues $\pm\sqrt{2}$. So $(1, 1)$ is a saddle. At $(2, 0)$, the Jacobian is

$$\begin{pmatrix} -2 & -2 \\ 0 & -2 \end{pmatrix},$$

which has a double eigenvalue of -2. Therefore, $(2, 0)$ is a sink.

(b)

17. **(a)** The Jacobian matrix is

$$\begin{pmatrix} -3x^2 & 0 \\ 0 & -1+2y \end{pmatrix}$$

so the linearized system at $(0, 0)$ is

$$\frac{dx}{dt} = 0$$

$$\frac{dy}{dt} = -y.$$

(b) The eigenvalues are 0 and -1. Any vector on the x-axis is an eigenvector for the eigenvalue 0, and any vector on the y-axis is an eigenvector for the eigenvalue -1. Hence, the linearized system has a line of equilibrium points along the x-axis. Every other solution moves vertically toward the x-axis.

(c) The linearized system at $(0, 1)$ is

$$\frac{dx}{dt} = 0$$
$$\frac{dy}{dt} = y.$$

(d) The eigenvalues of this system are 0 and 1. Any vector on the x-axis is an eigenvector for the eigenvalue 0, and any vector on the y-axis is an eigenvector for the eigenvalue 1. Hence, the linearized system has a line of equilibrium points along the x-axis. Every other solution moves vertically away from the x-axis. It is important to remember that the origin for the linearized system corresponds to the equilibrium point $(0, 1)$ for the nonlinear system.

(e) The phase portrait is essentially a "combination" of two phase lines. The x-phase line has a sink at the origin. The y-phase line has a sink at the origin and a source at $y = 1$. Hence, the full phase portrait has a sink at $(0, 0)$, and the equilibrium point at $(0, 1)$ looks like a saddle.

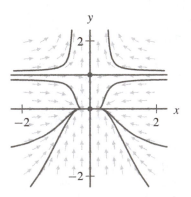

(f) The reason the linearizations and the nonlinear system look so different is that the equation for dx/dt contains only higher-order terms (just x^3 in this case). Since the equilibrium points occur along the y-axis ($x = 0$), the linearization has an entire line of equilibria in the x-direction.

19. (a)

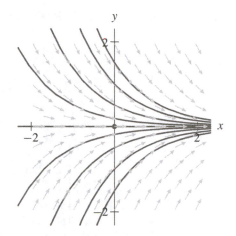

(b) The linearization of the equilibrium point at the origin has the coefficient matrix

$$\begin{pmatrix} 0 & 0 \\ 0 & -1 \end{pmatrix},$$

which has eigenvalues -1 and 0. So for the linearized system, the x-axis is a line of equilibria and solutions tend to zero in the y-direction. The nonlinear terms make solutions tend to zero in the x-direction for initial conditions with $x < 0$ and away from zero in the x-direction for initial conditions with $x > 0$.

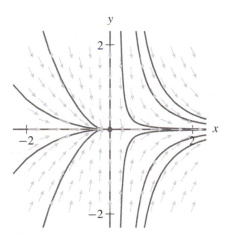

(c) The equilibria are $(\pm 1, 0)$. The coefficient matrix of the linearization at $(1, 0)$ is

$$\begin{pmatrix} 2 & 0 \\ 0 & -2 \end{pmatrix}.$$

The eigenvalues are 2 and -2, thus $(1, 0)$ is a saddle. The coefficient matrix of the linearization at $(-1, 0)$ is

$$\begin{pmatrix} -2 & 0 \\ 0 & -2 \end{pmatrix},$$

which has -2 as a repeated eigenvalue. So, $(-1, 0)$ is a sink.

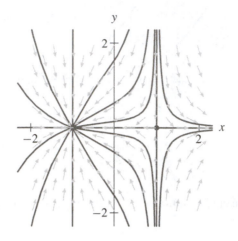

21. (a) The only equilibrium points occur if $a = 0$. Then all points on the curve $y = x^2$ are equilibrium points.

 (b) The bifurcation occurs at $a = 0$.

 (c) If $a < 0$, all solutions decrease in the y-direction since $dy/dt < 0$. If $a > 0$, all solutions increase in the y-direction since $dy/dt > 0$. If $a = 0$, there is a curve of equilibrium points located along $y = x^2$, and all solutions move horizontally.

Phase portrait for $a < 0$ Phase portrait for $a = 0$ Phase portrait for $a > 0$

23. (a) The equilibrium points are $(0, 0)$, $(\pm 1/\sqrt{a}, \pm 1/\sqrt{a})$, so there is only one equilibrium point if $a \leq 0$, and three equilibrium points if $a > 0$.

 (b) A bifurcation occurs at $a = 0$.

 (c) If $a < 0$, there are is only one equilibrium point at the origin, and this equilibrium point is a spiral source. If $a > 0$, the system has two additional equilibrium points, at $(\pm 1/\sqrt{a}, \pm 1/\sqrt{a})$. These equilibrium points come from infinity as a increases through 0.

Phase portrait for $a < 0$ Phase portrait for $a = 0$ Phase portrait for $a > 0$

25. **(a)** The equilibrium points are $(\pm\sqrt{a/2}, -a/2)$, so there are no equilibrium points if $a < 0$, one equilibrium if $a = 0$, and two equilibrium points if $a > 0$

(b) A bifurcation occurs at $a = 0$.

(c) If $a < 0$, there are no equilibrium points and all solutions come from and go to infinity. If $a = 0$, an equilibrium point appears at the origin. This equilibrium point has eigenvalues 0 and 1 and is a node. If $a > 0$, the system has two equilibrium points, at $(\pm\sqrt{a/2}, -a/2)$.

Phase portrait for $a < 0$ Phase portrait for $a = 0$ Phase portrait for $a > 0$

27. **(a)** The fact that $(0, 0)$ is an equilibrium point says that, if both X and Y are absent from the island, then neither will ever migrate to the island. However, it may be possible for one species to migrate if the other is already on the island.

(b) If a small population consisting solely of one of the species reproduces rapidly, then we expect both $\partial f/\partial x$ and $\partial g/\partial y$ to be positive and large at $(0, 0)$. We expect this because these partials are the coefficients of x and y in the linearization at $(0, 0)$.

(c) Since the species compete, an increase in y decreases dx/dt and an increase in x decreases dy/dt. Hence, both $\partial f/\partial y$ and $\partial g/\partial x$ are negative at $(0, 0)$ since $\partial f/\partial y$ is the coefficient of y in the dx/dt equation and $\partial g/\partial x$ is the coefficient for x in the dy/dt equation for the linearization at the origin.

(d) Suppose the coefficient matrix of the linearized system is

$$\begin{pmatrix} a & b \\ c & d \end{pmatrix},$$

with a and d positive and large and b and c negative. The eigenvalues are

$$\frac{(a+d) \pm \sqrt{(a-d)^2 + 4bc}}{2}.$$

If b and c are near zero, then $(0, 0)$ is a source. If b and c are very negative, then $(0, 0)$ is a saddle.

It is also possible to have 0 as an eigenvalue of the linearized system in which case the linearization fails to determine the behavior of the nonlinear system near $(0, 0)$.

(e) For the linearized system, note that $dx/dt < 0$ along the positive y-axis and $dy/dt < 0$ along the positive x-axis. If the origin is a saddle, the eigenvectors for the negative eigenvalue must be in the first and third quadrants, and a typical solution near the origin starting in the first quadrant has one of the species going extinct. If the origin is a source, then a typical solution near the origin has one or the other of the species going extinct except for one curve of solutions in the first quadrant.

  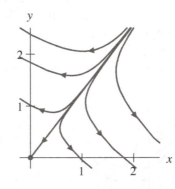

29.   **(a)** At $(0, 0)$, $\partial f/\partial x$ is positive and large, and $\partial g/\partial y$ is positive and small.

   (b) At $(0, 0)$, $\partial f/\partial y$ is negative with a large absolute value and $\partial g/\partial x = 0$.

   (c) With these assumptions, the Jacobian matrix is

$$\begin{pmatrix} a & b \\ 0 & d \end{pmatrix},$$

where $a > 0$, $b < 0$, and $d > 0$ is much smaller than a. The eigenvalues of this matrix are a and d, so $(0, 0)$ is a source.

   (d) Note that for $y = 0$, $dy/dt = 0$, and the eigenvector for a is in the x-direction.

EXERCISES FOR SECTION 5.2

1. For x- and y-nullclines, $dx/dt = 0$, and $dy/dt = 0$ respectively. Then, we obtain $y = -x + 2$ for the x-nullcline and $y = x^2$ for the y-nullcline. To find intersections, we set $-x + 2 = x^2$, or $(x + 2)(x - 1) = 0$. Solving this for x yields $x = 1, -2$. For $x = 1$, $y = 1$, and for $x = -2$, $y = 4$. So the equilibrium points are $(1, 1)$ and $(-2, 4)$.

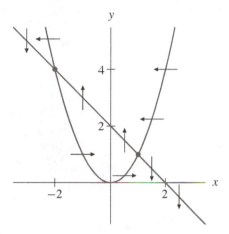

The solution for (a) is in the left-down region, and therefore, it eventually enters the region where $y < -x + 2$ and $y < x^2$. Once the solution enters this region, it stays there because the vector field on the boundaries never points out. Solutions for (b) and (c) start in this same region. Hence, all three solutions will go down and to the right without bound.

3. For the x-nullcline, $x(x - 1) = 0$, or $x = 0$ and $x = 1$, and for the y-nullcline, $y = x^2$. The equilibrium points are the intersection points of the x- and y-nullclines. They are $(0, 0)$ and $(1, 1)$.

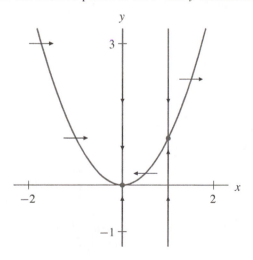

The initial conditions (a) and (b) are in right-up and left-up region respectively. Therefore, their solution curves eventually enter the region where $y > x^2$ and $x \leq 1$, and tend toward the equilibrium point at $(0, 0)$. The initial condition (c) is on the x-nullcline $x = 1$, and therefore its solution curve tends to the equilibrium point at $(1, 1)$.

5. **(a)** The x-nullcline is made up of the lines $x = 0$ and $y = -x/3 + 50$. The y-nullcline is made up of the lines $y = 0$ and $y = -2x + 100$.

(b)

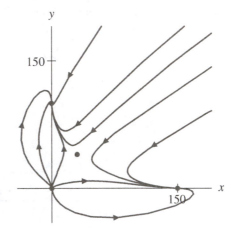

(c) Most solutions tend toward one of the equilibrium points $(0, 100)$ or $(150, 0)$. One curve of solutions divides these two behaviors. On this curve, solutions tend toward the saddle equilibrium at $(30, 40)$.

7. **(a)** The x-nullcline consists of the two lines $x = 0$ and $y = -x/2 + 50$. The y-nullcline consists of the two lines $y = 0$ and $y = -x/6 + 25$.

(b)

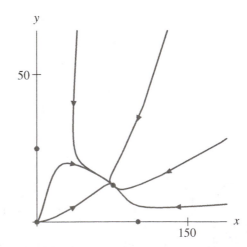

(c) All solutions off the axes tend toward the sink at $(75, 25/2)$. On the x-axis, solutions tend to the saddle at $(100, 0)$. On the y-axis, solutions tend to the saddle at $(0, 25)$.

9. **(a)** The x-nullcline is given by the two lines $x = 0$ and $y = -x + 40$. The y-nullcline is given by the line $y = 0$ and the circle $x^2 + y^2 = 50^2$.

(b)

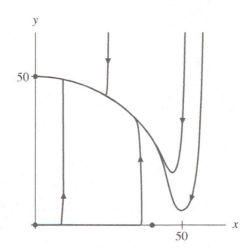

(c) Solutions off the x-axis tend toward the sink at $(0, 50)$. Solutions on the x-axis tend toward the saddle at $(40, 0)$.

11. **(a)** The x-nullcline is given by the line $x = 0$ and the line $y = -4x/3 + 80$. The y-nullcline is given by the line $y = 0$ and the circle $x^2 + y^2 = 50^2$. (Recall that we are only interested in the first quadrant).

(b)

(c) Most solutions tend toward the sink at $(30, 40)$ or the sink at $(60, 0)$. The curve dividing these two behaviors is a curve of solutions tending toward the saddle at $(234/5, 88/5)$. Solutions on the y-axis tend toward the saddle at $(0, 50)$.

13. (a) The x-nullcline is given by the lines $x = 0$ and $y = -x + 2$. The y-nullcline is given by the lines $y = 0$ and $y = x$.

(b)

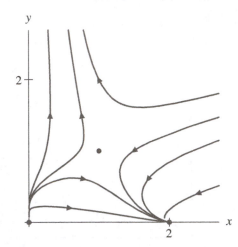

(c) Most solutions tend toward either the sink at $(2, 0)$ or toward infinity in the y-direction (with $x < 1$). The curve separating these two behaviors is a curve of solutions that tend toward the saddle at $(1, 1)$.

15. (a) Since the species are cooperative, an increase in y results in an increase in x and vice versa. Therefore, one needs to change the signs in front of B and D from $-$ to $+$.

(b) The x-nullcline is given by $x = 0$ or $-Ax + By + C = 0$. The y-nullcline is given by $y = 0$ or $Dx - Ey + F = 0$. The origin is always an equilibrium point. Also, $x = 0, y = F/E$ and $x = C/A, y = 0$ are equilibrium points. Equilibrium points with both x and y positive arise from solutions of

$$\begin{cases} -Ax + By + C = 0 \\ Dx - Ey + F = 0 \end{cases}$$

In matrix notation, we obtain

$$\begin{pmatrix} -A & B \\ D & -E \end{pmatrix} \begin{pmatrix} x \\ y \end{pmatrix} = \begin{pmatrix} -C \\ -F \end{pmatrix}.$$

In order for a unique solution to exist, $AE - BD \neq 0$. Then, the solution is

$$\begin{pmatrix} x \\ y \end{pmatrix} = \frac{1}{AE - BD} \begin{pmatrix} CE + BF \\ CD + AF \end{pmatrix}$$

Since A through F are all positive, we must have $AE - BD > 0$ for the solution to be in the first quadrant.

If $AE - BD = 0$, then $-Ax + By$ must be a negative multiple of $Dx - Ey$, so there are no solutions with both x and y positive.

17. **(a)** For the a-nullcline, $da/dt = 0$, so $2 - ab/2 = 0$, or $ab = 4$. For the b-nullcline, $db/dt = 0$, so $ab = 3$. Both nullclines are hyperbolas, and the curve of $ab = 4$ is above the one of $ab = 3$. Therefore, the direction of vector field on $ab = 4$ is vertical and downward, and the one on $ab = 3$ is horizontal and to the right.

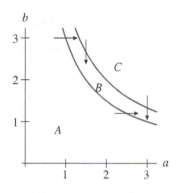

(b) Below and above $ab = 3$, $db/dt > 0$ and $db/dt < 0$ respectively. Below and above $ab = 3$, $da/dt > 0$ and $da/dt < 0$ respectively. Therefore, in region A, the vector field points up and to the right, in region B, the vector field points down and to the right, and in region C, the vector field points down and to the left.

(c) On the boundaries of B, the direction of the vector field never points out of B. Therefore, as time increases, these solutions are asymptotic to the positive x-axis from above.

19. **(a)** For the a-nullcline, $da/dt = 0$, so $b^2/6 - ab/2 + 2 = 0$. Multiplying by 6 yields the implicit equation $b^2 - 3ab + 12 = 0$, which can be solved using the quadratic formula.

For the b-nullcline, $db/dt = 0$, so $-b^2/3 - ab/2 + 3/2 = 0$. We multiply by -6 and obtain the implicit equation $2b^2 + 3ab - 9 = 0$, which can also be solved using the quadratic formula.

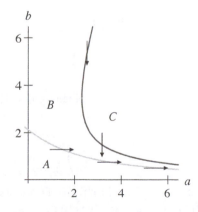

The a-nullcline is above the b-nullcline. Along the a-nullcline the vector field points down, that is, $db/dt < 0$. Along the b-nullcline the vector field points to the right, that is, $da/dt > 0$.

(b) In region A, both da/dt and db/dt are positive. In region B, $da/dt > 0$ and $db/dt < 0$. In region C, $da/dt < 0$, and $db/dt < 0$. Therefore, in region A, the vector field points up and to the right; in region B, the vector field points down and to the right; and in region C, the vector field points down and to the left.

(c) Solutions that start in region A eventually enter region B and head off to infinity in the a-direction. Solutions that start in region C head down and to the left until they enter region B and head off to infinity in the a-direction. Some solutions in region B can enter region C once. If they do, they eventually reenter region B and head off to infinity in the a-direction. Any solution in region B with $a > 2.3$ and $b < 3.2$ must stay in region B and heads off to infinity in the a-direction. (The values 2.3 and 3.2 are approximations.)

21. For the x-nullcline, $dx/dt = 0$; thus, $y = 0$. For the y-nullcline, $dy/dt = 0$; thus, $x(1 - x) = 0$. The line $y = 0$ is x-nullcline, and the lines $x = 0$ and $x = 1$ are y-nullclines. Since $dx/dt = y$, x increases for $y > 0$ and decreases for $y < 0$. Similarly, $dy/dt > 0$ for $0 < x < 1$ and $dy/dt < 0$ for $x < 0$ and $x > 1$. So, the function y increases for $0 < x < 1$ and decreases for $x < 0$ and $x > 1$.

23. The Jacobian matrix of the vector field is

$$\begin{pmatrix} 0 & 1 \\ 1 - 2x & 0 \end{pmatrix}.$$

The coefficient matrix of the linearization at $(0, 0)$ is

$$\begin{pmatrix} 0 & 1 \\ 1 & 0 \end{pmatrix}$$

which has eigenvalues ± 1 and hence is a saddle with eigenvectors $(1, 1)$ (for 1) and $(1, -1)$ (for -1). The coefficient matrix of the linearization at $(1, 0)$ is

$$\begin{pmatrix} 0 & 1 \\ -1 & 0 \end{pmatrix}$$

which has eigenvalues $\pm i$ and is hence a center. From this information we can conclude that solutions with initial conditions near $(0, 0)$ move away from the origin in the $(1, 1)$ or $(-1, -1)$ direction while solutions near $(1, 0)$ tend to rotate around $(1, 0)$.

What we can not tell is what solutions do over the long-term. For example, do solutions spiral toward or away from $(1, 0)$? The higher order terms could cause either behavior (or all solutions could be periodic). Also, what is the behavior of the stable and unstable separatrices associated with the saddle? It turns out that this information is available for this system using techniques of Section 5.3

EXERCISES FOR SECTION 5.3

1. **(a)** We compute that

$$\frac{\partial H}{\partial x} = x - x^3$$

and so

$$\frac{dy}{dt} = -\frac{\partial H}{\partial x}.$$

Also,

$$\frac{\partial H}{\partial y} = y = \frac{dx}{dt}.$$

Hence, this is a Hamiltonian system with Hamiltonian function H.

(b) Note that $(0, 0)$ is a local minimum and $(\pm 1, 0)$ are saddle points.

(c) The equilibrium point $(0, 0)$ is a center and $(\pm 1, 0)$ are saddles. The saddles are connected by separatrix solutions.

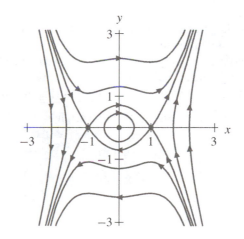

3. **(a)** If $H(x, y) = x \cos y + y^2$, then

$$\frac{\partial H}{\partial x} = \cos y$$

and so

$$\frac{dy}{dt} = -\frac{\partial H}{\partial x}.$$

Similarly,

$$\frac{\partial H}{\partial y} = -x \sin y + 2y = \frac{dx}{dt}.$$

(b)

(c) The equilibrium points occur at points of the form $((1 - 4n)\pi, (2n - \frac{1}{2})\pi)$ and $((1 + 4n)\pi, (2n + \frac{1}{2})\pi)$ where n is an integer.

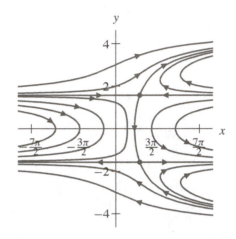

5. A large amplitude swing will take θ near $\pm\pi$, $v = 0$, the equilibrium point corresponding to the pendulum being balanced straight up. Near equilibrium points the vector field is very short, so solutions move very slowly. A solution passing close to $\pm\pi$, $v = 0$ must move slowly and hence, take a long time to make one complete swing. Hence, very high swings have long period. We must also be careful not to let the pendulum "swing over the top".

7. **(a)** The linearization at the origin is

$$\frac{d\theta}{dt} = v$$

$$\frac{dv}{dt} = -\frac{g}{l}\theta.$$

The eigenvalues of this system are $\pm i\sqrt{g/l}$, so the natural period is $2\pi/(\sqrt{g/l})$, which can also be written as $2\pi\sqrt{l}/\sqrt{g}$. Doubling the arm length corresponds to replacing l with $2l$, but the computations above stay the same. The natural period for arm length $2l$ is $2\pi\sqrt{2l}/\sqrt{g}$. Doubling the arm length multiplies the natural period by $\sqrt{2}$.

(b) Compute

$$\frac{d(2\pi\sqrt{l}/\sqrt{g})}{dl} = \frac{\pi}{\sqrt{gl}}.$$

9. We know that the equilibrium points of a Hamiltonian system cannot be sources or sinks. Phase portrait (b) has a spiral source, so it is not Hamiltonian. Phase portrait (c) has a sink and a source, so it is not Hamiltonian. Phase portraits (a) and (d) might come from Hamiltonian systems. (Try to imagine a function which has the solution curves as level sets.)

11. First note that

$$\frac{\partial(x - 3y^2)}{\partial x} = 1 = -\frac{\partial(-y)}{\partial y}.$$

Hence, the system is Hamiltonian. Integrating dx/dt with respect to y yields

$$H(x, y) = xy - y^3 + c(x).$$

If we differentiate $H(x, y)$ with respect to x, we get

$$y + c'(x),$$

which we want to be the negative of $dy/dt = -y$. Hence $c'(x) = 0$, and we pick the antiderivative $c(x) = 0$. A Hamiltonian function is

$$H(x, y) = xy - y^3.$$

13. First we check to see if the partial derivative with respect to x of the first component of the vector field is the negative of the partial derivative with respect to y of the second component. We have

$$\frac{\partial(x \cos y)}{\partial x} = \cos y$$

while

$$-\frac{\partial(-y \cos x)}{\partial y} = \cos x.$$

Since these two are not equal, the system is not Hamiltonian.

15. (a) We note that

$$\frac{\partial(1 - y^2)}{\partial x} = 0 \neq -\frac{\partial(x(1 + y^2))}{\partial y} = -2xy.$$

Hence, the system is not Hamiltonian.

(b)

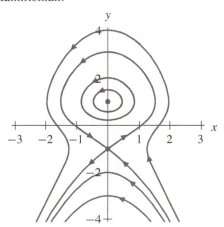

This phase portrait is consistent with the existence of a conserved quantity, but it is difficult to produce such a quantity from the figure alone.

(c) We note that

$$\frac{\partial((1 - y^2)/(1 + y^2))}{\partial x} = 0 = \frac{\partial(x)}{\partial y},$$

so the system is Hamiltonian. The Hamiltonian function is

$$H(x, y) = \int \frac{1 - y^2}{1 + y^2}\, dy - \frac{x^2}{2},$$

which is equal to

$$H(x, y) = -y + 2\arctan y - \frac{x^2}{2}.$$

(d) Multiplying the vector field by the function $f(x, y)$ changes the lengths of the vectors in the vector field, but it does not change their directions. Hence, the direction fields for the two systems agree. Consequently, the phase portraits are the same for the two systems, but solutions move at different speeds along the solution curves in the phase plane. Since H is a conserved quantity for the Hamiltonian system, the solution curves lie on level sets of H, and H is also a conserved quantity for the original system.

We can check that H is a conserved quantity by hand by computing the derivative with respect to t of $H(x(t), y(t))$, where $(x(t), y(t))$ is a solution of the original system. From the Chain Rule, we have

$$\frac{dH(x(t), y(t))}{dt} = \frac{\partial H}{\partial x}\frac{dx}{dt} + \frac{\partial H}{\partial y}\frac{dy}{dt}$$

$$= -x(1 - y^2) + \frac{1 - y^2}{1 + y^2}(x(1 + y^2))$$

$$= 0.$$

(e) See the answer to part (d).

17. Using the technique of Exercise 15, we we multiply the vector field by $1/(2 - y)$. As in Exercise 14, the resulting system

$$\frac{dx}{dt} = \frac{1 - y^2}{2 - y}$$

$$\frac{dy}{dt} = x$$

is Hamiltonian. The Hamiltonian is

$$H(x, y) = -\frac{x^2}{2} + \int \frac{y^2 - 1}{y - 2}\, dy$$

$$= -\frac{x^2}{2} + \int 2 + y + \frac{3}{y - 2}\, dy$$

$$= -\frac{x^2}{2} + 2y + \frac{y^2}{2} + 3\ln|y - 2|.$$

The function

$$H(x, y) = -\frac{x^2}{2} + 2y + \frac{y^2}{2} + 3\ln|y - 2|$$

is a conserved quantity for the original system. However, it is not defined on the line $y = 2$. From the system, we see that this line is a single solution curve that separates the two half-planes, $y < 2$ and $y > 2$.

19. First note that this system is Hamiltonian for every value of a. The Hamiltonian function depends on a and is given by

$$H(x, y) = x^2 y + xy^2 - ax.$$

If $a > 0$, then the system has two saddle equilibrium points on the y-axis at $(0, \pm\sqrt{a})$. If $a = 0$, then system has only one equilibrium point at $(0, 0)$. If $a < 0$, the system again has two saddles, but they are now located at $(\pm 2\sqrt{-3a}/3, \mp\sqrt{-3a}/3)$. This corresponds to a change in shape of the graph of H.

EXERCISES FOR SECTION 5.4

1. (a) Let $(x(t), y(t))$ be a solution of the system. Then

$$\frac{d}{dt}(L(x(t), y(t))) = \frac{\partial L}{\partial x}\frac{dx}{dt} + \frac{\partial L}{\partial y}\frac{dy}{dt}$$

$$= x(-x^3) + y(-y^3)$$

$$= -x^4 - y^4.$$

This is negative except at $x = y = 0$, which is an equilibrium point of the system.

(b) The level sets are circles centered at the origin.

(c) Since L decreases along solutions, every solution must approach the origin as t increases.

3. (a) We compute the eigenvalues. The characteristic polynomial is $\lambda^2 + 0.1\lambda + 4$, so the eigenvalues are

$$\lambda = \frac{-0.1 \pm i\sqrt{15.99}}{2}.$$

Hence the origin is a spiral sink.

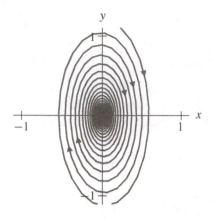

(b) Let $(x(t), y(t))$ be a solution. Then

$$\frac{d}{dt}(L(x(t), y(t))) = 2x(y) + 2y(-4x - 0.1y)$$

$$= -6xy - 0.2y^2.$$

In order for L to be a Lyapunov function, this quantity must be less than or equal to zero for all (x, y). However, if we take the solution with $x(0) = -1$, $y(0) = 1$ then

$$\frac{d}{dt}(L(x(t), y(t)))|_{t=0} = 5.8$$

which is positive, so L is not a Lyapunov function.

(c) Again let $(x(t), y(t))$ be a solution and compute

$$\frac{d}{dt}(K(x(t), y(t))) = 4x(y) + y(-4x - 0.1y)$$

$$= -0.1y^2$$

and this quantity is negative except at $y = 0$. (Solutions cross the $y = 0$ line at a discrete set of times. This is a technical point needed for the definition of Lyapunov functions.) So K is a Lyapunov function for the system.

5. We assume that the swings are small, hence we consider solutions near $\theta = v = 0$. The linearized system at the origin has Jacobian

$$\begin{pmatrix} 0 & 1 \\ -g/l & -b/m \end{pmatrix}$$

which has eigenvalues

$$\frac{-b/m \pm \sqrt{(b/m)^2 - 4g/l}}{2}.$$

When b/m is small, the eigenvalues are complex and the natural period of the oscillations is

$$2\pi / \sqrt{\frac{g}{l} - \frac{b^2}{4m^2}}$$

or

$$2\pi \left((g/l) - b^2 m^{-2}/4\right)^{-1/2}.$$

If we differentiate this expression with respect to m we obtain

$$\frac{d}{dm}\left(2\pi \left(\frac{g}{l} - \frac{b^2}{4}m^{-2}\right)^{-1/2}\right) = -\pi \left(\frac{g}{l} - \frac{b^2}{4}m^{-2}\right)^{-3/2}\left(-\frac{b^2}{4}\left(-2m^{-3}\right)\right)$$

which is less than zero. Hence, as the mass m increases, the natural period of the linearization near $(0, 0)$ decreases and the clock runs fast.

7. In order to be usable as a clock, the pendulum arm must swing back and forth. That is, the equilibrium point at the origin must be a spiral sink or center so that solutions oscillate. So we study the equilibrium point at the origin. The linearized system at the origin has Jacobian

$$\begin{pmatrix} 0 & 1 \\ -g/l & -b/m \end{pmatrix},$$

which has eigenvalues

$$\frac{-b/m \pm \sqrt{(b/m)^2 - 4g/l}}{2}.$$

To have oscillating solutions we must have complex eigenvalues, that is we must have

$$\frac{b^2}{m^2} - \frac{4g}{l} < 0.$$

This holds if

$$m > \frac{b\sqrt{l}}{2\sqrt{g}}.$$

(We could also have $m < -b\sqrt{l}/(2\sqrt{g})$, but negative mass is "unphysical".)

9. The linearized system at the origin has complex eigenvalues (when b/m is small) that have real part equal to $-b/(2m)$. So near $\theta = v = 0$ the amplitude of the oscillations decrease at the same rate as

$$e^{-tb/(2m)}.$$

In order for the clock to keep accurate time, we need θ and v to remain close to zero. An absolutely "worst case" for the size of θ is $|\theta| < \pi$. So we can get a rough approximation of how long it takes for the amplitude of an initial swing to decay to 0.1 by finding t such that

$$\pi e^{-tb/(2m)} = 0.1.$$

This occurs when

$$t = -\frac{2m}{b}\ln\left(\frac{0.1}{\pi}\right).$$

Pendulum clocks need to be wound because energy must be added to the system since it is dissipative.

11. (a) The linearized system at the origin has Jacobian

$$\begin{pmatrix} 0 & 1 \\ -g/l & -b/m \end{pmatrix},$$

which has eigenvalues

$$\frac{-b/m \pm \sqrt{(b/m)^2 - 4g/l}}{2}.$$

If b/m is small and negative, these eigenvalues have a positive real part and the origin is a spiral source. The linearized system at $(\pi, 0)$ has Jacobian

$$\begin{pmatrix} 0 & 1 \\ g/l & -b/m \end{pmatrix},$$

which has eigenvalues

$$\frac{-b/m \pm \sqrt{(b/m)^2 + 4g/l}}{2}.$$

If b/m is small (positive or negative), the two eigenvalues differ in sign, so this equilibrium point is a saddle. The equilibrium points $(\pm 2n\pi, 0)$ are just translates of the origin while the equilibrium points $(\pm(2n + 1)\pi, 0)$ are translates of $(\pi, 0)$.

(b)

(c) A solution that has initial point near $(0, 0)$ will spiral away from the origin. Eventually, θ will become larger than π or smaller than $-\pi$ and then θ will increase or decrease monotonically toward infinity. This corresponds to the pendulum arm swinging with higher and higher amplitude until it "passes over the top" either clockwise or counter clockwise. After it passes over the top once, then it will continue to rotate in the same direction, accelerating with each rotation.

13. (a) We have $\nabla G(x, y) = (2x, -2y)$, so

$$\frac{dx}{dt} = 2x$$
$$\frac{dy}{dt} = -2y.$$

(b) The system is linear and has eigenvalues 2 and -2. Hence the origin is a saddle.

(c) The graph of G is a saddle surface turning up in the x-direction and down in the y-direction.

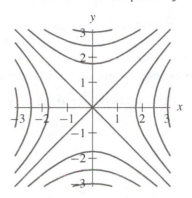

(d) The line of eigenvectors for eigenvalue 2 is the x-axis, the line of eigenvectors for eigenvalue -2 is the y-axis (see Chapter 3).

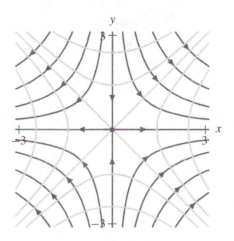

Phase portrait shown with level sets of G in gray.

15. (a) The Jacobian matrix is

$$\begin{pmatrix} 1 - 3x^2 & 0 \\ 0 & -1 \end{pmatrix}$$

At the origin the coefficient matrix of the linearization is

$$\begin{pmatrix} 1 & 0 \\ 0 & -1 \end{pmatrix}$$

which has eigenvalues 1 and -1. Hence, the origin is a saddle.

(b) At $(1, 0)$ the Jacobian matrix is

$$\begin{pmatrix} -2 & 0 \\ 0 & -1 \end{pmatrix}$$

which has eigenvalues -2 and -1. Hence, the point $(1, 0)$ is a sink.

(c) Since the eigenvalue -1 has eigenvectors along the y-axis, (and $-2 < -1$), solutions which approach $(1, 0)$ as t tends to infinity do so in the y direction (the only exception being solutions on the x-axis).

(d) The Jacobian matrix at $(-1, 0)$ is the same as at $(1, 0)$.

17. **(a)** The system is formed by taking the gradient of S. Hence, the system is

$$\frac{dx}{dt} = 2x - x^3 - 6xy^2$$
$$\frac{dy}{dt} = 2y - y^3 - 6x^2y.$$

(b)

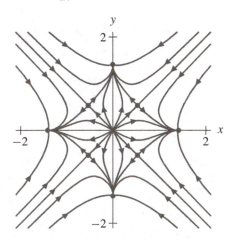

(c) From the phase portrait, we see that there are four sinks. Hence, there are four dead fish.

(d)

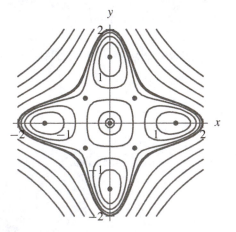

(e) For $|x|$ or $|y|$ large, S is negative, which is not physically reasonable.

19. **(a)** Since $f = \partial G/\partial x$ and $g = \partial G/\partial y$,

$$\frac{\partial f}{\partial y} = \frac{\partial^2 G}{\partial y \partial x} = \frac{\partial^2 G}{\partial x \partial y} = \frac{\partial g}{\partial x}.$$

(b) We compute

$$\frac{\partial f}{\partial y} = 3x$$

$$\frac{\partial g}{\partial x} = 2,$$

and these partials are not equal. By part (a), the system is not a gradient system.

21. Let $(y(t), v(t))$ be a solution of the "damped" system (with $k > 0$). Compute

$$\frac{d}{dt}(H(y(t), v(t))) = \frac{\partial H}{\partial y}\frac{dy}{dt} + \frac{\partial H}{\partial v}\frac{dv}{dt}$$

$$= \frac{dV}{dy}v + v\left(-\frac{dV}{dy} - kv\right)$$

$$= -kv^2.$$

Hence, for this system, H is a Lyapunov function.

23. Let $(x_1(t), x_2(t), p_1(t), p_2(t))$ be a solution and differentiate

$$H(x_1(t), x_2(t), p_1(t), p_2(t))$$

with respect to t using the Chain Rule. That is,

$$\frac{dH}{dt} = \frac{\partial H}{\partial x_1}\frac{dx_1}{dt} + \frac{\partial H}{\partial x_2}\frac{dx_2}{dt} + \frac{\partial H}{\partial p_1}\frac{dp_1}{dt} + \frac{\partial H}{\partial p_2}\frac{dp_2}{dt}.$$

Note that

$$\frac{\partial H}{\partial x_1} = k_1(x_1 - L_1) - k_2(x_2 - x_1 - L_2)$$

$$\frac{\partial H}{\partial x_2} = k_2(x_2 - x_1 - L_2)$$

$$\frac{\partial H}{\partial p_1} = \frac{p_1}{m_1}$$

$$\frac{\partial H}{\partial p_2} = \frac{p_2}{m_2}.$$

Using these partials along with the formulas for dx_1/dt, dx_2/dt, dp_1/dt, and dp_2/dt given by the system, we see that

$$\frac{dH}{dt} = \frac{\partial H}{\partial x_1}\frac{dx_1}{dt} + \frac{\partial H}{\partial x_2}\frac{dx_2}{dt} + \frac{\partial H}{\partial p_1}\frac{dp_1}{dt} + \frac{\partial H}{\partial p_2}\frac{dp_2}{dt} = 0.$$

EXERCISES FOR SECTION 5.5

1. **(a)** The Jacobian matrix for the system is

$$\begin{pmatrix} 1 - 2x - y & -x & 0 \\ y & 1 - 2y + x - z & -y \\ 0 & z & 1 - 2z + y \end{pmatrix}$$

so at $(1, 0, 0)$ the Jacobian matrix is

$$\begin{pmatrix} -1 & -1 & 0 \\ 0 & 2 & 0 \\ 0 & 0 & 1 \end{pmatrix}.$$

(b) The characteristic polynomial is $(1 - \lambda)(2 - \lambda)(-1 - \lambda)$, and the eigenvalues are $1, 2$ and -1. An eigenvector for eigenvalue 1 is $(0, 0, 1)$, an eigenvector for eigenvalue -1 is $(1, 0, 0)$ and an eigenvector for eigenvalue 2 is $(1, -3, 0)$.

(c) The equilibrium point is a saddle with two eigenvalues positive and one negative. Solutions tend toward the equilibrium point along the x-axis as t increases. Solutions in the plane formed by the two vectors $(0, 0, 1)$ and $(1, -3, 0)$ through the equilibrium point tend toward the equilibrium point as t decreases.

(d)

(e) If the initial condition satisfies $y = z = 0$ and x is close to 1, then the corresponding solution tends toward the equilibrium point as t increases. If y or z is small and positive and x is close to 1, then the solution may move closer to the equilibrium point if y and z are small, but eventually moves away from the equilibrium point.

3. **(a)** The Jacobian matrix at $(0, 0, 1)$ is

$$\begin{pmatrix} 1 & 0 & 0 \\ 0 & 0 & 0 \\ 0 & 1 & -1 \end{pmatrix}.$$

(b) The characteristic polynomial is $(1 - \lambda)\lambda(-1 - \lambda)$, so the eigenvalues are $-1, 0$ and 1. Corresponding eigenvectors are $(0, 0, 1)$, $(0, 1, 1)$ and $(1, 0, 0)$ respectively. (We can see this by computation, or use the fact that x decouples from y and z in the linearized system to reduce the dimension of the problem.)

(c) The linearized system has solutions that tend toward the original equilibrium point as t increases, solutions that tend toward the original equilibrium point as t decreases, and a line of equilibrium points through the original equilibrium point.

(d)

(e) For the linearized system, solutions with initial conditions with $x = 0$ tend in the z-direction toward an equilibrium point on the line of equilibrium points through the original equilibrium point. The behavior of solutions for the nonlinear system on the $x = 0$ plane is not so clear because nonlinear terms could cause solutions to tend toward or away from the equilibrium point. Off the plane $x = 0$, solutions move away from the equilibrium point in the x-direction.

5. **(a)** When $y = 0$ we have $dy/dt = 0$ and the equations for the rate of change of x and z are

$$\frac{dx}{dt} = x(1 - x)$$

$$\frac{dz}{dt} = z(1 - z).$$

These are both logistic equations with sinks at $x = 1$ and $z = 1$. Hence, $(1, 0, 1)$ is an equilibrium point. When $x = 0$ we have $dx/dt = 0$ and

$$\frac{dy}{dt} = y(1 - y) - yz$$

$$\frac{dz}{dt} = z(1 - z) + yz$$

and $y = z = 1$ is not an equilibrium point for the yz system so $(0, 1, 1)$ is not an equilibrium point for the three-dimensional system. The argument for $(1, 1, 0)$ is very similar.

(b) Having $y = 0$ corresponds to the moose being extinct. The food chain then has no connection between the wolves and the trees. Hence, both wolf and tree populations will tend toward their carrying capacity. When $x = 0$ the trees are extinct. The moose and wolfs then form a two dimensional predator-prey system. We do not expect that both moose and wolf populations to be maintained at their carrying capacitys because wolves eat moose. The argument when the wolves are extinct is similar.

7. **(a)** We must solve

$$x_0(1 - x_0) - x_0 y_0 = 0$$
$$y_0(1 - y_0) + x_0 y_0 - y_0 z_0 = 0$$
$$\zeta z_0(1 - z_0) + y_0 z_0 = 0.$$

Since we are looking for the solution where none of the coordinates is zero, we can divide the first equation by x_0, the second by y_0 and the third by z_0 to obtain

$$1 - x_0 - y_0 = 0$$
$$1 - y_0 + x_0 - z_0 = 0$$
$$\zeta - \zeta z_0 + y_0 = 0.$$

The first equation gives $y_0 = 1 - x_0$ which when plugged into the second equation gives $z_0 = 2x_0$ and these in the third equation give

$$x_0 = \frac{\zeta + 1}{2\zeta + 1}.$$

Hence, $y_0 = \zeta/(2\zeta + 1)$ and $z_0 = (2\zeta + 2)/(2\zeta + 1)$.

(b) We can compute the rate of change of the position of this equilibrium point as ζ changes as follows:

$$\frac{dx_0}{d\zeta} = \frac{-1}{(2\zeta + 1)^2}$$

$$\frac{dy_0}{d\zeta} = \frac{1}{(2\zeta + 1)^2}$$

$$\frac{dz_0}{d\zeta} = \frac{-2}{(2\zeta + 1)^2}.$$

Note that $dz_0/d\zeta$ is negative. Hence, a decrease in ζ, which we claimed corresponded to a decrease in the growth rate of the wolves, yields an increase in z_0. Similarly, a decrease in ζ gives a decrease in y_0 and an increase in x_0 because $dy_0/d\zeta$ is positive and $dx_0/d\zeta$ is negative (remember we are decreasing ζ).

(c) Note that the dz/dt equation is

$$\frac{dz}{dt} = \zeta z(1 - z) + yz = \zeta z - \zeta z^2 + yz$$

so ζ is multiplied onto the z and the z^2 terms. If we compare this to the modification of the equations given in the section we see that the change must be causes by the fact that ζ is multiplied onto the z^2 term. It does seem "unphysical" that a decrease in the growth-rate parameter could cause an increase in the equilibrium population, but the parameter ζ gives both the growth rate for small z (the ζz term) and the proportionality constant for the effect of large populations (the ζz^2) term. This observation is interesting and deserves further study.

EXERCISES FOR SECTION 5.6

1. Comparing the y values at the indicated points with the y values of the $y(t)$-graphs for $t = 2n\pi$ shows this return map corresponds to graph (ii). The solution oscillates with increasing and decreasing amplitude around $y = 1$.

3. Comparing the y values at the indicated points with the y values of the $y(t)$-graphs for $t = 2n\pi$ shows this return map corresponds to graph (i). The solution oscillates in a regular, but fairly complicated way.

5. (a) This return map corresponds to graph (iii).

(b) The values of the θ-coordinate of the indicated points are decreasing monotonically and are all less than one. Hence, graph (iii) is the only one that fits.

(c) The solution should oscillate with rising and falling amplitude which corresponds to the pendulum arm swinging higher and lower, but never swinging "over the top".

7. (a) This return map corresponds to graph (iv).

(b) The θ-coordinates at the indicated points are bounded below by minus two, are all nonpositive and decrease, then increase. The graph with the corresponding behavior is graph (iv).

(c) The pendulum swings back and forth, never "going over the top" with amplitude and period oscillating in a regular way.

9. (a) The second-order differential equation is

$$\frac{d^2 y}{dt^2} + 3y = 0.2\sin t.$$

The general solution for the homogeneous equation is

$$y(t) = k_1 \sin\sqrt{3}\,t + k_2 \cos\sqrt{3}\,t.$$

Suppose $y(t) = a\sin t$. Differentiation yields

$$\frac{d^2 y}{dt^2} + 3y = 2a\sin t = 0.2\sin t.$$

Then, $a = 1/10$ and the general solution is

$$y(t) = k_1 \sin\sqrt{3}\,t + k_2 \cos\sqrt{3}\,t + \tfrac{1}{10}\sin t.$$

By the initial condition, $y(0) = 1$ and $v(0) = 0$, one obtains

$$\begin{cases} k_2 = 1 \\ \sqrt{3}\,k_1 + \tfrac{1}{10} = 0 \end{cases}$$

or $k_1 = -\sqrt{3}/30$ and $k_2 = 1$. The solution is

$$y(t) = -\frac{\sqrt{3}}{30}\sin\sqrt{3}\,t + \cos\sqrt{3}\,t + \tfrac{1}{10}\sin t.$$

(b) By the solution,

$$y(t) = -\frac{\sqrt{3}}{30}\sin\sqrt{3}\,t + \cos\sqrt{3}\,t + \tfrac{1}{10}\sin t.$$

After 2π in time, $0.1\sin(t + 2\pi)$ gives the same value. Therefore, the Poincaré map is due to the first two terms and it moves on an ellipse.

(c) The change in the initial condition affects only constants, k_1 and k_2. Therefore, Poincaré map traces an ellipse and the scale of the ellipse depends on k_1 and k_2.

(d) The forced harmonic oscillator system is linear and the solution is the sum of the general solution for the homogeneous equation and the particular solution for the nonhomogeneous equation. In this case, the Poincaré map is due to the homogeneous solution as we saw in (c). On the other hand, the forced pendulum system is nonlinear and therefore, the solution becomes more complicated.

REVIEW EXERCISES FOR CHAPTER 5

1. Since the equilibrium point is at the origin and the system has only polynomial terms, the linearized system is just the linear terms in dx/dt and dy/dt, that is,

$$\frac{dx}{dt} = x$$
$$\frac{dy}{dt} = -2y.$$

3. The Jacobian matrix for this system is

$$\begin{pmatrix} 2x + 3\cos 3x & 0 \\ -y\cos xy & 2 - x\cos xy \end{pmatrix},$$

and evaluating at $(0, 0)$, we get

$$\begin{pmatrix} 3 & 0 \\ 0 & 2 \end{pmatrix}.$$

So the linearized system at the origin is

$$\frac{dx}{dt} = 3x$$
$$\frac{dy}{dt} = 2y.$$

5. The x-nullcline is where $dx/dt = 0$, that is, the line $y = x$. The y-nullcline is where $dy/dt = 0$, that is, the circle $x^2 + y^2 = 2$.

Along the x-nullcline, $dy/dt < 0$ if and only if $-\sqrt{2} < x < \sqrt{2}$. Along the y-nullcline, $dx/dt < 0$ if and only if $y > x$.

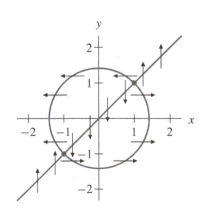

7. This system is not a gradient system. If it were, then we would have

$$\frac{\partial G}{\partial x} = \frac{dx}{dt} \quad \text{and} \quad \frac{\partial G}{\partial y} = \frac{dy}{dt}$$

for some function $G(x, y)$. In that case, equality of mixed partials would imply that

$$\frac{\partial}{\partial y}\left(\frac{dx}{dt}\right) = \frac{\partial}{\partial x}\left(\frac{dy}{dt}\right).$$

For this system, we have

$$\frac{\partial}{\partial y}\left(\frac{dx}{dt}\right) = 2x + 2y \quad \text{and} \quad \frac{\partial}{\partial x}\left(\frac{dy}{dt}\right) = 2x.$$

Since these two partials do not agree, no such function $G(x, y)$ exists.

9. If the system is a linear system, then all nonequilibrium solutions tend to infinity as t increases, that is, $|\mathbf{Y}(t)| \to \infty$ as $t \to \infty$.

 If the system is not linear, it is possible for a solution to spiral toward a periodic solution. For example, consider the Van der Pol equation discussed in Section 5.1. (These two behaviors are the only possibilities.)

11. True. The x-nullcline is where $dx/dt = 0$ and the y-nullcline is where $dy/dt = 0$, so any point in common must be an equilibrium point.

13. False. These two numbers are the diagonal entries of the Jacobian matrix. The other two entries of the Jacobian matrix also affect the eigenvalues.

15. **(a)** Setting $dx/dt = 0$ and $dy/dt = 0$, we obtain the simultaneous equations

$$\begin{cases} x - 3y^2 = 0 \\ x - 3y - 6 = 0. \end{cases}$$

Solving for x and y yields the equilibrium points $(12, 2)$ and $(3, -1)$.

 To determine the type of an equilibrium point, we compute the Jacobian matrix. We get

$$\begin{pmatrix} 1 & -6y \\ 1 & -3 \end{pmatrix}.$$

At $(12, 2)$, the Jacobian is

$$\begin{pmatrix} 1 & -12 \\ 1 & -3 \end{pmatrix},$$

and its eigenvalues are $-1 \pm 2\sqrt{2}\,i$. Hence, $(12, 2)$ is a spriral sink.

 At $(3, -1)$, the Jacobian matrix is

$$\begin{pmatrix} 1 & 6 \\ 1 & -3 \end{pmatrix},$$

and the eigenvalues are $-1 \pm \sqrt{10}$. So $(3, -1)$ is a saddle.

(b) The x-nullcline is $x = 3y^2$, and the y-nullcline is $x = 3y + 6$. We compute the direction of the vector field by computing the sign of dy/dt on the x-nullcline and the sign of dx/dt on the y-nullcline.

(c)

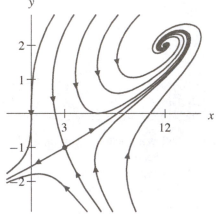

17. **(a)** To find the equilibria, we solve the system of equations

$$\begin{cases} 4x - x^2 - xy = 0 \\ 6y - 2y^2 - xy = 0. \end{cases}$$

We obtain the four equilibrium points $(0, 0)$, $(0, 3)$, $(4, 0)$, and $(2, 2)$.

To classify the equilibria, we compute the Jacobian

$$\begin{pmatrix} 4 - 2x - y & -x \\ -y & 6 - 4y - x \end{pmatrix},$$

evaluate at each equilibrium point, and compute the eigenvalues.

At $(0, 0)$, the Jacobian matrix is

$$\begin{pmatrix} 4 & 0 \\ 0 & 6 \end{pmatrix}.$$

The eigenvalues are 4 and 6, so the origin is a source.

At $(0, 3)$, the Jacobian matrix is

$$\begin{pmatrix} 1 & 0 \\ -3 & -6 \end{pmatrix}.$$

The eigenvalues are 1 and -6, so $(0, 3)$ is a saddle.

At $(4, 0)$, the Jacobian matrix is

$$\begin{pmatrix} -4 & -4 \\ 0 & 2 \end{pmatrix}.$$

The eigenvalues are -4 and 2, so $(4, 0)$ is a saddle.

Finally, at $(2, 2)$ the Jacobian matrix is

$$\begin{pmatrix} -6 & -2 \\ -2 & -4 \end{pmatrix}.$$

The eigenvalues are $-5 \pm \sqrt{5}$. Both are negative, so $(2, 2)$ is a sink.

(b) The x-nullcline satisfies the equation $4x - x^2 - xy = 0$, which can be rewritten as

$$x(4 - x - y) = 0.$$

We get two lines, $x = 0$ (the y-axis) and $y = 4 - x$.

The y-nullcline satisfies the equation $6y - 2y^2 - xy = 0$, which can be rewritten as

$$y(6 - 2y - x) = 0.$$

We get two lines, $y = 0$ (the x-axis) and $x + 2y = 6$.

We compute the direction of the vector field by computing the sign of dy/dt on the x-nullcline and the sign of dx/dt on the y-nullcline.

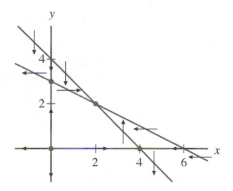

(c) The phase portrait is

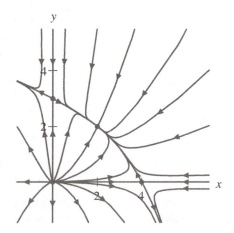

19. To find the equilibrium points, we solve the simultaneous equations

$$\begin{cases} x^2 - a = 0 \\ y^2 - b = 0 \end{cases}$$

and obtain the four points $(\sqrt{a}, \sqrt{b})$, $(-\sqrt{a}, \sqrt{b})$, $(\sqrt{a}, -\sqrt{b})$, and $(-\sqrt{a}, -\sqrt{b})$.

The Jacobian matrix for this system is

$$\begin{pmatrix} 2x & 0 \\ 0 & 2y \end{pmatrix}$$

which is diagonal. Hence, the eigenvalues are simply $2x$ and $2y$.

At $(\sqrt{a}, \sqrt{b})$, both eigenvalues are positive, so this equilibrium point is a source.

At $(-\sqrt{a}, \sqrt{b})$, one eigenvalue is negative and the other is positive. This equilibrium point is a saddle. The same is true for the equilibrium point $(\sqrt{a}, -\sqrt{b})$.

At $(-\sqrt{a}, -\sqrt{b})$, both eigenvalues are negative. This equilibrium point is a sink.

21. If $a = 0$, the equilibria are $(0, \sqrt{b})$ and $(0, -\sqrt{b})$.

The Jacobian matrix is

$$\begin{pmatrix} 2x & 0 \\ 0 & 2y \end{pmatrix},$$

and its eigenvalues are $2x$ and $2y$.

At $(0, \sqrt{b})$, the eigenvalues are 0 and $2\sqrt{b}$. Solutions of the nonlinear system approach the equilibrium along the negative x-direction, and solutions tend away from the equilibrium along the positive x-direction and along the y-axis.

At $(0, -\sqrt{b})$, the eigenvalues are 0 and $-2\sqrt{b}$. Solutions of the nonlinear system approach the equilibrium along the negative x-direction and along the y-axis and solutions tend away from the equilibrium in the positive x-direction.

The x-nullcline is the line $x = 0$, and the y-nullcline consists of the two lines $y = \pm\sqrt{b}$.

In the following figures, the nullclines are on the left and the phase portrait is on the right.

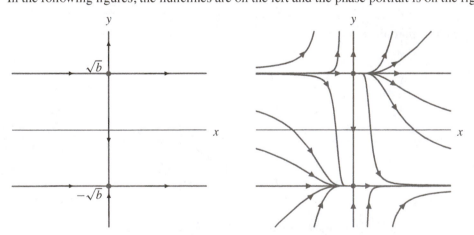

As $a \to 0$ from above, the x-nullclines merge. The pair of equilibrium points $(\sqrt{a}, \sqrt{b})$ and $(-\sqrt{a}, \sqrt{b})$ coalesce at $(0, \sqrt{b})$, and the pair of equilibrium points $(\sqrt{a}, -\sqrt{b})$ and $(-\sqrt{a}, -\sqrt{b})$ coalesce at $(0, -\sqrt{b})$.

23. The equilibrium points are given by $dx/dt = x^2 = 0$ and $dy/dt = y^2 = 0$, so $(0,0)$ is the only equilibrium point. The Jacobian at this point is the zero matrix, that is, all four entries are zero.

Solutions on the negative x- and y-axes approach the origin, and solutions on the positive x- and y-axes move away from the origin as t increases. Hence, solutions in the third quadrant approach the origin as t increases, and solutions in all other quadrants tend to infinity as t increases.

The x-nullcline is the y-axis, and the y-nullcline is the x-axis.

In the following figures, the nullclines are on the left and the phase portrait is on the right.

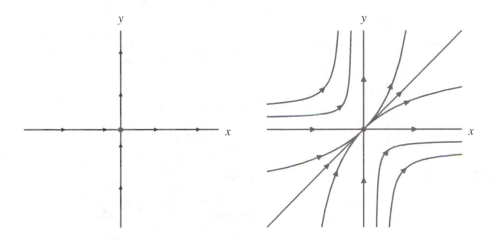

25. **(a)** The equilibrium points are the solutions of

$$\begin{cases} y^2 - x^2 - 1 = 0 \\ 2xy = 0, \end{cases}$$

that is, $(0, \pm 1)$.

The Jacobian matrix is

$$\begin{pmatrix} -2x & 2y \\ 2y & 2x \end{pmatrix}.$$

At $(0, 1)$, the Jacobian is

$$\begin{pmatrix} 0 & 2 \\ 2 & 0 \end{pmatrix}.$$

Its characteristic polynomial is $\lambda^2 - 4$, so its eigenvalues are $\lambda = \pm 2$. The equilibrium point is a saddle.

At $(0, -1)$, the Jacobian is

$$\begin{pmatrix} 0 & -2 \\ -2 & 0 \end{pmatrix}.$$

Its characteristic polynomial is $\lambda^2 - 4$, so its eigenvalues are $\lambda = \pm 2$. The equilibrium point is a saddle.

(b) The x-nullcline is the hyperbola $y^2 - x^2 = 1$, and the y-nullclines are the x- and y-axes. In the following figures, the nullclines are on the left and the phase portrait is on the right.

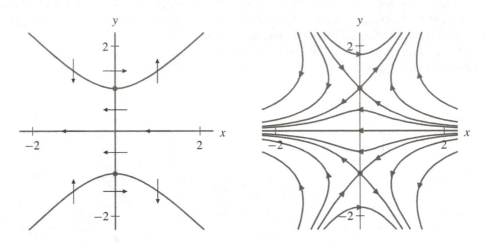

(c) To see if the system is Hamiltonian, we compute

$$\frac{\partial(y^2 - x^2 - 1)}{\partial x} = -2x \quad \text{and} \quad -\frac{\partial(2xy)}{\partial y} = -2x.$$

Since these partials agree, the system is Hamiltonian.

The Hamiltonian is a function $H(x, y)$ such that

$$\frac{\partial H}{\partial y} = \frac{dx}{dt} = y^2 - x^2 - 1 \quad \text{and} \quad \frac{\partial H}{\partial x} = -\frac{dy}{dt} = -2xy.$$

We integrate the second equation with respect to x to see that

$$H(x, y) = -x^2 y + \phi(y),$$

where $\phi(y)$ represents the terms whose derivative with respect to x are zero. Using this expression for $H(x, y)$ in the first equation, we obtain

$$-x^2 + \phi'(y) = y^2 - x^2 - 1.$$

Hence, $\phi'(y) = y^2 - 1$, and we can take $\phi(y) = \frac{1}{3}y^3 - y$. The function

$$H(x, y) = -x^2 y + \frac{y^3}{3} - y$$

is a Hamiltonian function for this system.

(d) To see if the system is a gradient system, we compute

$$\frac{\partial(y^2 - x^2 - 1)}{\partial y} = 2y \quad \text{and} \quad \frac{\partial(2xy)}{\partial x} = 2y.$$

Since these partials agree, the system is a gradient system.

We must now find a function $G(x, y)$ such that

$$\frac{\partial G}{\partial x} = \frac{dx}{dt} = y^2 - x^2 - 1 \quad \text{and} \quad \frac{\partial G}{\partial y} = \frac{dy}{dt} = 2xy.$$

Integrating the second equation with respect to y, we obtain

$$G(x, y) = xy^2 + h(x),$$

where $h(x)$ represents the terms whose derivative with respect to y are zero. Using this expression for $G(x, y)$ in the first equation, we obtain

$$y^2 + h'(x) = y^2 - x^2 - 1.$$

Hence, $h'(x) = -x^2 - 1$, and we can take $h(x) = -\frac{1}{3}x^3 - x$. The function

$$G(x, y) = xy^2 - \frac{x^3}{3} - x$$

is the required function.

27. To see if the system is Hamiltonian, we compute

$$\frac{\partial(-3x + 10y)}{\partial x} = -3 \quad \text{and} \quad -\frac{\partial(-x + 3y)}{\partial y} = -3.$$

Since these partials agree, the system is Hamiltonian.

To find the Hamiltonian function, we use the fact that

$$\frac{\partial H}{\partial y} = \frac{dx}{dt} = -3x + 10y.$$

Integrating with respect to y gives

$$H(x, y) = -3xy + 5y^2 + \phi(x),$$

where $\phi(x)$ represents the terms whose derivative with respect to y are zero. Differentiating this expression for $H(x, y)$ with respect to x gives

$$-3y + \phi'(x) = -\frac{dy}{dt} = x - 3y.$$

We choose $\phi(x) = \frac{1}{2}x^2$ and obtain the Hamiltonian function

$$H(x, y) = -3xy + 5y^2 + \frac{x^2}{2}.$$

We know that the solution curves of a Hamiltonian system remain on the level sets of the Hamiltonian function. Hence, solutions of this system satisfy the equation

$$-3xy + 5y^2 + \frac{x^2}{2} = h$$

for some constant h. Multiplying through by 2 yields the equation

$$x^2 - 6xy + 10y^2 = k$$

where $k = 2h$ is a constant.

29.   **(a)** Since θ represents an angle in this model, we restrict θ to the interval $-\pi < \theta < \pi$. The equilibria must satisfy the equations

$$\begin{cases} \cos\theta = s^2 \\ \sin\theta = -s^2. \end{cases}$$

Therefore,

$$\tan\theta = \frac{\sin\theta}{\cos\theta} = \frac{-s^2}{s^2} = -1,$$

and consequently, $\theta = -\arctan 1 = -\pi/4$.

To find s, we note that $s^2 = \cos(-\pi/4) = 1/\sqrt{2}$. Hence, $s = 1/\sqrt[4]{2}$, and the only equilibrium point is

$$(\theta, s) = \left(-\frac{\pi}{4}, \frac{1}{\sqrt[4]{2}}\right).$$

(b) The Jacobian matrix for this system is

$$\begin{pmatrix} \dfrac{\sin\theta}{s} & 1 + \dfrac{\cos\theta}{s^2} \\ -\cos\theta & -2s \end{pmatrix}.$$

Evaluating at the equilibrium point, we get

$$\begin{pmatrix} -2^{-3/4} & 2 \\ -2^{-1/2} & -2^{3/4} \end{pmatrix}.$$

The characteristic polynomial of this matrix is

$$\lambda^2 + \frac{1 + 2\sqrt{2}}{2^{3/4}}\lambda + (1 + \sqrt{2}).$$

Since

$$\left(\frac{1 + 2\sqrt{2}}{2^{3/4}}\right)^2 - 4(1 + \sqrt{2}) < 0,$$

the eigenvalues are complex. Their real part is

$$-\frac{1 + 2\sqrt{2}}{2^{7/4}},$$

which is negative. Consequently, the equilibrium point is a spiral sink.

CHAPTER 6

Laplace Transforms

EXERCISES FOR SECTION 6.1

1. We have

$$\mathcal{L}[3] = \int_0^\infty 3e^{-st}\, dt$$

$$= \lim_{b\to\infty} \int_0^b 3e^{-st}\, dt$$

$$= \lim_{b\to\infty} \left(\frac{-3}{s} e^{-st} \Big|_0^b \right)$$

$$= \lim_{b\to\infty} -\frac{3}{s} \left(e^{-sb} - e^0 \right)$$

$$= \frac{3}{s} \quad \text{if } s > 0,$$

since $\lim\limits_{b\to\infty} e^{-sb} = \lim\limits_{b\to\infty} 1/e^{sb} = 0$ if $s > 0$.

3. We use the fact that $\mathcal{L}[df/dt] = s\mathcal{L}[f] - f(0)$. Letting $f(t) = t^2$ we have $f(0) = 0$ and

$$\mathcal{L}[2t] = s\mathcal{L}[t^2] - 0$$

or

$$2\mathcal{L}[t] = s\mathcal{L}[t^2]$$

using the fact that the Laplace transform is linear. Then since $\mathcal{L}[t] = 1/s^2$ (by the previous exercise), we have

$$\mathcal{L}[-5t^2] = -5\mathcal{L}[t^2] = -5 \left(\frac{2\mathcal{L}[t]}{s} \right) = -\frac{10}{s^3}.$$

5. To show a rule by induction, we need two steps. First, we need to show the rule is true for $n = 1$. Then, we need to show that if the rule holds for n, then it holds for $n + 1$.

   (a) $n = 1$. We need to show that $\mathcal{L}[t] = 1/s^2$.

$$\mathcal{L}[t] = \int_0^\infty te^{-st}\, dt.$$

Using integration by parts with $u = t$ and $dv = e^{-st}\, dt$, we find

$$\mathcal{L}[t] = \frac{te^{-st}}{-s} \Big|_0^\infty + \int_0^\infty \frac{e^{-st}}{s}\, dt$$

$$= \lim_{b\to\infty} \left[\frac{te^{-st}}{-s} \Big|_0^b \right] + \int_0^\infty \frac{e^{-st}}{s}\, dt$$

$$= \int_0^\infty \frac{e^{-st}}{s}\, dt$$

$$= -\frac{e^{-st}}{s^2} \Big|_0^\infty$$

$$= \frac{1}{s^2} \quad (s > 0).$$

(b) Now we assume that the rule holds for n, that is, that $\mathcal{L}[t^n] = n!/s^{n+1}$, and show it holds true for $n + 1$, that is, $\mathcal{L}[t^{n+1}] = (n + 1)!/s^{n+2}$. There are two different methods to do so:

(i)
$$\mathcal{L}[t^{n+1}] = \int_0^\infty t^{n+1} e^{-st} dt$$

Using integration by parts with $u = t^{n+1}$ and $dv = e^{-st} dt$, we find

$$\mathcal{L}[t^{n+1}] = -\frac{t^{n+1} e^{-st}}{s} \Big|_0^\infty + \int_0^\infty \frac{n+1}{s} t^n e^{-st} dt.$$

Now,

$$-\frac{t^{n+1} e^{-st}}{s} \Big|_0^\infty = \lim_{b \to \infty} \left[-\frac{t^{n+1} e^{-st}}{s} \Big|_0^b \right]$$

$$= \lim_{b \to \infty} \frac{-b^{n+1} e^{-sb}}{s} + 0$$

$$= 0 \quad (s > 0).$$

So

$$\mathcal{L}[t^{n+1}] = \int_0^\infty \frac{n+1}{s} t^n e^{-st} dt$$

$$= \frac{n+1}{s} \int_0^\infty t^n e^{-st} dt$$

$$= \frac{n+1}{s} \mathcal{L}[t^n].$$

Since we assumed that $\mathcal{L}[t^n] = n!/s^{n+1}$, we get that

$$\mathcal{L}[t^{n+1}] = \frac{n+1}{s} \cdot \frac{n!}{s^{n+1}} = \frac{(n+1)!}{s^{n+2}}$$

which is what we wanted to show.

(ii) We use the fact that $\mathcal{L}[df/dt] = s\mathcal{L}[f] - f(0)$. Letting $f(t) = t^{n+1}$ we have $f(0) = 0$ and

$$\mathcal{L}[(n+1)t^n] = s\mathcal{L}[t^{n+1}] - 0$$

or

$$(n+1)\mathcal{L}[t^n] = s\mathcal{L}[t^{n+1}]$$

using the fact that the Laplace transform is linear. Since we assumed $\mathcal{L}[t^n] = n!/s^{n+1}$, we have

$$\mathcal{L}[t^{n+1}] = \frac{n+1}{s} \mathcal{L}[t^n] = \frac{n+1}{s} \cdot \frac{n!}{s^{n+1}} = \frac{(n+1)!}{s^{n+2}},$$

which is what we wanted to show.

7. Since we know that $\mathcal{L}[e^{at}] = 1/(s - a)$, we have $\mathcal{L}[e^{3t}] = 1/(s - 3)$, and therefore,

$$\mathcal{L}^{-1}\left[\frac{1}{s - 3} \right] = e^{3t}.$$

9. We see that

$$\frac{2}{3s+5} = \frac{2}{3} \cdot \frac{1}{s+5/3},$$

so

$$\mathcal{L}^{-1}\left[\frac{2}{3s+5}\right] = \frac{2}{3}e^{-\frac{5}{3}t}.$$

11. Using the method of partial fractions, we write

$$\frac{4}{s(s+3)} = \frac{A}{s} + \frac{B}{s+3}.$$

Putting the right-hand side over a common denominator gives $A(s+3) + B(s) = 4$, which can be written as $(A+B)s + 3A = 4$. Thus, $A + B = 0$, and $3A = 4$. This gives $A = 4/3$ and $B = -4/3$, so

$$\mathcal{L}^{-1}\left[\frac{4}{s(s+3)}\right] = \mathcal{L}^{-1}\left[\frac{4/3}{s} - \frac{4/3}{s+3}\right].$$

Hence,

$$\mathcal{L}^{-1}\left[\frac{4}{s(s+3)}\right] = \frac{4}{3} - \frac{4}{3}e^{-3t}.$$

13. Using the method of partial fractions, we have

$$\frac{2s+1}{(s-1)(s-2)} = \frac{A}{s-1} + \frac{B}{s-2}.$$

Putting the right-hand side over a common denominator gives $A(s-2) + B(s-1) = 2s+1$, which can be written as $(A+B)s + (-2A-B) = 2s+1$. So, $A + B = 2$, and $-2A - B = 1$. Thus $A = -3$ and $B = 5$, which gives

$$\mathcal{L}^{-1}\left[\frac{2s+1}{(s-1)(s-2)}\right] = \mathcal{L}^{-1}\left[\frac{5}{s-2} - \frac{3}{s-1}\right].$$

Finally,

$$\mathcal{L}^{-1}\left[\frac{2s+1}{(s-1)(s-2)}\right] = 5e^{2t} - 3e^t.$$

15.   (a) We have

$$\mathcal{L}\left[\frac{dy}{dt}\right] = s\mathcal{L}[y] - y(0)$$

and

$$\mathcal{L}[-y + e^{-2t}] = \mathcal{L}[-y] + \mathcal{L}[e^{-2t}] = -\mathcal{L}[y] + \frac{1}{s+2}$$

using linearity of the Laplace transform and the formula $\mathcal{L}[e^{at}] = 1/(s-a)$ from the text.

(b) Substituting the initial condition yields

$$s\mathcal{L}[y] - 2 = -\mathcal{L}[y] + \frac{1}{s+2}$$

so that

$$(s+1)\mathcal{L}[y] = 2 + \frac{1}{s+2}$$

which gives

$$\mathcal{L}[y] = \frac{1}{(s+1)(s+2)} + \frac{2}{s+1} = \frac{2s+5}{(s+1)(s+2)}.$$

(c) Using the method of partial fractions,

$$\frac{2s+5}{(s+1)(s+2)} = \frac{A}{s+1} + \frac{B}{s+2}.$$

Putting the right-hand side over a common denominator gives $A(s+2) + B(s+1) = 2s+5$, which can be written as $(A+B)s + (2A+B) = 2s+5$. So we have $A+B = 2$, and $2A+B = 5$. Thus, $A = 3$ and $B = -1$, and

$$\mathcal{L}[y] = \frac{3}{s+1} - \frac{1}{s+2}.$$

Therefore, $y(t) = 3e^{-t} - e^{-2t}$ is the desired function.

(d) Since $y(0) = 3e^0 - e^0 = 2$, $y(t)$ satisfies the given initial condition. Also,

$$\frac{dy}{dt} = -3e^{-t} + 2e^{-2t}$$

and

$$-y + e^{-2t} = -3e^{-t} + e^{-2t} + e^{-2t} = -3e^{-t} + 2e^{-2t},$$

so our solution also satisfies the differential equation.

17. **(a)** Taking Laplace transforms of both sides of the equation and simplifying gives

$$\mathcal{L}\left[\frac{dy}{dt}\right] + 7\mathcal{L}[y] = \mathcal{L}[1]$$

so

$$s\mathcal{L}[y] - y(0) + 7\mathcal{L}[y] = \frac{1}{s}$$

and $y(0) = 3$ gives

$$s\mathcal{L}[y] - 3 + 7\mathcal{L}[y] = \frac{1}{s}.$$

(b) Solving for $\mathcal{L}[y]$ gives

$$\mathcal{L}[y] = \frac{3}{s+7} + \frac{1}{s(s+7)} = \frac{3s+1}{s(s+7)}.$$

(c) Using the method of partial fractions, we get

$$\frac{3s+1}{s(s+7)} = \frac{A}{s} + \frac{B}{s+7}.$$

Putting the right-hand side over a common denominator gives $A(s+7) + Bs = 3s+1$, which can be written as $(A+B)s + 7A = 3s+1$. So $A+B = 3$, and $7A = 1$. Hence, $A = 1/7$ and $B = 20/7$, and we have

$$\mathcal{L}[y] = \frac{1/7}{s} + \frac{20/7}{s+7}.$$

Thus,

$$y(t) = \frac{20}{7}e^{-7t} + \frac{1}{7}.$$

(d) To check, we compute

$$\frac{dy}{dt} + 7y = -20e^{-7t} + 7\left(\frac{20}{7}e^{-7t} + \frac{1}{7}\right) = 1,$$

and $y(0) = 20/7 + 1/7 = 3$, so our solution satisfies the initial-value problem.

19. **(a)** Taking Laplace transforms of both sides of the equation and simplifying gives

$$\mathcal{L}\left[\frac{dy}{dt}\right] + 9\mathcal{L}[y] = \mathcal{L}[2]$$

so

$$s\mathcal{L}[y] - y(0) + 9\mathcal{L}[y] = \frac{2}{s}$$

and $y(0) = -2$ gives

$$s\mathcal{L}[y] + 2 + 9\mathcal{L}[y] = \frac{2}{s}.$$

(b) Solving for $\mathcal{L}[y]$ gives

$$\mathcal{L}[y] = -\frac{2}{s+9} + \frac{2}{s(s+9)} = \frac{-2s+2}{s(s+9)}.$$

(c) Using the method of partial fractions,

$$\frac{-2s+2}{s(s+9)} = \frac{A}{s} + \frac{B}{s+9}.$$

Putting the right-hand side over a common denominator gives $A(s+9) + Bs = -2s+2$, which can be written as $(A+B)s + 9A = -2s+2$. So $A+B = -2$ and $9A = 2$. Hence, $A = 2/9$ and $B = -20/9$, which gives us

$$\mathcal{L}[y] = \frac{2/9}{s} - \frac{20/9}{s+9}.$$

Finally,

$$y(t) = -\frac{20}{9}e^{-9t} + \frac{2}{9}.$$

(d) To check, we compute

$$\frac{dy}{dt} + 9y = 20e^{-9t} + 9\left(\frac{-20}{9}e^{-9t} + \frac{2}{9}\right) = 2,$$

and $y(0) = -20/9 + 2/9 = -2$, so our solution satisfies the initial-value problem.

21. **(a)** Putting the equation in the form

$$\frac{dy}{dt} + y = e^{-2t},$$

taking Laplace transforms of both sides of the equation and simplifying gives

$$\mathcal{L}\left[\frac{dy}{dt}\right] + \mathcal{L}[y] = \mathcal{L}[e^{-2t}]$$

so

$$s\mathcal{L}[y] - y(0) + \mathcal{L}[y] = \frac{1}{s+2}$$

and $y(0) = 1$ gives

$$s\mathcal{L}[y] - 1 + \mathcal{L}[y] = \frac{1}{s+2}.$$

(b) Solving for $\mathcal{L}[y]$ gives

$$\mathcal{L}[y] = \frac{1}{s+1} + \frac{1}{(s+1)(s+2)} = \frac{s+3}{(s+1)(s+2)}.$$

(c) Using partial fractions,

$$\frac{s+3}{(s+1)(s+2)} = \frac{A}{s+1} + \frac{B}{s+2}.$$

Putting the right-hand side over a common denominator gives $A(s+2) + B(s+1) = s+3$, which can be written as $(A+B)s + (2A+B) = s+3$. Thus, $A+B = 1$, and $2A+B = 3$. So $A = 2$ and $B = -1$, which gives us

$$\mathcal{L}[y] = \frac{2}{s+1} - \frac{1}{s+2}.$$

Hence,

$$y(t) = 2e^{-t} - e^{-2t}.$$

(d) To check, we compute

$$\frac{dy}{dt} + y = -2e^{-t} + 2e^{-2t} + \left(2e^{-t} - e^{-2t}\right) = e^{-2t},$$

and $y(0) = 2 - 1 = 1$, so our solution satisfies the initial-value problem.

23. **(a)** We have

$$\mathcal{L}\left[\frac{dy}{dt}\right] = s\mathcal{L}[y] - y(0)$$

and

$$\mathcal{L}[-y + t^2] = \mathcal{L}[-y] + \mathcal{L}[t^2] = -\mathcal{L}[y] + \frac{2}{s^3}$$

using linearity of the Laplace transform and the formula $\mathcal{L}[t^n] = n!/s^{n+1}$ from Exercise 5.

(b) Substituting the initial condition yields

$$s\mathcal{L}[y] - 1 = -\mathcal{L}[y] + \frac{2}{s^3}$$

so that

$$\mathcal{L}[y] = \frac{2/s^3 + 1}{1 + s} = \frac{2 + s^3}{s^3(s + 1)}.$$

(c) The best way to deal with this problem is with partial fractions. We seek constants $A, B, C,$ and D such that

$$\frac{A}{s} + \frac{B}{s^2} + \frac{C}{s^3} + \frac{D}{s+1} = \frac{2 + s^3}{s^3(s+1)}.$$

Multiplying through by $s^3(s+1)$ and equating like terms yields the system of equations

$$\begin{cases} A + D = 1 \\ A + B = 0 \\ B + C = 0 \\ C = 2. \end{cases}$$

Solving simultaneously gives us $A = 2, B = -2, C = 2,$ and $D = -1$. Therefore we seek a function $y(t)$ whose Laplace transform is

$$\frac{2}{s} - \frac{2}{s^2} + \frac{2}{s^3} - \frac{1}{s+1}.$$

We have $\mathcal{L}[e^{-t}] = 1/(s+1)$ so that

$$\mathcal{L}[-e^{-t}] = -\mathcal{L}[e^{-t}] = -\frac{1}{s+1}.$$

Also, using the formula from Exercise 5, we have

$$\mathcal{L}[t^2] = \frac{2}{s^3}, \quad \mathcal{L}[t] = \frac{1}{s^2}, \quad \text{and} \quad \mathcal{L}[1] = \frac{1}{s}$$

so that

$$\mathcal{L}[t^2 - 2t + 2] = \mathcal{L}[t^2] - 2\mathcal{L}[t] + 2\mathcal{L}[1] = \frac{2}{s^3} - \frac{2}{s^2} + \frac{2}{s}.$$

Therefore, $y(t) = t^2 - 2t + 2 - e^{-t}$ is the desired function.

(d) Since $y(0) = 2 - e^0 = 1$, $y(t)$ satisfies the given initial condition. Also,

$$\frac{dy}{dt} = 2t - 2 + e^{-t}$$

and

$$-y + t^2 = -t^2 + 2t - 2 + e^{-t} + t^2 = 2t - 2 + e^{-t}$$

so our solution also satisfies the differential equation.

25. First take Laplace transforms of both sides of the equation

$$\mathcal{L}\left[\frac{dy}{dt}\right] = 2\mathcal{L}[y] + 2\mathcal{L}[e^{-3t}]$$

and use the rules to simplify, obtaining

$$s\mathcal{L}[y] - y(0) = 2\mathcal{L}[y] + \frac{2}{s+3}$$

$$(s-2)\mathcal{L}[y] = y(0) + \frac{2}{s+3}$$

$$\mathcal{L}[y] = \frac{y(0)}{s-2} + \frac{2}{(s-2)(s+3)}.$$

Next note that

$$\mathcal{L}[y(0)e^{2t}] = y(0)/(s-2).$$

For the other summand, first simplify using partial fractions,

$$\frac{2}{(s-2)(s+3)} = \frac{A}{s-2} + \frac{B}{s+3}.$$

Putting the right-hand side over a common denominator gives $A(s+3) + B(s-2) = 2$, which can be written as $(A+B)s + (3A-2B) = 2$. This yields $A + B = 0$ and $3A - 2B = 2$. Hence $B = -2/5$ and $A = 2/5$, and

$$\frac{2}{(s-2)(s+3)} = \frac{2/5}{s-2} - \frac{2/5}{s+3}.$$

Now, $\mathcal{L}[e^{2t}] = 1/(s-2)$ and $\mathcal{L}[e^{-3t}] = 1/(s+3)$ so

$$\mathcal{L}[y] = \frac{y(0)}{s-2} + \frac{2}{5}\frac{1}{s-2} - \frac{2}{5}\frac{1}{s+3}.$$

Hence,

$$y(t) = y(0)e^{2t} + \frac{2}{5}e^{2t} - \frac{2}{5}e^{-3t}.$$

The first two terms can be combined into one, giving

$$y(t) = ce^{2t} - \frac{2}{5}e^{-3t},$$

where $c = y(0) + 2/5$.

27. As always the first step must be to take Laplace transform of both sides of the differential equation, giving

$$\mathcal{L}\left[\frac{dy}{dt}\right] = \mathcal{L}[y^2].$$

Simplifying, we obtain

$$s\mathcal{L}[y] - 1 = \mathcal{L}[y^2].$$

To solve for $\mathcal{L}[y]$ we must come up with an expression for $\mathcal{L}[y^2]$ in terms of $\mathcal{L}[y]$. This is not so easy! In particular, there is no easy way to simplify

$$\mathcal{L}[y^2] = \int_0^\infty y^2 e^{-st}\, dt$$

since we do not have a rule for the Laplace transform of a product.

EXERCISES FOR SECTION 6.2

1. **(a)** The function $g_a(t) = 1$ precisely when $u_a(t) = 0$, and $g_a(t) = 0$ precisely when $u_a(t) = 1$, so

$$g_a(t) = 1 - u_a(t).$$

(b) We can compute the Laplace transform of $g_a(t)$ from the definition

$$\mathcal{L}[g_a] = \int_0^a 1e^{-st}\, dt = -\frac{e^{-as}}{s} + \frac{e^{-0s}}{s} = \frac{1}{s} - \frac{e^{-as}}{s}.$$

Alternately, we can use the table

$$\mathcal{L}[g_a] = \mathcal{L}[1 - u_a(t)] = \frac{1}{s} - \frac{e^{-as}}{s}.$$

3.

$$\mathcal{L}[g_a(t)] = \int_0^\infty g_a(t)\, e^{-st}\, dt$$

$$= \int_0^a \frac{t}{a} e^{-st}\, dt + \int_a^\infty e^{-st}\, dt$$

Using integration by parts with $u = t$ and $dv = e^{-st}dt$, we have $du = dt$, $v = -e^{-st}/s$ and

$$\int_0^a \frac{t}{a} e^{-st}\, dt = \frac{1}{a} \int_0^a t e^{-st}\, dt$$

$$= \frac{1}{a}\left(-\frac{te^{-st}}{s}\bigg|_0^a - \int_0^a -\frac{e^{-st}}{s}\, dt \right)$$

$$= \frac{1}{a}\left(-\frac{ae^{-as}}{s} - \frac{1}{s^2}e^{-st}\bigg|_0^a \right)$$

$$= \frac{1}{a}\left(-\frac{ae^{-as}}{s} - \frac{1}{s^2}(e^{-as} - 1) \right)$$

$$= -\frac{e^{-as}}{s} - \frac{1}{as^2}\left(e^{-as} - 1\right).$$

Also,

$$\int_a^\infty e^{-st}\, dt = \lim_{b\to\infty} \int_a^b e^{-st}\, dt$$

$$= \lim_{b\to\infty} -\frac{1}{s}e^{-st}\bigg|_a^b$$

$$= \lim_{b\to\infty} -\frac{1}{s}\left(e^{-sb} - e^{-as}\right)$$

$$= \frac{1}{s}e^{-as}.$$

Therefore,

$$\mathcal{L}[g_a(t)] = -\frac{e^{-as}}{s} - \frac{1}{as^2}\left(e^{-as} - 1\right) + \frac{1}{s}e^{-as}$$

$$= \frac{1}{as^2}\left(1 - e^{-as}\right).$$

5. First use partial fractions to write

$$\frac{1}{(s-1)(s-2)} = \frac{A}{s-1} + \frac{B}{s-2}.$$

Putting the right-hand side over a common denominator yields $As - 2A + Bs - B = 1$ which can be written as $(A + B)s + (-2A - B) = 1$. Thus, $A + B = 0$, and $-2A - B = 1$. Solving for A and B yields $A = -1$ and $B = 1$, so

$$\frac{1}{(s-1)(s-2)} = \frac{-1}{s-1} + \frac{1}{s-2}.$$

Now, as above

$$\mathcal{L}[u_3(t)e^{2(t-3)}] = \frac{e^{-3s}}{s-2}$$

and

$$\mathcal{L}[u_3(t)e^{t-3}] = \frac{e^{-3s}}{s-1}$$

and the desired function is

$$u_3(t)\left(e^{2(t-3)} - e^{(t-3)}\right).$$

7. Using partial fractions, we get

$$\frac{14}{(3s+2)(s-4)} = \frac{A}{3s+2} + \frac{B}{s-4}.$$

Hence, we must have $As - 4A + 3Bs + 2B = 14$, which can be written as

$$(A + 3B)s + (-4A + 2B) = 14.$$

Therefore, $A + 3B = 0$, and $-4A + 2B = 14$. Solving for A and B yields $A = -3$ and $B = 1$, so

$$\frac{14}{(3s+2)(s-4)} = \frac{1}{s-4} - \frac{3}{3s+2} = \frac{1}{s-4} - \frac{1}{s+2/3}.$$

Applying the rules

$$\mathcal{L}[u_1(t)e^{4(t-1)}] = \frac{e^{-s}}{s-4}$$

and

$$\mathcal{L}[u_1(t)e^{-\frac{2}{3}(t-1)}] = \frac{e^{-s}}{s+2/3},$$

the desired function is

$$y(t) = u_1(t)\left(e^{4(t-1)} - e^{-\frac{2}{3}(t-1)}\right).$$

9. Taking the Laplace transform of both sides of the equation, we have

$$\mathcal{L}\left[\frac{dy}{dt}\right] + 9\mathcal{L}[y] = \mathcal{L}[u_5(t)],$$

which is equivalent to

$$s\mathcal{L}[y] - y(0) + 9\mathcal{L}[y] = \frac{e^{-5s}}{s}.$$

Since $y(0) = -2$, we have

$$s\mathcal{L}[y] + 2 + 9\mathcal{L}[y] = \frac{e^{-5s}}{s},$$

which yields

$$\mathcal{L}[y] = \frac{-2}{s+9} + \frac{e^{-5s}}{s(s+9)}.$$

Using the partial fractions decomposition

$$\frac{1}{s(s+9)} = \frac{1/9}{s} - \frac{1/9}{s+9},$$

we see that

$$\mathcal{L}[y] = \frac{-2}{s+9} + \frac{1}{9}\left(\frac{e^{-5s}}{s}\right) - \frac{1}{9}\left(\frac{e^{-5s}}{s+9}\right).$$

Taking the inverse of the Laplace transform, we obtain

$$y(t) = -2e^{-9t} + \frac{1}{9}u_5(t) - \frac{1}{9}u_5(t)e^{-9(t-5)}$$

$$= -2e^{-9t} + \frac{1}{9}u_5(t)\left(1 - e^{-9(t-5)}\right).$$

To check our answer, we compute

$$\frac{dy}{dt} = 18e^{-9t} + \frac{1}{9}\frac{du_5}{dt}\left(1 - e^{-9(t-5)}\right) + \frac{1}{9}u_5(t)\left(9e^{-9(t-5)}\right),$$

and since $du_5/dt = 0$ except at $t = 5$ (where it is undefined),

$$\frac{dy}{dt} + 9y = 18e^{-9t} + u_5(t)e^{-9(t-5)} + 9\left(-2e^{-9t} + \frac{1}{9}u_5(t)\left(1 - e^{-9(t-5)}\right)\right)$$

$$= u_5(t).$$

Hence, our $y(t)$ satisfies the differential equation except when $t = 5$. (We cannot expect $y(t)$ to satisfy the differential equation at $t = 5$ because the differential equation is not continuous there.) Note that $y(t)$ also satisfies the initial condition $y(0) = -2$.

11. Taking the Laplace transform of both sides of the equation, we obtain

$$\mathcal{L}\left[\frac{dy}{dt}\right] = \mathcal{L}[-y] + \mathcal{L}[u_2(t)e^{-2(t-2)}],$$

which is equivalent to

$$s\mathcal{L}[y] - y(0) = -\mathcal{L}[y] + \frac{e^{-2s}}{s+2}$$

(using linearity of the Laplace transform and the formula

If $\mathcal{L}[f] = F(s)$ then $\mathcal{L}[u_a(t)f(t-a)] = e^{-as}F(s)$

where $f(t) = e^{-2t}$ and $a = 2$.)
Substituting the initial condition yields

$$s\mathcal{L}[y] - 1 = -\mathcal{L}[y] + \frac{e^{-2s}}{s+2}$$

so that

$$\mathcal{L}[y] = \frac{1}{s+1} + \frac{e^{-2s}}{(s+1)(s+2)}.$$

By partial fractions, we know that

$$\frac{1}{s+1} - \frac{1}{s+2} = \frac{1}{(s+1)(s+2)},$$

so we have

$$\frac{e^{-2s}}{(s+1)(s+2)} = e^{-2s}\left(\frac{1}{s+1} - \frac{1}{s+2}\right) = \frac{e^{-2s}}{s+1} - \frac{e^{-2s}}{s+2}.$$

Taking the inverse of the Laplace transform yields

$$y(t) = e^{-t} + u_2(t)e^{-(t-2)} - u_2(t)e^{-2(t-2)}$$

$$= e^{-t} + u_2(t)\left(e^{-(t-2)} - e^{-2(t-2)}\right).$$

To check our answer, we compute

$$\frac{dy}{dt} = -e^{-t} + \frac{du_2}{dt}\left(e^{-(t-2)} - e^{-2(t-2)}\right) + u_2(t)\left(-e^{-(t-2)} + 2e^{-2(t-2)}\right),$$

and since $du_2/dt = 0$ except at $t = 2$ (where it is undefined),

$$\frac{dy}{dt} + y = -e^{-t} + u_2(t)\left(-e^{-(t-2)} + 2e^{-2(t-2)}\right) + e^{-t} + u_2(t)\left(e^{-(t-2)} - e^{-2(t-2)}\right)$$

$$= u_2(t)e^{-2(t-2)}.$$

Hence, our $y(t)$ satisfies the differential equation except when $t = 2$. (We cannot expect $y(t)$ to satisfy the differential equation at $t = 2$ because the differential equation is not continuous there.) Note that $y(t)$ also satisfies the initial condition $y(0) = 1$.

13. Taking the Laplace transform of both sides of the equation, we obtain

$$\mathcal{L}\left[\frac{dy}{dt}\right] = -\mathcal{L}[y] + \mathcal{L}[u_1(t)(t-1)],$$

which is equivalent to

$$s\mathcal{L}[y] - y(0) = -\mathcal{L}[y] + \frac{e^{-s}}{s^2}.$$

Substituting the initial condition yields

$$s\mathcal{L}[y] - 2 = -\mathcal{L}[y] + \frac{e^{-s}}{s^2}$$

so that

$$\mathcal{L}[y] = \frac{e^{-s}}{s^2(s+1)} + \frac{2}{s+1}.$$

Using the technique of partial fractions, we write

$$\frac{1}{s^2(s+1)} = \frac{A}{s} + \frac{B}{s^2} + \frac{C}{s+1}.$$

Putting the right-hand side over a common denominator gives us $As(s+1) + B(s+1) + Cs^2 = 1$ which can be written as $(A+C)s^2 + (A+B)s + B = 1$. So $A+C = 0$, $A+B = 0$, and $B = 1$. Thus $A = -1$ and $C = 1$, and

$$\frac{1}{s^2(s+1)} = \frac{-1}{s} + \frac{1}{s^2} + \frac{1}{s+1}.$$

Taking the inverse of the Laplace transform gives us

$$\begin{aligned}
y(t) &= \mathcal{L}^{-1}\left[\frac{e^{-s}}{s^2(s+1)}\right] + \mathcal{L}^{-1}\left[\frac{2}{s+1}\right] \\
&= -\mathcal{L}^{-1}\left[\frac{e^{-s}}{s}\right] + \mathcal{L}^{-1}\left[\frac{e^{-s}}{s^2}\right] + \mathcal{L}^{-1}\left[\frac{e^{-s}}{s+1}\right] + \mathcal{L}^{-1}\left[\frac{2}{s+1}\right] \\
&= -u_1(t) + u_1(t)(t-1) + u_1(t)e^{-(t-1)} + 2e^{-t} \\
&= u_1(t)\left((t-2) + e^{-(t-1)}\right) + 2e^{-t}.
\end{aligned}$$

To check our answer, we compute

$$\frac{dy}{dt} = \frac{du_1}{dt}\left((t-2) + e^{-(t-1)}\right) + u_1(t)\left(1 - e^{-(t-1)}\right) - 2e^{-t},$$

and since $du_1/dt = 0$ except at $t = 1$ (where it is undefined),

$$\frac{dy}{dt} + y = u_1(t)\left(1 - e^{-(t-1)}\right) - 2e^{-t} + u_1(t)\left((t-2) + e^{-(t-1)}\right) + 2e^{-t}$$

$$= u_1(t) + u_1(t)(t-2)$$

$$= u_1(t)(t-1).$$

Hence, our $y(t)$ satisfies the differential equation except when $t = 1$. (We cannot expect $y(t)$ to satisfy the differential equation at $t = 1$ because the differential equation is not continuous there.) Note that $y(t)$ also satisfies the initial condition $y(0) = 2$.

15. Taking the Laplace transform of both sides of the equation, we have

$$\mathcal{L}\left[\frac{dy}{dt}\right] = -\mathcal{L}[y] + \mathcal{L}[u_a(t)],$$

which is equivalent to

$$s\mathcal{L}[y] - y(0) = -\mathcal{L}[y] + \frac{e^{-as}}{s}.$$

Solving for $\mathcal{L}[y]$ yields

$$\mathcal{L}[y] = \frac{e^{-as}}{s(s+1)} + \frac{y(0)}{s+1}.$$

Using the partial fractions decomposition

$$\frac{1}{s(s+1)} = \frac{1}{s} - \frac{1}{s+1},$$

we get

$$\mathcal{L}[y] = \frac{e^{-as}}{s} - \frac{e^{-as}}{s+1} + \frac{y(0)}{s+1}.$$

Taking the inverse Laplace transform, we obtain

$$y(t) = u_a(t) - u_a(t)e^{-(t-a)} + y(0)e^{-t}$$

$$= u_a(t)\left(1 - e^{-(t-a)}\right) + y(0)e^{-t}.$$

To check our answer, we compute

$$\frac{dy}{dt} = \frac{du_a}{dt}\left(1 - e^{-(t-a)}\right) + u_a(t)e^{-(t-a)} - y(0)e^{-t}$$

and since $du_a/dt = 0$ except at $t = a$ (where it is undefined),

$$\frac{dy}{dt} + y = u_a(t)e^{-(t-a)} - y(0)e^{-t} + u_a(t)\left(1 - e^{-(t-a)}\right) + y(0)e^{-t}$$

$$= u_a(t).$$

Hence, our $y(t)$ satisfies the differential equation except when $t = a$. (We cannot expect $y(t)$ to satisfy the differential equation at $t = a$ because the differential equation is not continuous there.)

17. From the formula in Exercise 16, we see that we need only compute the integral $\int_0^2 w(t)\,e^{-st}\,dt$. We write

$$\int_0^2 w(t)\,e^{-st}\,dt = \int_0^1 e^{-st}\,dt - \int_1^2 e^{-st}\,dt$$

$$= \frac{e^{-st}}{-s}\bigg|_0^1 - \frac{e^{-st}}{-s}\bigg|_1^2$$

$$= \left(-\frac{e^{-s}}{s} + \frac{1}{s}\right) - \left(-\frac{e^{-2s}}{s} + \frac{e^{-s}}{s}\right)$$

$$= \frac{1}{s} - 2\frac{e^{-s}}{s} + \frac{e^{-2s}}{s}$$

$$= \frac{1}{s}(1 - 2e^{-s} + e^{-2s})$$

$$= \frac{1}{s}(1 - e^{-s})^2.$$

From the formula of Exercise 16, we get

$$\mathcal{L}[w] = \frac{1}{1 - e^{-2s}} \left(\frac{1}{s}(1 - e^{-s})^2\right)$$

$$= \frac{(1 - e^{-s})^2}{s(1 - e^{-s})(1 + e^{-s})}$$

$$= \frac{1 - e^{-s}}{s(1 + e^{-s})}.$$

19. (a) Transforming both sides of the equation, we have $\mathcal{L}[dy/dt] = -\mathcal{L}[y] + \mathcal{L}[w(t)]$, and using the result of Exercise 17, we get

$$s\mathcal{L}[y] - y(0) = -\mathcal{L}[y] + \frac{1 - e^{-s}}{s(1 + e^{-s})}.$$

Solving for $\mathcal{L}[y]$ using the fact that $y(0) = 0$, we obtain

$$\mathcal{L}[y] = \frac{1 - e^{-s}}{s(s + 1)(1 + e^{-s})}.$$

(b) The function $w(t)$ is alternatively 1 and -1. While $w(t) = 1$, the solution decays exponentially toward $y = 1$. When $w(t)$ changes to -1, the solution then decays toward $y = -1$.
In fact, there is a periodic solution with initial condition $y(0) = (1 - e)/(1 + e) \approx -0.462$, and our solution tends toward this periodic solution as $t \to \infty$.

EXERCISES FOR SECTION 6.3

1. We use integration by parts twice to compute

$$\mathcal{L}[\sin \omega t] = \int_0^\infty \sin \omega t \, e^{-st} \, dt.$$

First, letting $u = \sin \omega t$ and $dv = e^{-st}\, dt$, we get

$$\mathcal{L}[\sin \omega t] = \sin \omega t \left.\frac{e^{-st}}{-s}\right|_0^\infty - \int_0^\infty \frac{e^{-st}}{-s}\, \omega \cos \omega t\, dt$$

$$= \lim_{b \to \infty} \left[\frac{e^{-st}}{-s} \sin \omega t \Big|_0^b\right] + \frac{\omega}{s} \int_0^\infty e^{-st} \cos \omega t\, dt$$

$$= \frac{\omega}{s} \int_0^\infty e^{-st} \cos \omega t\, dt,$$

since the limit of $e^{-sb} \sin \omega b$ is 0 as $b \to \infty$ and $s > 0$.

Using integration by parts on

$$\int_0^\infty e^{-st} \cos \omega t\, dt,$$

with $u = \cos \omega t$ and $dv = e^{-st}\, dt$, we get

$$\int_0^\infty e^{-st} \cos \omega t\, dt = \left.\frac{e^{-st}}{-s} \cos \omega t\right|_0^\infty - \int_0^\infty \frac{e^{-st}}{-s}\, (-\omega \sin \omega t)\, dt$$

$$= \lim_{b \to \infty} \left[\frac{e^{-st}}{-s} \cos \omega t \Big|_0^b\right] - \frac{\omega}{s} \int_0^\infty e^{-st} \sin \omega t\, dt$$

$$= \lim_{b \to \infty} \left[\frac{e^{-sb}}{-s} \cos \omega b\right] + \frac{1}{s} - \frac{\omega}{s} \int_0^\infty e^{-st} \sin \omega t\, dt$$

$$= \frac{1}{s} - \frac{\omega}{s} \int_0^\infty e^{-st} \sin \omega t\, dt,$$

since the limit of $e^{-sb} \cos \omega b$ is 0 as $b \to \infty$ and $s > 0$.

Thus,

$$\int_0^\infty \sin \omega t\, e^{-st}\, dt = \frac{\omega}{s} \int_0^\infty e^{-st} \cos \omega t\, dt$$

$$= \frac{\omega}{s} \left(\frac{1}{s} - \frac{\omega}{s} \int_0^\infty \sin \omega t\, e^{-st}\, dt\right)$$

$$= \frac{\omega}{s^2} - \frac{\omega^2}{s^2} \int_0^\infty \sin \omega t\, e^{-st}\, dt.$$

So

$$\frac{s^2 + \omega^2}{s^2} \int_0^\infty \sin \omega t\, e^{-st}\, dt = \frac{\omega}{s^2},$$

and

$$\int_0^\infty \sin \omega t\, e^{-st}\, dt = \frac{\omega}{s^2 + \omega^2}.$$

3. We need to compute

$$\mathcal{L}[e^{at}\sin\omega t] = \int_0^\infty e^{at}\sin\omega t\, e^{-st}\, dt = \int_0^\infty \sin\omega t\, e^{-(s-a)t}\, dt.$$

We can do this using integration by parts twice and ending up with $\mathcal{L}[e^{at}\sin\omega t]$ on both sides of the equation. Alternately, if we let $r = s - a$, then

$$\int_0^\infty \sin\omega t\, e^{-(s-a)t}\, dt = \int_0^\infty \sin\omega t\, e^{-rt}\, dt$$

The integral on the right is the Laplace transform of $\sin\omega t$ with r as the new independent variable. From Exercise 1, we know

$$\int_0^\infty \sin\omega t\, e^{-rt}\, dt = \frac{\omega}{r^2 + \omega^2}.$$

Substituting back we have

$$\mathcal{L}[e^{at}\sin\omega t] = \frac{\omega}{(s-a)^2 + \omega^2}.$$

5. Using the formula

$$\mathcal{L}\left[\frac{d^2y}{dt^2}\right] = s^2\mathcal{L}[y] - y'(0) - sy(0),$$

and the linearity of the Laplace transform, we get that

$$s^2\mathcal{L}[y] - y'(0) - sy(0) + \omega^2\mathcal{L}[y] = 0.$$

Substituting the initial conditions and solving for $\mathcal{L}[y]$ gives

$$\mathcal{L}[y] = \frac{s}{s^2 + \omega^2}.$$

7. Since

$$\mathcal{L}[\sin\omega t] = \frac{\omega}{s^2 + \omega^2},$$

we can compute that

$$\frac{d}{d\omega}\mathcal{L}[\sin\omega t] = \frac{s^2 - \omega^2}{(s^2 + \omega^2)^2},$$

but

$$\frac{d}{d\omega}\mathcal{L}[\sin\omega t] = \mathcal{L}\left[\frac{d}{d\omega}\sin\omega t\right] = \mathcal{L}[t\cos\omega t].$$

So

$$\mathcal{L}[t\cos\omega t] = \frac{s^2 - \omega^2}{(s^2 + \omega^2)^2}.$$

9. From Exercise 10, we know that

$$\mathcal{L}[te^{at}] = \frac{1}{(s-a)^2}.$$

Differentiating both sides of this formula with respect to a gives

$$\frac{d}{da}\mathcal{L}[te^{at}] = \mathcal{L}\left[\frac{d}{da}te^{ta}\right] = \mathcal{L}[t^2e^{at}]$$

while

$$\frac{d}{da}\frac{1}{(s-a)^2} = \frac{2}{(s-a)^3}.$$

Hence,

$$\mathcal{L}[t^2e^{at}] = \frac{2}{(s-a)^3}.$$

11. In this case, $b = 2$, and $(s+b/2)^2 = (s+1)^2 = s^2 + 2s + 1$, so $s^2 + 2s + 10 = (s+1)^2 + 3^2$.

13. In this case, $b = 1$, and $(s+b/2)^2 = (s+1/2)^2 = s^2 + s + 1/4$, so $s^2 + s + 1 = (s+1/2)^2 + 3/4 = (s+1/2)^2 + (\sqrt{3}/2)^2$.

15. In Exercise 11, we completed the square and obtained $s^2 + 2s + 10 = (s+1)^2 + 3^2$, so

$$\mathcal{L}^{-1}\left[\frac{1}{s^2+2s+10}\right] = \mathcal{L}^{-1}\left[\frac{1}{(s+1)^2+3^2}\right]$$

$$= \frac{1}{3}\mathcal{L}^{-1}\left[\frac{3}{(s+1)^2+3^2}\right]$$

$$= \frac{1}{3}e^{-t}\sin 3t.$$

17. In Exercise 13, we completed the square and obtained $s^2 + s + 1 = (s+1/2)^2 + (\sqrt{3}/2)^2$, so

$$\frac{2s+3}{s^2+s+1} = \frac{2s+3}{(s+1/2)^2+(\sqrt{3}/2)^2}.$$

We want to put this fraction in the right form so that we can use the formulas for $\mathcal{L}[e^{at}\cos\omega t]$ and $\mathcal{L}[e^{at}\sin\omega t]$. We see that

$$\frac{2s+3}{(s+1/2)^2+(\sqrt{3}/2)^2} = \frac{2s+1}{(s+1/2)^2+(\sqrt{3}/2)^2} + \frac{2}{(s+1/2)^2+(\sqrt{3}/2)^2}$$

$$= \frac{2(s+1/2)}{(s+1/2)^2+(\sqrt{3}/2)^2} + \frac{(4/\sqrt{3})(\sqrt{3}/2)}{(s+1/2)^2+(\sqrt{3}/2)^2}.$$

So

$$\mathcal{L}^{-1}\left[\frac{2s+3}{s^2+s+1}\right] = 2\mathcal{L}^{-1}\left[\frac{(s+1/2)}{(s+1/2)^2+(\sqrt{3}/2)^2}\right] + \frac{4}{\sqrt{3}}\mathcal{L}^{-1}\left[\frac{\sqrt{3}/2}{(s+1/2)^2+(\sqrt{3}/2)^2}\right]$$

$$= 2e^{-t/2}\cos\left(\frac{\sqrt{3}}{2}t\right) + \frac{4}{\sqrt{3}}e^{-t/2}\sin\left(\frac{\sqrt{3}}{2}t\right).$$

19. We compute

$$\mathcal{L}\left[e^{(a+ib)t}\right] = \int_0^\infty e^{(a+ib)t} e^{-st} \, dt$$

$$= \int_0^\infty e^{-(s-(a+ib))t} \, dt$$

$$= -\frac{1}{s-(a+ib)} \left(\lim_{u\to\infty} \left[e^{-(s-a)u} e^{-ibu} \right] - 1 \right).$$

The limit is zero as long as $s > a$. Hence,

$$\mathcal{L}\left[e^{(a+ib)t}\right] = \frac{1}{s-(a+ib)}$$

if $s > a$ and undefined otherwise. This is the same formula as for real exponentials. It can also be written

$$\mathcal{L}\left[e^{(a+ib)t}\right] = \frac{s-a+ib}{(s-a)^2 + b^2}.$$

21. We recall that

$$e^{at} \cos \omega t = \text{Re}(e^{(a+ib)t}).$$

So

$$\mathcal{L}[e^{at} \cos \omega t] = \text{Re}(\mathcal{L}[e^{(a+ib)t}])$$

$$= \text{Re}\left(\frac{s-a+i\omega}{(s-a)^2 + \omega^2} \right)$$

$$= \frac{s-a}{(s-a)^2 + \omega^2}.$$

Similarly,

$$\mathcal{L}[e^{at} \sin \omega t] = \text{Im}(\mathcal{L}[e^{(a+ib)t}])$$

$$= \text{Im}\left(\frac{s-a+i\omega}{(s-a)^2 + \omega^2} \right)$$

$$= \frac{\omega}{(s-a)^2 + \omega^2}.$$

23. Using the quadratic formula, we see that the roots of $s^2 + 2s + 10 = 0$ are $s = -1 \pm 3i$. Thus $s^2 + 2s + 10 = (s + 1 + 3i)(s + 1 - 3i)$. So we want to find A and B so that

$$\frac{1}{s^2 + 2s + 10} = \frac{A}{s + 1 + 3i} + \frac{B}{s + 1 - 3i}.$$

So, finding common denominators (that is, usual partial fractions only with complex numbers) gives

$$\begin{cases} A + B = 0 \\ A + B + 3i(-A + B) = 1. \end{cases}$$

Solving, we get $A = i/6$ and $B = -i/6$, so

$$\frac{1}{s^2 + 2s + 10} = \frac{i/6}{s + 1 + 3i} + \frac{-i/6}{s + 1 - 3i}.$$

Thus

$$\mathcal{L}^{-1}\left[\frac{1}{s^2 + 2s + 10}\right] = \mathcal{L}^{-1}\left[\frac{i/6}{s + 1 + 3i} + \frac{-i/6}{s + 1 - 3i}\right]$$

$$= \frac{i}{6}e^{(-1-3i)t} - \frac{i}{6}e^{(-1+3i)t}$$

$$= \frac{i}{6}\left(e^{-t}\cos(-3t) + ie^{-t}\sin(-3t)\right) - \frac{i}{6}\left(e^{-t}\cos 3t + ie^{-t}\sin 3t\right)$$

$$= -\frac{i}{6}\left(2ie^{-t}\sin 3t\right)$$

$$= \frac{1}{3}e^{-t}\sin 3t.$$

25. Using the quadratic formula, the roots of the denominator are $(-1 \pm i\sqrt{3})/2$. Hence, we can factor the denominator into

$$\left(s - \left(\tfrac{-1+i\sqrt{3}}{2}\right)\right)\left(s - \left(\tfrac{-1-i\sqrt{3}}{2}\right)\right).$$

We then do the partial fractions decomposition

$$\frac{2s + 3}{s^2 + s + 1} = \frac{A}{s - \left(\tfrac{-1+i\sqrt{3}}{2}\right)} + \frac{B}{s - \left(\tfrac{-1-i\sqrt{3}}{2}\right)},$$

which gives rise to the equations

$$\begin{cases} A + B = 2 \\ \left(\tfrac{1+i\sqrt{3}}{2}\right)A + \left(\tfrac{1-i\sqrt{3}}{2}\right)B = 3. \end{cases}$$

Solving yields $A = 1 - \frac{2}{\sqrt{3}}i$ and $B = 1 + \frac{2}{\sqrt{3}}i$. So

$$\frac{2s + 3}{s^2 + s + 1} = \frac{1 - \frac{2}{\sqrt{3}}i}{s - \left(\tfrac{-1+i\sqrt{3}}{2}\right)} + \frac{1 + \frac{2}{\sqrt{3}}i}{s - \left(\tfrac{-1-i\sqrt{3}}{2}\right)}.$$

Taking inverse Laplace transforms of the right-hand side gives

$$\left(1 - \tfrac{2}{\sqrt{3}}i\right)e^{(-1+i\sqrt{3})t/2} + \left(1 + \tfrac{2}{\sqrt{3}}i\right)e^{(-1-i\sqrt{3})t/2}.$$

Using Euler's formula to replace the complex exponentials and simplifying yields

$$2e^{-t/2}\cos\left(\tfrac{\sqrt{3}}{2}t\right) + \tfrac{4}{\sqrt{3}}e^{-t/2}\sin\left(\tfrac{\sqrt{3}}{2}t\right).$$

27. **(a)** Taking the Laplace transform of both sides of the equation, we obtain

$$\mathcal{L}\left[\frac{d^2y}{dt^2}\right] + 4\mathcal{L}[y] = \frac{8}{s},$$

and using the fact that $\mathcal{L}[d^2y/dt^2] = s^2\mathcal{L}[y] - sy(0) - y'(0)$, we have

$$(s^2 + 4)\mathcal{L}[y] - sy(0) - y'(0) = \frac{8}{s}.$$

(b) Substituting the initial conditions yields

$$(s^2 + 4)\mathcal{L}[y] - 11s - 5 = \frac{8}{s},$$

and solving for $\mathcal{L}[y]$ we get

$$\mathcal{L}[y] = \frac{11s + 5}{s^2 + 4} + \frac{8}{s(s^2 + 4)}.$$

The partial fractions decomposition of $8/(s(s^2 + 4))$ is

$$\frac{8}{s(s^2 + 4)} = \frac{A}{s} + \frac{Bs + C}{s^2 + 4}.$$

Putting the right-hand side over a common denominator gives us

$$(A + B)s^2 + Cs + 4A = 8,$$

and consequently, $A = 2$, $B = -2$, and $C = 0$. In other words,

$$\frac{8}{s(s^2 + 4)} = \frac{2}{s} + \frac{-2s}{s^2 + 4}.$$

We obtain

$$\mathcal{L}[y] = \frac{2}{s} + \frac{9s + 5}{s^2 + 4}.$$

(c) To take the inverse Laplace transform, we rewrite $\mathcal{L}[y]$ in the form

$$\mathcal{L}[y] = \frac{2}{s} + 9\left(\frac{s}{s^2 + 4}\right) + \frac{5}{2}\left(\frac{2}{s^2 + 4}\right).$$

Therefore, $y(t) = 2 + 9\cos 2t + \tfrac{5}{2}\sin 2t$.

29. **(a)** Taking the Laplace transform of both sides of the equation, we obtain

$$\mathcal{L}\left[\frac{d^2y}{dt^2}\right] - 4\mathcal{L}\left[\frac{dy}{dt}\right] + 5\mathcal{L}[y] = \frac{2}{s-1},$$

and using the formulas for $\mathcal{L}[dy/dt]$ and $\mathcal{L}[d^2y/dt^2]$ in terms of $\mathcal{L}[y]$, we have

$$(s^2 - 4s + 5)\mathcal{L}[y] - sy(0) - y'(0) + 4y(0) = \frac{2}{s-1}.$$

(b) Substituting the initial conditions yields

$$(s^2 - 4s + 5)\mathcal{L}[y] - 3s + 11 = \frac{2}{s-2},$$

and solving for $\mathcal{L}[y]$ we get

$$\mathcal{L}[y] = \frac{3s-11}{s^2-4s+5} + \frac{2}{(s-1)(s^2-4s+5)}.$$

Using the partial fractions decomposition

$$\frac{2}{(s-1)(s^2-4s+5)} = \frac{1}{s-1} + \frac{-s+3}{s^2-4s+5},$$

we obtain

$$\mathcal{L}[y] = \frac{1}{s-1} + \frac{2s-8}{s^2-4s+5}.$$

(c) In order to compute the inverse Laplace transform, we first write

$$s^2 - 4s + 5 = (s-2)^2 + 1$$

by completing the square, and then we write

$$\frac{2s-8}{s^2-4s+5} = \frac{2(s-2)}{(s-2)^2+1} - \frac{4}{(s-2)^2+1}.$$

Taking the inverse Laplace transform, we have

$$y(t) = e^t + 2e^{2t}\cos t - 4e^{2t}\sin t.$$

31. **(a)** Note that this is resonant forcing of an undamped oscillator. We take the Laplace transform of both sides

$$\mathcal{L}\left[\frac{d^2y}{dt^2}\right] + 4\mathcal{L}[y] = \mathcal{L}[\cos 2t]$$

and obtain

$$s^2\mathcal{L}[y] + 2s + 4\mathcal{L}[y] = \frac{s}{s^2+4}.$$

(b) Solving for $\mathcal{L}[y]$, we get

$$\mathcal{L}[y] = -\frac{2s}{s^2+4} + \frac{s}{(s^2+4)^2}.$$

(c) To take the inverse Laplace transform, we note that

$$\mathcal{L}^{-1}\left[-\frac{2s}{s^2+4}\right] = -2\cos 2t$$

and

$$\mathcal{L}^{-1}\left[\frac{s}{(s^2+4)^2}\right] = \frac{1}{4}\mathcal{L}^{-1}\left[\frac{4s}{(s^2+4)^2}\right] = \frac{t}{4}\sin 2t.$$

So

$$y(t) = -2\cos 2t + \frac{t}{4}\sin 2t,$$

which is of the form we would expect for a resonant response.

33. (a) We take Laplace transform of both sides to obtain

$$\mathcal{L}\left[\frac{d^2y}{dt^2}\right] + 4\mathcal{L}\left[\frac{dy}{dt}\right] + 9\mathcal{L}[y] = \mathcal{L}[20u_2(t)\sin(t-2)],$$

which is equivalent to

$$s^2\mathcal{L}[y] - y(0)s - y'(0) + 4\left(s\mathcal{L}[y] - y(0)\right) + 9\mathcal{L}[y] = 20e^{-2s}\left(\frac{1}{s^2+1}\right).$$

Using the given initial conditions, we have

$$(s^2+4s+9)\mathcal{L}[y] - s - 6 = e^{-2s}\left(\frac{20}{s^2+1}\right).$$

(b) Solving for $\mathcal{L}[y]$ gives

$$\mathcal{L}[y] = \frac{s+6}{s^2+4s+9} + e^{-2s}\left(\frac{20}{(s^2+4s+9)(s^2+1)}\right).$$

(c) Now to find the inverse Laplace transform, we first note that

$$\frac{s+6}{s^2+4s+9} = \frac{s+2}{(s+2)^2+5} + \frac{4}{(s+2)^2+5}.$$

Therefore,

$$\mathcal{L}^{-1}\left[\frac{s+6}{s^2+4s+9}\right] = \mathcal{L}^{-1}\left[\frac{s+2}{(s+2)^2+5}\right] + \frac{4}{\sqrt{5}}\mathcal{L}^{-1}\left[\frac{\sqrt{5}}{(s+2)^2+5}\right]$$

$$= e^{-2t}\cos\sqrt{5}\,t + \frac{4}{\sqrt{5}}e^{-2t}\sin\sqrt{5}\,t.$$

For the second term, we use partial fractions to write

$$\frac{20}{(s^2+4s+9)(s^2+1)} = \frac{s+2}{s^2+4s+9} - \frac{s-2}{s^2+1}.$$

We have already seen that

$$\mathcal{L}^{-1}\left[\frac{s+2}{s^2+4s+9}\right] = e^{-2t}\cos\sqrt{5}\,t.$$

Moreover,

$$\mathcal{L}^{-1}\left[\frac{s-2}{s^2+1}\right] = \mathcal{L}^{-1}\left[\frac{s}{s^2+1}\right] - 2\mathcal{L}^{-1}\left[\frac{1}{s^2+1}\right] = \cos t - 2\sin t.$$

Therefore, the inverse Laplace transform

$$\mathcal{L}^{-1}\left[e^{-2s}\left(\frac{s+2}{s^2+4s+9} - \frac{s-2}{s^2+1}\right)\right]$$

is

$$u_2(t)\left(e^{-2(t-2)}\cos(\sqrt{5}(t-2)) - \cos(t-2) + 2\sin(t-2)\right).$$

Combining the two results, we obtain the solution of the initial-value problem as

$$y(t) = e^{-2t}\cos\sqrt{5}\,t + \frac{4}{\sqrt{5}}e^{-2t}\sin\sqrt{5}\,t +$$

$$u_2(t)\left(e^{-2(t-2)}\cos(\sqrt{5}\,(t-2)) - \cos(t-2) + 2\sin(t-2)\right).$$

35. **(a)** Consider

$$\mathcal{L}[f] = F(s) = \int_0^\infty f(t)\,e^{-st}\,dt.$$

We can calculate dF/ds by differentiating under the integral sign. That is,

$$\frac{dF}{ds} = \int_0^\infty \frac{\partial}{\partial s}\left(f(t)e^{-st}\right)dt$$

$$= \int_0^\infty f(t)(-t)e^{-st}\,dt$$

$$= -\mathcal{L}[tf(t)].$$

(b) If we apply this result to

$$\mathcal{L}[\sin\omega t] = \frac{\omega}{s^2+\omega^2} = \omega(s^2+\omega^2)^{-1},$$

we obtain

$$\mathcal{L}[t\sin\omega t] = -\omega(-1)(s^2+\omega^2)^{-2}(2s)$$

$$= \frac{2\omega s}{(s^2+\omega^2)^2}.$$

Compare this result with the result of Exercise 6.

EXERCISES FOR SECTION 6.4

1. This is the $\frac{0}{0}$ case of L'Hôpital's Rule. Differentiating numerator and denominator with respect to Δt, we obtain

$$\frac{se^{s\Delta t} - (-s)e^{-s\Delta t}}{2},$$

which simplifies to

$$\frac{s(e^{s\Delta t} + e^{-s\Delta t})}{2}.$$

Since both $e^{s\Delta t}$ and $e^{-s\Delta t}$ tend to 1 as $\Delta t \to 0$, the desired limit is s.

3. Applying the Laplace transform to both sides, using the rules, and the fact that $\mathcal{L}[\delta_3] = e^{-3s}$, we get

$$s^2 \mathcal{L}[y] - sy(0) - y'(0) + 2s\mathcal{L}[y] - 2y(0) + 5\mathcal{L}[y] = e^{-3s}.$$

Substituting the given initial conditions, we have

$$\mathcal{L}[y] = \frac{s+3}{s^2 + 2s + 5} + \frac{e^{-3s}}{s^2 + 2s + 5}.$$

Using the fact that $s^2 + 2s + 5 = (s+1)^2 + 4$, we obtain

$$\mathcal{L}[y] = \frac{s+1}{(s+1)^2 + 4} + \frac{2}{(s+1)^2 + 4} + \frac{1}{2}e^{-3s}\frac{2}{(s+1)^2 + 4}.$$

Therefore,

$$y(t) = e^{-t}\cos 2t + e^{-t}\sin 2t + \frac{1}{2}u_3(t)e^{-(t-3)}\sin(2(t-3)).$$

5. Applying Laplace transform to both sides, using the rules, and the fact that $\mathcal{L}[\delta_a] = e^{-as}$, we get

$$s^2 \mathcal{L}[y] - sy(0) - y'(0) + 2s\mathcal{L}[y] - 2y(0) + 3\mathcal{L}[y] = e^{-s} - 3e^{-4s}.$$

Substituting the initial conditions gives us

$$\mathcal{L}[y] = \frac{e^{-s}}{s^2 + 2s + 3} - \frac{3e^{-4s}}{s^2 + 2s + 3}.$$

Now, using that $s^2 + 2s + 3 = (s+1)^2 + 2$, we have

$$\mathcal{L}[y] = \frac{1}{\sqrt{2}}e^{-s}\frac{\sqrt{2}}{(s+1)^2 + 2} + \frac{3}{\sqrt{2}}e^{-4s}\frac{\sqrt{2}}{(s+1)^2 + 2}.$$

So,

$$y(t) = \frac{1}{\sqrt{2}}u_1(t)e^{-(t-1)}\sin(\sqrt{2}(t-1)) - \frac{3}{\sqrt{2}}u_4(t)e^{-(t-4)}\sin(\sqrt{2}(t-4)).$$

7. **(a)** From the table

$$\mathcal{L}[\delta_a] = e^{-as}$$

$$s\mathcal{L}[u_a] - u_a(0) = s\frac{e^{-as}}{s} - 0 = e^{-as}.$$

(b) The formula for the Laplace transform of a derivative is

$$\mathcal{L}\left[\frac{dy}{dt}\right] = s\mathcal{L}[y] - y(0)$$

and this is exactly the relationship between the Laplace transforms of $u_a(t)$ and $\delta_a(t)$. Hence, it is tempting to think of the Dirac delta function as the derivative of the Heaviside function.

(c) We can think of the Heaviside function $u_a(t)$ as a limit of piecewise linear functions equal to zero for t less than $a - \Delta t$, equal to one for t greater than $a + \Delta t$ and a straight line for t between $a - \Delta t$ and $a + \Delta t$. The derivative of this function is precisely the function $g_{\Delta t}$ used to define the Dirac delta function. This is still just an informal relationship until we specify in what sense we are taking the limit.

9. **(a)** To compute the Laplace transform of the infinite sum on the right-hand side of the equation, we can either sum the geometric series that results from the fact that $\mathcal{L}[\delta_n] = e^{-ns}$ or use Exercise 16 in Section 6.2. Either way, we get

$$\mathcal{L}\left[\sum_{n=1}^{\infty}\delta_n(t)\right] = \frac{e^{-s}}{1 - e^{-s}} = \frac{1}{e^s - 1}.$$

For our purposes, it is actually better to leave the Laplace transform of the right-hand side as

$$\mathcal{L}\left[\sum_{n=1}^{\infty}\delta_n(t)\right] = \sum_{n=1}^{\infty}e^{-ns}.$$

Since $y(0) = 0$ and $y'(0) = 0$, the transformed equation is

$$s^2\mathcal{L}[y] + 2\mathcal{L}[y] = \sum_{n=1}^{\infty}e^{-ns},$$

which simplifies to

$$\mathcal{L}[y] = \frac{1}{s^2 + 2}\sum_{n=1}^{\infty}e^{-ns} = \sum_{n=1}^{\infty}\frac{e^{-ns}}{s^2 + 2}.$$

(b) Since

$$\mathcal{L}^{-1}\left[\frac{e^{-ns}}{s^2 + 2}\right] = \frac{1}{\sqrt{2}}u_n(t)\sin(\sqrt{2}\,(t - n)),$$

we have

$$y(t) = \frac{1}{\sqrt{2}}\sum_{n=1}^{\infty}u_n(t)\sin(\sqrt{2}\,(t - n)).$$

(c) The period of the forcing is different from the natural period of the unforced oscillator. Hence, the solution oscillates but not periodically.

EXERCISES FOR SECTION 6.5

1. Using the definition of the convolution with f and g, we see that

$$(f * g)(t) = \int_0^t 1 \cdot e^{-u} \, du$$

$$= \int_0^t e^{-u} \, du$$

$$= -e^{-u} \Big|_0^t$$

$$= 1 - e^{-t}.$$

Checking the convolution property ($\mathcal{L}[f * g] = \mathcal{L}[f] \cdot \mathcal{L}[g]$) for Laplace transforms, we have

$$\mathcal{L}[f] = \frac{1}{s}, \quad \mathcal{L}[g] = \frac{1}{s+1},$$

and

$$\mathcal{L}[f * g] = \frac{1}{s} - \frac{1}{s+1}$$

$$= \frac{s+1-s}{s(s+1)}$$

$$= \frac{1}{s(s+1)}.$$

So, $\mathcal{L}[f] \cdot \mathcal{L}[g] = \mathcal{L}[f * g]$.

3. Using the definition of the convolution with f and g, we see that

$$(f * g)(t) = \int_0^t \cos(t - v) u_2(v) \, dv.$$

(We're using v as the integrating variable instead of u so as not to confuse it with the Heaviside function.) First, notice that if $0 < v < 2$, then $u_2(v) = 0$. Thus, if $t < 2$, then the function $u_2(v)$ is always 0, which means the integral is 0. Now, if $t \geq 2$, then

$$\int_0^t \cos(t - v) u_2(v) \, dv = \int_0^2 \cos(t - v) \, 0 \, dv + \int_2^t \cos(t - v) \, 1 \, dv$$

$$= \int_2^t \cos(t - v) \, dv.$$

So,

$$\int_0^t \cos(t - v) u_2(v) \, dv = \begin{cases} 0, & \text{if } t < 2, \\ \int_2^t \cos(t - v) \, dv, & \text{if } t \geq 2. \end{cases}$$

Evaluating the second integral, we get

$$\int_2^t \cos(t - v)\, dv = -\sin(t - v)\Big|_2^t = \sin(t - 2).$$

We have a function that is 0 for $t < 2$ and equal to $\sin(t - 2)$ for $t \geq 2$, so our function is

$$u_2(t) \sin(t - 2).$$

Checking the convolution property ($\mathcal{L}[f * g] = \mathcal{L}[f] \cdot \mathcal{L}[g]$) for Laplace transforms, we have

$$\mathcal{L}[f] = \frac{s}{s^2 + 1}, \quad \mathcal{L}[g] = \frac{e^{-2s}}{s},$$

and

$$\mathcal{L}[f * g] = \frac{e^{-2s}}{s^2 + 1}.$$

Hence, $\mathcal{L}[f] \cdot \mathcal{L}[g] = \mathcal{L}[f * g]$.

5. Using the definition of the convolution with f and g, we see that

$$(f * g)(t) = \int_0^t 3 \sin(t - u) \cos(2u)\, du.$$

We will use four trigonometric identities to evaluate this integral:

$$\sin(t - u) = \sin t \cos u - \cos t \sin u$$

$$\sin(mt) \sin(nt) = \tfrac{1}{2}\left[\cos((m - n)t) - \cos((m + n)t)\right]$$

$$\cos(mt) \cos(nt) = \tfrac{1}{2}\left[\cos((m + n)t) + \cos((m - n)t)\right]$$

$$\sin(mt) \cos(nt) = \tfrac{1}{2}\left[\sin((m + n)t) + \sin((m - n)t)\right].$$

So

$$\int_0^t 3 \sin(t - u) \cos(2u)\, du$$

$$= \int_0^t [3 \cos 2u \cos u \sin t - \cos 2u \sin u \cos t]\, du$$

$$= \int_0^t \left[\tfrac{3}{2}(\cos 3u + \cos u) \sin t - \tfrac{3}{2}(\sin 3u - \sin u) \cos t\right] du$$

$$= \sin t \left[\tfrac{1}{2} \sin 3u + \tfrac{3}{2} \sin u\right]_0^t + \cos t \left[\tfrac{1}{2} \cos 3u - \tfrac{3}{2} \cos u\right]_0^t$$

$$= \sin t \left(\tfrac{1}{2} \sin 3t + \tfrac{3}{2} \sin t\right) + \cos t \left(\tfrac{1}{2} \cos 3t - \tfrac{3}{2} \cos t + 1\right)$$

$$= \tfrac{1}{2} \sin 3t \sin t + \tfrac{3}{2} \sin^2 t + \tfrac{1}{2} \cos 3t \cos t - \tfrac{3}{2} \cos^2 t + \cos t$$

$$= \tfrac{1}{4}\left(\cos 2t - \cos 4t\right) + \tfrac{3}{2}\sin^2 t + \tfrac{1}{4}\left(\cos 4t + \cos 2t\right) - \tfrac{3}{2}\cos^2 t + \cos t$$

$$= \tfrac{1}{2}\cos 2t + \tfrac{3}{2}\left(\sin^2 t - \cos^2 t\right) + \cos t$$

$$= \tfrac{1}{2}\cos 2t - \tfrac{3}{2}\cos 2t + \cos t$$

$$= \cos t - \cos 2t,$$

which is the same answer obtained in the text using the technique of Laplace transforms.

7. Taking Laplace transform of both sides of the equation and solving for $\mathcal{L}[\zeta]$ (see page 613), we obtain

$$\mathcal{L}[\zeta] = \frac{1}{s^2 + ps + q}.$$

Hence, if we let

$$z(s) = s^2 + ps + q,$$

we have that $z(0) = 5$ and $z(2) = 17$. Now $z(0) = 5$ implies $q = 5$. Using $z(2) = 17 = 2^2 + 2p + 5$, we see that $p = 4$.

9. (a) Since ζ solves the initial-value problem above, we know that

$$\frac{d\zeta^2}{dt^2} + p\frac{d\zeta}{dt} + q\zeta = \delta_0(t), \quad \zeta(0) = \zeta'(0) = 0^-.$$

Taking Laplace transforms of both sides, and substituting initial conditions gives us

$$s^2\mathcal{L}[\zeta] + ps\mathcal{L}[\zeta] + q\mathcal{L}[\zeta] = 1,$$

which yields

$$\mathcal{L}[\zeta] = \frac{1}{s^2 + ps + q}.$$

Now, taking Laplace transforms of both sides of

$$\frac{d^2 y}{dt^2} + p\frac{dy}{dt} + qy = 0, \quad y(0) = a, \, y'(0) = 0$$

gives us

$$s^2\mathcal{L}[y] - sa + ps\mathcal{L}[y] - pa + q\mathcal{L}[y] = 0.$$

Solving for $\mathcal{L}[y]$ gives

$$\mathcal{L}[y] = \frac{a(s + p)}{s^2 + ps + q},$$

so

$$\mathcal{L}[y] = a(s + p)\mathcal{L}[\zeta].$$

(b) Taking Laplace transforms of both sides of

$$\frac{d^2 y}{dt^2} + p\frac{dy}{dt} + qy = 0, \quad y(0) = 0, \, y'(0) = b$$

gives us

$$s^2 \mathcal{L}[y] - b + ps\mathcal{L}[y] + q\mathcal{L}[y] = 0.$$

Solving for $\mathcal{L}[y]$ gives

$$\mathcal{L}[y] = \frac{b}{s^2 + ps + q},$$

so

$$\mathcal{L}[y] = b\mathcal{L}[\zeta].$$

(c) Taking Laplace transforms of both sides of

$$\frac{d^2 y}{dt^2} + p\frac{dy}{dt} + qy = f(t), \quad y(0) = a, y'(0) = b$$

gives us

$$s^2 \mathcal{L}[y] - sa - b + ps\mathcal{L}[y] - pa + q\mathcal{L}[y] = \mathcal{L}[f].$$

Solving for $\mathcal{L}[y]$ gives

$$\mathcal{L}[y] = \frac{\mathcal{L}[f] + a(s + p) + b}{s^2 + ps + q},$$

so

$$\mathcal{L}[y] = (\mathcal{L}[f] + a(s + p) + b)\, \mathcal{L}[\zeta].$$

11. (a) We know that

$$\frac{d^2 y_1}{dt^2} + p\frac{dy_1}{dt} + qy_1 = f_1(t).$$

Taking Laplace transforms of both sides and solving for $\mathcal{L}[y_1]$ gives

$$\mathcal{L}[y_1] = \frac{\mathcal{L}[f_1]}{s^2 + ps + q}.$$

(b) As above, we know that

$$\frac{d^2 y_2}{dt^2} + p\frac{dy_2}{dt} + qy_2 = f_2(t),$$

so

$$\mathcal{L}[y_2] = \frac{\mathcal{L}[f_2]}{s^2 + ps + q}.$$

Now we see that

$$\frac{\mathcal{L}[f_1]}{\mathcal{L}[y_1]} = s^2 + ps + q,$$

and

$$\frac{\mathcal{L}[f_2]}{\mathcal{L}[y_2]} = s^2 + ps + q,$$

so

$$\frac{\mathcal{L}[f_1]}{\mathcal{L}[y_1]} = \frac{\mathcal{L}[f_2]}{\mathcal{L}[y_2]}.$$

(c) Solving for $\mathcal{L}[y_2]$ gives us

$$\mathcal{L}[y_2] = \mathcal{L}[f_2]\frac{\mathcal{L}[y_1]}{\mathcal{L}[f_1]}.$$

EXERCISES FOR SECTION 6.6

1. **(a)** Taking Laplace transform of both sides and substituting the initial conditions yields

$$s^2 \mathcal{L}[y] - 2s + 2 + 2s\mathcal{L}[y] - 4 + 2\mathcal{L}[y] = \frac{4}{(s+2)^2 + 16}.$$

Hence,

$$\mathcal{L}[y] = \frac{2s+2}{s^2 + 2s + 2} + \frac{4}{(s^2 + 4s + 20)(s^2 + 2s + 2)}.$$

(b) The poles are the roots of $s^2 + 2s + 2$ and $s^2 + 4s + 20$, or $-1 \pm i$ and $-2 \pm 4i$.

(c) Since all poles have negative real parts, the solution tends to zero at an exponential rate. The real part closest to 0 is -1 so solutions tend to zero at a rate of e^{-t}. Since the poles are complex, the solutions oscillate. The oscillations with period $2\pi/4 = \pi/2$ decay at the rate e^{-2t} while the oscillations with period 2π decay at the rate e^{-t}.

3. **(a)** Taking the Laplace transform of both sides and plugging in the initial conditions gives

$$s^2 \mathcal{L}[y] + s\mathcal{L}[y] + 8\mathcal{L}[y] = \mathcal{L}[\cos(t-4)] - \mathcal{L}[u_4(t)\cos(t-4)].$$

To take the Laplace transform of the first term on the right we recall that

$$\cos(t-4) = \cos(4)\cos(t) + \sin(4)\sin(t),$$

so

$$(s^2 + s + 8)\mathcal{L}[y] = \cos(4)\frac{s}{s^2 + 1} + \sin(4)\frac{1}{s^2 + 1} - \frac{se^{-4s}}{s^2 + 1}.$$

Therefore,

$$\mathcal{L}[y] = \frac{\cos(4)s + \sin(4) - se^{-4s}}{(s^2 + s + 8)(s^2 + 1)}.$$

(b) The poles are given by $s^2 + 1 = 0$ and $s^2 + s + 8 = 0$ or $s = \pm i$ and $s = -1/2 \pm i\sqrt{31}/2$.

(c) Note that the forcing "turns off" at time $t = 4$. Hence up until time $t = 4$ there is a forced response with period 2π. Since $t = 4$ is only about half a period for the forced response, this is not very significant. The natural response is an oscillation with period $4\pi/\sqrt{31}$ which decays like $e^{-t/2}$, and this is the long-term behavior of the solution.

5. **(a)** Computing the Laplace transform of both sides of the equation and using the initial conditions gives

$$s^2 \mathcal{L}[y] - s - 1 + 16\mathcal{L}[y] = 0,$$

so

$$\mathcal{L}[y] = \frac{s+1}{s^2 + 16}.$$

(b) The poles are the solutions of $s^2 + 16 = 0$ or $s = \pm 4i$.

(c) If we find the inverse Laplace transform, we see that

$$y(t) = \cos(4t) + (1/4)\sin(4t).$$

So a reasonable conjecture would be that when the poles of the Laplace transform are on the imaginary axis, the solution is periodic and does not decay.

7. **(a)** As usual, we take the Laplace transform of both sides of the equation and substitute the initial conditions to get

$$s^2 \mathcal{L}[y] - s - 2 + 2s\mathcal{L}[y] - 2 + \mathcal{L}[y] = 0.$$

Hence,

$$\mathcal{L}[y] = \frac{s+4}{s^2 + 2s + 1}.$$

(b) The poles are the roots of $s^2 + 2s + 1 = (s+1)^2$. Hence, there is a "double pole" at $s = -1$.

(c) For homogeneous second-order equations, double poles play the same role as double eigenvalues. In the case of a double pole on the real axis, the solution is a critically damped oscillator.

9. **(a)** As usual, we compute

$$s^2\mathcal{L}[y] - sy(0) - y'(0) + 20s\mathcal{L}[y] - 20y(0) + 200\mathcal{L}[y] = \mathcal{L}[w(t)]$$

and using the initial conditions and Exercise 17 of Section 6.2, we have that

$$(s^2 + 20s + 200)\mathcal{L}[y] - s - 20 = \frac{1 - e^{-s}}{s(1 + e^{-s})}.$$

Hence,

$$\mathcal{L}[y] = \frac{s+20}{s^2 + 20s + 200} + \frac{1 - e^{-s}}{s(1 + e^{-s})(s^2 + 20s + 200)}.$$

(b) The poles are the roots of $s^2 + 20s + 200 = 0$ ($s = -10 \pm 10i$) and the zeros of $s(1 + e^{-s}) = 0$. One zero is $s = 0$, and using Euler's formula, we obtain the other zeros $s = (2n+1)i\pi$ for $n = 0, \pm 1, \pm 2, \ldots$.

(c) The natural response corresponds to the poles $s = -10 \pm 10i$, so it decays like e^{-10t} (very rapidly), while oscillating with a period of $\pi/5$. The remainder of the poles correspond to the forcing and indicate a forced response, which is an oscillation between $\pm 1/200$. Because the forcing is discontinuous, solutions settle (quickly) toward $1/200$ for $0 < t < 1$. When the forcing switches, they tend (quickly) to $-1/200$. (The function $y(t) = 1/200$ is a particular solution of the equation

$$\frac{d^2 y}{dt^2} + 20\frac{dy}{dt} + 200y = 1,$$

and $y(t) = -1/200$ is a solution for the same equation if the forcing is -1 rather than 1.)

REVIEW EXERCISES FOR CHAPTER 6

1. Note that

$$\frac{4}{s^2 - 1} = \frac{2}{s-1} - \frac{2}{s+1}.$$

Since

$$\mathcal{L}^{-1}\left[\frac{1}{s-1}\right] = e^t \quad \text{and} \quad \mathcal{L}^{-1}\left[\frac{1}{s+1}\right] = e^{-t},$$

we have

$$\mathcal{L}^{-1}\left[\frac{4}{s^2 - 1}\right] = 2(e^t - e^{-t}).$$

3. We know that the general solution to this differential equation is $y(t) = \alpha \cos \sqrt{5}\,t + \beta \sin \sqrt{5}\,t$. Therefore, the solution to the given initial-value problem is $y(t) = \cos \sqrt{5}\,t$. Its Laplace transform is

$$\frac{s}{s^2 + 5}.$$

5. Note that $y(t) = 1 - u_6(t)$. Therefore,

$$\mathcal{L}[y(t)] = \mathcal{L}[1 - u_6(t)] = \mathcal{L}[1] - \mathcal{L}[u_6(t)] = \frac{1}{s} - \frac{e^{-6s}}{s} = \frac{1 - e^{-6s}}{s}.$$

7. This improper integral is the Laplace transform of $\delta_2(t)$ evaluated at $s = 2$. Since $\mathcal{L}[\delta_2(t)] = e^{-2s}$,

$$\int_0^\infty \delta_2(t) e^{-2t}\, dt = e^{-4}.$$

9. We can use the formula in Exercise 16 from Section 6.2. However, the result of Review Exercise 8 can be generalized to yield the fact that $u_n(t)$ is the solution to the initial-value problem

$$\frac{dy}{dt} = \delta_n(t), \quad y(0) = 0.$$

If we do not worry about the convergence of any of the infinite sums involved, we can use the linearity of the differential equation to produce the solution

$$y(t) = u_1(t) + u_2(t) + u_3(t) + \dots.$$

(See the result of Exercise 19 in Section 1.8).

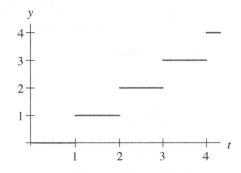

11.   **(a)** The transform $\mathcal{L}[\sin 2t] = 2/(s^2 + 4)$ is second nature by now. Hence, this transform is transform (iii).

   (b) We know that

$$\mathcal{L}[e^{2t} \cos 2t] = \frac{s - 2}{(s - 2)^2 + 4} = \frac{s - 2}{s^2 - 4s + 8}.$$

Hence, this transform is transform (xi).

(c) We have

$$\mathcal{L}[e^{2t} - e^{-2t}] = \frac{1}{s-2} - \frac{1}{s+2} = \frac{4}{s^2 - 4}.$$

This transform is transform (i).

(d) We know that $\sin(t - \pi) = -\sin t$. Therefore,

$$\mathcal{L}[\sin(t - \pi)] = \mathcal{L}[-\sin t] = \frac{-1}{s^2 + 1},$$

and this transform is transform (vi).

(e) The transform $\mathcal{L}[\cos 2t] = s/(s^2 + 4)$ is also second nature. Hence, this transform is transform (vii).

(f) We have

$$\mathcal{L}[u_2(t)e^{3(t-2)}] = e^{-2s}\mathcal{L}[e^{3t}] = \frac{e^{-2s}}{s-3}.$$

Hence, this transform is transform (ix).

13. True. The Laplace transform of $\delta_1(t)$ is e^{-s}, which is an infinitely differentiable function in s. (The Laplace transform is a "smoothing" operator. The transformed function $Y(s) = \mathcal{L}[y(t)]$ tends to be more differentiable than the original function $y(t)$.)

15. True. We use the linearity of the equation and solve it for the forcing functions $\delta_\pi(t), \delta_{2\pi}(t), \delta_{4\pi}(t)$, ... individually. For example, for the initial-value problem

$$\frac{d^2 y}{dt^2} + 4y = \delta_\pi(t), \quad y(0) = 0, \quad y'(0) = 0,$$

we get

$$\mathcal{L}[y] = \frac{e^{-\pi s}}{s^2 + 4},$$

and therefore, $y(t) = u_\pi(t) \sin 2(t - \pi)$.

Next we solve

$$\frac{d^2 y}{dt^2} + 4y = \delta_{2\pi}(t), \quad y(0) = 0, \quad y'(0) = 0,$$

and we obtain $y(t) = u_{2\pi}(t) \sin 2(t - 2\pi)$. Continuing in this manner, we obtain

$$y(t) = u_\pi(t) \sin 2(t - \pi) + u_{2\pi}(t) \sin 2(t - 2\pi) + u_{4\pi}(t) \sin 2(t - 4\pi) + \ldots$$
$$= u_\pi(t) \sin 2t + u_{2\pi}(t) \sin 2t + u_{4\pi}(t) \sin 2t + \ldots$$
$$= (u_\pi(t) + u_{2\pi}(t) + u_{4\pi}(t) + \ldots) \sin 2t,$$

as the solution to the original initial-value problem. This solution oscillates with increasing and unbounded amplitude.

17. False. Since $s^4 + 6s^2 + 9 = (s^2 + 3)^2$, the fraction

$$\frac{3s + 5}{s^4 + 6s^2 + 9}$$

is already in its partial fractions form. Furthermore, the denominator of $(s^2 + 3)^2$ indicates that the inverse Laplace transform contains terms that involve $t \sin \sqrt{3}\, t$ and $t \cos \sqrt{3}\, t$ (see Exercise 35 in Section 6.3). These functions oscillate with amplitudes that grow linearly. If t is large enough, $|y(t)| > 15$.

19. First we note that $s^2 + 2s - 8 = (s - 2)(s + 4)$. Hence, the partial fractions decomposition is

$$\frac{s + 16}{s^2 + 2s - 8} = \frac{A}{s - 2} + \frac{B}{s + 4} = \frac{A(s + 4) + B(s - 2)}{s^2 + 2s + 8}.$$

We solve the equations

$$\begin{cases} A + B = 1 \\ 4A - 2B = 16 \end{cases}$$

and obtain $A = 3$ and $B = -2$. Therefore,

$$\frac{s + 16}{s^2 + 2s - 8} = \frac{3}{s - 2} - \frac{2}{s + 4}.$$

The inverse Laplace transform is

$$\mathcal{L}^{-1}\left[\frac{s + 16}{s^2 + 2s - 8}\right] = 3\mathcal{L}^{-1}\left[\frac{1}{s - 2}\right] - 2\mathcal{L}^{-1}\left[\frac{1}{s + 4}\right] = 3e^{2t} - 2e^{-4t}.$$

21. First we compute the partial fractions decomposition of

$$\frac{5s - 12}{s^2 - 5s + 6}.$$

We note that $s^2 - 5s + 6 = (s - 2)(s - 3)$. Hence, the partial fractions decomposition is

$$\frac{5s - 12}{s^2 - 5s + 6} = \frac{A}{s - 2} + \frac{B}{s - 3} = \frac{A(s - 3) + B(s - 2)}{s^2 - 5s + 6}.$$

We solve the equations

$$\begin{cases} A + B = 5 \\ 3A + 2B = 12 \end{cases}$$

and obtain $A = 2$ and $B = 3$. Therefore,

$$\frac{5s - 12}{s^2 - 5s + 6} = \frac{2}{s - 2} + \frac{3}{s - 3}.$$

We conclude that

$$\mathcal{L}^{-1}\left[\frac{5s - 12}{s^2 - 5s + 6}\right] = 2e^{2t} + 3e^{3t},$$

and consequently,

$$\mathcal{L}^{-1}\left[\frac{(5s - 12)e^{-3s}}{s^2 - 5s + 6}\right] = u_3(t)\left(2e^{2(t-3)} + 3e^{3(t-3)}\right).$$

23. Note that $s^2 - 4s + 4 = (s-2)^2$, so $s = 2$ is a repeated root of the denominator. The fraction

$$\frac{1}{s^2 - 4s + 4}$$

cannot be further decomposed, and we must use the result of Exercise 35 in Section 6.3.
We know that

$$\mathcal{L}\left[e^{2t}\right] = \frac{1}{s-2},$$

so

$$\mathcal{L}\left[te^{2t}\right] = -\frac{d}{ds}\left(\frac{1}{s-2}\right) = \frac{1}{(s-2)^2}.$$

Consequently,

$$\mathcal{L}^{-1}\left[\frac{1}{(s-2)^2}\right] = te^{2t}.$$

25. **(a)** The characteristic polynomial of the unforced equation is

$$s^2 - 4s + 4,$$

which has $s = 2$ as a double root. Therefore, the natural guesses of $y_p(t) = ke^{2t}$ and $y_p(t) = kte^{2t}$ fail to be solutions of the forced equation because they are both solutions of the unforced equation.

So we guess $y_p(t) = kt^2 e^{2t}$. Substituting this guess into the left-hand side of the differential equation gives

$$\frac{d^2 y_p}{dt^2} - 4\frac{dy_p}{dt} + 4y_p = (2ke^{2t} + 8kte^{2t} + 4kt^2 e^{2t}) - 4(2kte^{2t} + 2kt^2 e^{2t}) + 4kt^2 e^{2t}$$

$$= 2ke^{2t}.$$

So $k = 1/2$ yields the solution

$$y_p(t) = \tfrac{1}{2}t^2 e^{2t}.$$

From the characteristic polynomial, we know that the general solution of the unforced equation is

$$k_1 e^{2t} + k_2 te^{2t}.$$

Consequently, the general solution of the forced equation is

$$y(t) = k_1 e^{2t} + k_2 te^{2t} + \tfrac{1}{2}t^2 e^{2t}.$$

To solve the initial-value problem, we note that $y(0) = 1$ implies that $k_1 = 1$. We find k_2 by differentiation. We get

$$y'(t) = 2e^{2t} + k_2(1 + 2t)e^{2t} + (t^2 + t)e^{2t}.$$

Since $y'(0) = 5$, we have $k_2 = 3$ and

$$y(t) = (1 + 3t + \tfrac{1}{2}t^2)e^{2t}.$$

(b) Transforming both sides of the differential equation, we have

$$s^2 \mathcal{L}[y] - s - 5 - 4(s\mathcal{L}[y] - 1) + 4\mathcal{L}[y] = \frac{1}{s-2}$$

$$(s^2 - 4s + 4)\mathcal{L}[y] - s - 1 = \frac{1}{s-2}$$

$$(s-2)^2 \mathcal{L}[y] = s + 1 + \frac{1}{s-2}.$$

So

$$\mathcal{L}[y] = \frac{s+1}{(s-2)^2} + \frac{1}{(s-2)^3}.$$

Note that

$$\frac{s+1}{(s-2)^2} = \frac{s-2}{(s-2)^2} + \frac{3}{(s-2)^2} = \frac{1}{s-2} + \frac{3}{(s-2)^2}.$$

Writing $\mathcal{L}[y]$ in terms of its partial fractions decomposition, we have

$$\frac{1}{s-2} + \frac{3}{(s-2)^2} + \frac{1}{(s-2)^3}.$$

The inverse Laplace transform of the first term is routine. The inverse transform of the second term can be deduced from the result of Exercise 23, and the inverse transform of the third term can also be obtained using the method of solution for Exercise 23. That is, we use the result of Exercise 35 in Section 6.3. In particular, we know that

$$\mathcal{L}[te^{2t}] = \frac{1}{(s-2)^2}$$

from Exercise 23. Consequently,

$$\mathcal{L}[t^2 e^{2t}] = -\frac{d}{ds}\left(\frac{1}{(s-2)^2}\right) = \frac{2}{(s-2)^3}.$$

Therefore,

$$\mathcal{L}^{-1}\left[\frac{1}{(s-2)^3}\right] = \tfrac{1}{2}t^2 e^{2t},$$

and the solution to the initial-value problem is

$$y(t) = (1 + 3t + \tfrac{1}{2}t^2)e^{2t}.$$

(c) The solutions in parts (a) and (b) were more or less equally complicated. One advantage of the solution in part (a) is that we obtained the general solution to the equation as part of the process.

27. We transform both sides of the differential equation and obtain

$$\mathcal{L}\left[\frac{dy}{dt}\right] - 4\mathcal{L}[y] = \mathcal{L}[50u_2(t)\sin(3(t-2))]$$

$$s\mathcal{L}[y] - y(0) - 4\mathcal{L}[y] = 50e^{-2s}\left(\frac{3}{s^2+9}\right)$$

$$(s-4)\mathcal{L}[y] - 5 = \frac{150e^{-2s}}{s^2+9}.$$

Solving for $\mathcal{L}[y]$, we get

$$\mathcal{L}[y] = \frac{5}{s-4} + \frac{150e^{-2s}}{(s^2+9)(s-4)}.$$

To find the inverse Laplace transform, we first note that

$$\mathcal{L}^{-1}\left[\frac{5}{s-4}\right] = 5e^{4t}.$$

To invert the second term, we need the partial fractions decomposition of

$$\frac{150}{(s^2+9)(s-4)}.$$

We set

$$\frac{150}{(s^2+9)(s-4)} = \frac{As+B}{s^2+9} + \frac{C}{s-4} = \frac{(As+B)(s-4)+C(s^2+9)}{(s^2+9)(s-4)},$$

and we get $A = -6$, $B = -24$, and $C = 6$. Therefore,

$$\mathcal{L}^{-1}\left[\frac{150}{(s^2+9)(s-4)}\right] = \mathcal{L}^{-1}\left[\frac{-6s}{s^2+9}\right] + \mathcal{L}^{-1}\left[\frac{-24}{s^2+9}\right] + \mathcal{L}^{-1}\left[\frac{6}{s-4}\right]$$

$$= -6\cos 3t - 8\sin 3t + 6e^{4t}.$$

Hence,

$$\mathcal{L}^{-1}\left[\frac{150e^{-2s}}{(s^2+9)(s-4)}\right] = u_2(t)\left(-6\cos(3(t-2)) - 8\sin(3(t-2)) + 6e^{4(t-2)}\right).$$

Combining the two results, we obtain the solution of the initial-value problem

$$y(t) = 5e^{4t} + u_2(t)\left(6e^{4(t-2)} - 6\cos(3(t-2)) - 8\sin(3(t-2))\right)$$

29. We transform both sides of the differential equation and obtain

$$s^2\mathcal{L}[y] - 2s - 1 + 2(s\mathcal{L}[y] - 2) + 3\mathcal{L}[y] = e^{-4s}.$$

Solving for $\mathcal{L}[y]$, we get

$$(s^2 + 2s + 3)\mathcal{L}[y] = 2s + 5 + e^{-4s}$$

$$\mathcal{L}[y] = \frac{2s+5}{s^2+2s+3} + \frac{e^{-4s}}{s^2+2s+3}.$$

To compute the inverse Laplace transform, we first note that $s^2 + 2s + 3 = (s+1)^2 + 2$. To invert the first term, we write it as

$$\frac{2s+5}{s^2+2s+3} = \frac{2(s+1)}{(s+1)^2+2} + \frac{3}{(s+1)^2+2}.$$

We see that

$$\mathcal{L}^{-1}\left[\frac{2s+5}{s^2+2s+3}\right] = 2e^{-t}\cos\sqrt{2}\,t + \frac{3}{\sqrt{2}}e^{-t}\sin\sqrt{2}\,t.$$

To invert the second term, we compute

$$\mathcal{L}^{-1}\left[\frac{1}{s^2+2s+3}\right] = \frac{1}{\sqrt{2}}e^{-t}\sin\sqrt{2}\,t.$$

Therefore,

$$\mathcal{L}^{-1}\left[\frac{e^{-4s}}{s^2+2s+3}\right] = \frac{u_4(t)}{\sqrt{2}}e^{-(t-4)}\sin\sqrt{2}\,(t-4).$$

Combining both results, we get the solution to the original initial-value problem

$$y(t) = 2e^{-t}\cos\sqrt{2}\,t + \frac{3}{\sqrt{2}}e^{-t}\sin\sqrt{2}\,t + \frac{u_4(t)}{\sqrt{2}}e^{-(t-4)}\sin\sqrt{2}\,(t-4).$$

CHAPTER 7

Numerical Methods

EXERCISES FOR SECTION 7.1

1. (a) The differential equation is separable. Therefore, one way to obtain the solution to the initial-value problem is to integrate

    $$\int y^{-2}\,dy = \int -2t\,dt.$$

    We obtain

    $$\frac{y^{-1}}{-1} = -t^2 + c$$

    $$y^{-1} = t^2 + k$$

    $$y = \frac{1}{t^2 + k}.$$

    We determine the value of k using the initial condition $y(0) = 1$. Hence, $k = 1$, and the solution to the given initial-value problem is $y(t) = 1/(t^2 + 1)$.

    (b) To calculate y_{20}, we must apply Euler's method 20 times. Table 7.1 contains the results of a number of intermediate calculations.

    (c) The total error e_{20} is the difference between the actual value $y(2) = 0.2$ and the approximate value $y_{20} = 0.193342$. Therefore, $e_{20} = 0.0066581$.

    (d) Table 7.2 contains the results of Euler's method and the corresponding total errors for $n = 1000$, $2000, \ldots, 6000$.

Table 7.1
Results of Euler's method

k	t_k	y_k	k	t_k	y_k
0	0	1.0	10	1.0	0.503642
1	0.1	1.0	$\vdots$	$\vdots$	$\vdots$
2	0.2	0.98	19	1.9	0.210119
3	0.3	0.941584	20	2.0	0.193342
4	0.4	0.888389			
5	0.5	0.82525			

Table 7.2
Results of Euler's method and the corresponding total errors

n	y_n	e_n
1000	0.199874	0.000125848
2000	0.199937	0.0000628901
3000	0.199958	0.0000419192
4000	0.199969	0.0000314366
5000	0.199975	0.0000251479
6000	0.199979	0.0000209558

(e) Table 7.3 gives values of e_n for some n that are intermediate to the ones above in case that you want to double check the ones you have computed. Also, the graph of e_n as a function of n for $100 \leq n \leq 6000$ is given.

Table 7.3
Selected total errors

n	e_n
100	0.00127095
200	0.000631985
1400	0.0000898639
3700	0.0000339862
5600	0.000022453

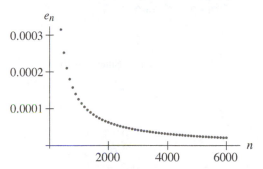

(f) Our computer math system fits the data to the function $0.126731/n$. The following figure includes both the data and the graph of this function.

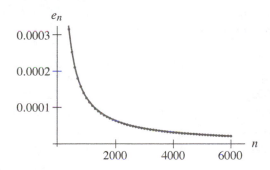

3. (a) The differential equation is both separable and linear. Therefore, one way to obtain the solution to the initial-value problem is to integrate

$$\int \frac{1}{y}\,dy = \int t\,dt.$$

We obtain

$$\ln|y| = \frac{t^2}{2} + c$$

$$y = ke^{t^2/2}.$$

We determine the value of k using the initial condition $y(0) = 1$. Hence, $k = 1$, and the solution to the given initial-value problem is $y(t) = e^{t^2/2}$.

(b) To calculate y_{20}, we must apply Euler's method 20 times. Table 7.4 contains the results of a number of intermediate calculations.

(c) The total error e_{20} is the difference between the actual value $y(\sqrt{2}) = e$ and the approximate value $y_{20} = 2.51066$. Therefore, $e_{20} = 0.20762$.

(d) Table 7.5 contains the results of Euler's method and the corresponding total errors for $n = 1000$, $2000, \ldots, 6000$.

Table 7.4

Results of Euler's method

k	t_k	y_k	k	t_k	y_k
0	0	1	10	0.707107	1.24797
1	0.0707107	1	$\vdots$	$\vdots$	$\vdots$
2	0.141421	1.005	19	1.3435	2.29284
3	0.212132	1.01505	20	1.41421	2.51066
4	0.282843	1.03028			
5	0.353553	1.05088			

Table 7.5

Results of Euler's method and the corresponding total errors

n	y_n	e_n
1000	2.71376	0.00452218
2000	2.71602	0.00226316
3000	2.71677	0.00150923
4000	2.71715	0.0011321
5000	2.71738	0.000905762
6000	2.71753	0.000754848

(e) Table 7.6 gives values of e_n for some n that are intermediate to the ones above in case that you want to double check the ones you have computed. Also, the graph of e_n as a function of n for $100 \leq n \leq 6000$ is given.

Table 7.6

Selected total errors

n	e_n
100	0.0444901
200	0.0224467
1400	0.00323182
3700	0.00122384
5600	0.000808748

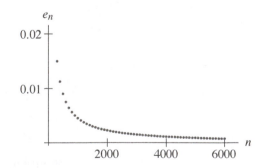

(f) Our computer math system fits the data to the function $4.47023/n$. The following figure includes both the data and the graph of this function.

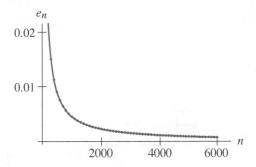

5. **(a)** The differential equation is linear, and we can use a guessing technique to find an analytic solution (see Section 1.8). First, we rewrite the equation as

$$\frac{dy}{dt} - 3y = 1 - t.$$

This form of the equation suggests the guess $y_p(t) = at + b$. Substituting this guess into the differential equation yields

$$a - 3(at + b) = 1 - t.$$

We obtain the equations $-3a = -1$ and $a - 3b = 1$. Then $y_p(t)$ is a solution if and only if $a = 1/3$ and $b = -2/9$.

The general solution of the associated homogeneous equation is ce^{3t}. Therefore, the general solution to this linear equation is

$$y(t) = \tfrac{1}{3}t - \tfrac{2}{9} + ce^{3t}.$$

We determine the value of c using the initial condition $y(0) = 1$. We get $c = 11/9$, and the solution to the given initial-value problem is

$$y(t) = \tfrac{11}{9}e^{3t} + \tfrac{1}{3}t - \tfrac{2}{9}.$$

(b) To calculate y_{20}, we must apply Euler's method 20 times. Table 7.7 contains the results of a number of intermediate calculations.

(c) The total error e_{20} is the difference between the actual value $y(1) = (11e^3 + 1)/9$ and the approximate value $y_{20} = 20.1147$. Therefore, $e_{20} = 4.54544$.

(d) Table 7.8 contains the results of Euler's method and the corresponding total errors for $n = 1000$, $2000, \ldots, 6000$.

Table 7.7
Results of Euler's method

k	t_k	y_k	k	t_k	y_k
0	0	1.	10	0.5	4.88902
1	0.05	1.2	⋮	⋮	⋮
2	0.1	1.4275	19	0.95	17.4888
3	0.15	1.68663	20	1.	20.1147
4	0.2	1.98212			
5	0.25	2.31944			

Table 7.8
Results of Euler's method and the corresponding total errors

n	y_n	e_n
1000	24.5501	0.110003
2000	24.605	0.0551181
3000	24.6233	0.0367714
4000	24.6325	0.0275883
5000	24.638	0.0220753
6000	24.6417	0.0183987

(e) Table 7.3 gives values of e_n for some n that are intermediate to the ones above in case that you want to double check the ones you have computed. Also, the graph of e_n as a function of n for $100 \le n \le 6000$ is given.

Table 7.9
Selected total errors

n	e_n
100	1.05955
200	0.540843
1400	0.0786686
3700	0.0298226
5600	0.0197119

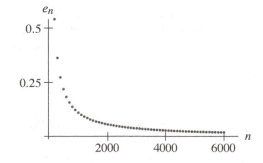

(f) Our computer math system fits the data to the function $107.125/n$. The following figure includes both the data and the graph of this function.

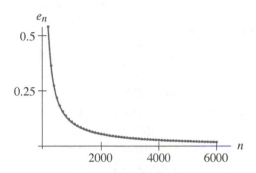

7. **(a)** The partial derivative $\partial f/\partial y$ of $f(t, y) = -2ty^2$ is $-4ty$. Thus, $M_2 = -4ty$.
The partial derivative $\partial f/\partial t$ of $f(t, y) = -2ty^2$ is $-2y^2$. Therefore,

$$M_1 = \frac{\partial f}{\partial t} + \frac{\partial f}{\partial y} f(t, y) = -2y^2 + (-4ty)(-2ty^2) = -2y^2 + 8t^2y^3.$$

(b) Since $f(t_0, y_0) = 0$, the result of the first step of Euler's method is the point $(t_1, y_1) = (0.02, 1)$. Consequently, to estimate the error, we compute the quantity M_1 at the point $(0.01, 1)$. We obtain $M_1 \approx -1.9992$.

Once we have M_1 at this point, we can estimate the error e_1 (which is the same as the truncation error) by computing $|M_1(\Delta t)^2/2|$. In this case, we obtain 0.00039984.

To calculate the actual error, we can compare the result y_1 of Euler's method with the value of the solution $y(0.2)$. Since we know that the solution is $1/(1+t^2)$, we have $y(0.2) = 0.9996$, and thus the actual error is 0.00039984. For this computation, note that the estimated error and the actual error essentially agree. (In order to see a difference in these two quantities, we had to do the calculations to 11 decimal places.)

(c) The second point (t_2, y_2) obtained from Euler's method is the point $(0.04, 0.9992)$. For the second step, the estimated error is no longer simply the truncation error, so we must compute both M_1 and M_2. Evaluating M_1 at the point $(0.03, 0.9996)$, we obtain $M_1 = -1.99121$. Evaluating M_2 at the point $(0.02, 0.9996)$, we obtain $M_2 = -0.079968$.

Now to estimate the error in the second step, we use the approximation

$$e_k \approx (1 + M_2\Delta t)\, e_{k-1} + M_1 \frac{(\Delta t)^2}{2}$$

and obtain $e_2 \approx 0.000797442$.

To compare this estimate to the actual error, we compute the value of the solution $y(0.04) = 0.998403$. Since $y_2 = 0.9992$, we see that the actual error e_2 is 0.000797444. Note that our estimate of e_2 and the true value of e_2 are very close.

(d) Table 7.10 gives the values of e_k and our estimates of e_k for $k = 10, 20, 30, \ldots, 100$ in case that you want to double check your computations.

Table 7.10
Selected total errors

k	t_k	e_k	estimated e_k
10	0.2	-0.00354477	-0.00354363
20	0.4	-0.00487796	-0.00486646
30	0.6	-0.00399684	-0.00396879
40	0.8	-0.00226733	-0.00223171
50	1.	-0.000714495	-0.000683793
60	1.2	0.000335713	0.000355439
70	1.4	0.000929043	0.000937796
80	1.6	0.00120617	0.00120657
90	1.8	0.00129161	0.0012866
100	2.	0.00127095	0.00126289

(e) Compare your plots with the Figures 7.5 and 7.6 in Section 7.1.

9. (a) The partial derivative $\partial f/\partial y$ of $f(t, y) = \sin ty$ is $t \cos ty$. Thus, $M_2 = t \cos ty$.
    The partial derivative $\partial f/\partial t$ of $f(t, y) = \sin ty$ is $y \cos ty$. Therefore,

    $$M_1 = \frac{\partial f}{\partial t} + \frac{\partial f}{\partial y} f(t, y) = (y + t \sin ty) \cos ty.$$

    (b) The result of the first step of Euler's method is the point $(t_1, y_1) = (0.03, 3.0)$. Consequently, to estimate the error, we compute the quantity M_1 at the point $(0.015, 3)$. We obtain $M_1 \approx 2.99764$.

    Once we have M_1 at this point, we can estimate the error e_1 (which is the same as the truncation error) by computing $|M_1(\Delta t)^2/2|$. In this case, we obtain 0.00134894.

    We cannot compare our estimate to the true error since we do not know how to calculate the true error for this differential equation. (We cannot find a closed-form solution.)

    (c) The second point (t_2, y_2) obtained from Euler's method is the point $(0.06, 3.0027)$. For the second step, the estimated error is no longer simply the truncation error, so we must compute both M_1 and M_2. Evaluating M_1 at the point $(0.045, 3.00135)$, we obtain $M_1 = 2.98002$. Evaluating M_2 at the point $(0.03, 3.00135)$, we obtain $M_2 = 0.0298785$.

    Now to estimate the error in the second step, we use the approximation

    $$e_k \approx (1 + M_2 \Delta t) e_{k-1} + M_1 \frac{(\Delta t)^2}{2}$$

and obtain $e_2 \approx 0.00269115$.

(d) The following table gives the values of our estimates of e_k for $k = 10, 20, 30, \ldots, 100$ in case that you want to double check your computations. We also plot our estimates of the error as a function of k.

Table 7.11
Selected error estimates

k	t_k	estimated e_k
10	0.3	0.0123521
20	0.6	0.0138193
30	0.9	0.00135629
40	1.2	0.0109331
50	1.5	0.0121639
60	1.8	0.0119345
70	2.1	0.0129518
80	2.4	0.0164855
90	2.7	0.0240115
100	3.	0.0343504

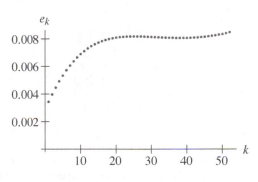

11. (a) The argument that justifies the inequality

$$e_1 \le M_1 \frac{(\Delta t)^2}{2}$$

is given on pages 637 and 638. In particular, the truncation error in the first step is given by Taylor's Theorem.

(b) The total error e_2 involved in the second step is discussed on pages 639 and 640. On the right-hand side of the inequality

$$e_2 \le e_1 + M_2 e_1 \Delta t + M_1 \frac{(\Delta t)^2}{2},$$

the first term is the error involved in the previous step. The second term measures the contribution to the error that arises from evaluating the right-hand side of the differential equation at the point (t_1, y_1) rather than at $(t_1, y(t_1))$. The third term measures the truncation error associated this step of the approximation (see Figure 7.4).

(c) The analysis of the error e_{k+1} is essentially identical to the analysis of the error e_2. That is,

$$e_{k+1} = |y(t_{k+1}) - y_{k+1}|$$

where $y(t_{k+1})$ is the actual value of the function and y_{k+1} is the value given by Euler's method. In other words,

$$y_{k+1} = y_k + f(t_k, y_k) \Delta t.$$

Applying Taylor's Theorem to $y(t)$ at the point $(t_k, y(t_k))$, we have

$$y(t_{k+1}) = y(t_k) + f(t_k, y(t_k)) + y''(\xi_k)\frac{(\Delta t)^2}{2}.$$

Thus,

$$e_{k+1} = |y(t_{k+1}) - y_{k+1}|$$

$$\leq |y(t_k) + f(t_k, y(t_k)) + y''(\xi_k)\frac{(\Delta t)^2}{2} - (y_k + f(t_k, y_k)\,\Delta t)|$$

$$\leq |y(t_k) - y_k| + |f(t_k, y(t_k)) - f(t_k, y_k)|\,\Delta t + |y''(\xi_k)|\frac{(\Delta t)^2}{2}$$

$$\leq e_k + |f(t_k, y(t_k)) - f(t_k, y_k)|\,\Delta t + |y''(\xi_k)|\frac{(\Delta t)^2}{2}.$$

The term $|f(t_k, y(t_k)) - f(t_k, y_k)|$ is bounded by the product of M_2 and e_k, and the third term (the truncation error) is bounded by $M_1(\Delta t)^2/2$. Hence, we have

$$e_{k+1} \leq e_k + (M_2)(e_k)(\Delta t) + M_1\frac{(\Delta t)^2}{2}$$

$$= (1 + M_2\Delta t)e_k + M_1\frac{(\Delta t)^2}{2}.$$

(d) We know that $e_1 \leq K_2$ because e_1 is the same as the truncation error. Using the result of part (b) (or part (c) with $k = 1$), we know that

$$e_2 \leq (1 + M_2\Delta t)e_1 + M_1\frac{(\Delta t)^2}{2} = (K_1)e_1 + K_2.$$

However, given that $e_1 \leq K_2$, we have

$$e_2 \leq (K_1 + 1)K_2.$$

Finally, from part (c), we know that

$$e_3 \leq (K_1)e_2 + K_2 \leq (K_1)(K_1 + 1)K_2 + K_2 = (K_1^2 + K_1 + 1)K_2.$$

(e) We can verify this assertion by induction. In fact, we can use the result of part (d) as the first step. Then we assume the inductive hypothesis that

$$e_{n-1} \leq \left(K_1^{n-2} + K_1^{n-3} + \cdots + K_1 + 1 \right) K_2.$$

Since part (c) says that $e_n \leq (K_1)(e_{n-1}) + K_2$, we can use the inductive hypothesis to obtain

$$e_n \leq (K_1) \left(K_1^{n-2} + K_1^{n-3} + \cdots + K_1 + 1 \right) K_2 + K_2$$

$$= \left(K_1^{n-1} + K_1^{n-2} + \cdots + K_1 + 1 \right) K_2.$$

(f) We can verify the formula

$$K_1^{n-1} + K_1^{n-2} + \cdots + K_1 + 1 = \frac{K_1^n - 1}{K_1 - 1}$$

using induction. However, it is probably easier to see why it holds if we multiply both sides by the factor $(K_1 - 1)$ and cancel on the left-hand side.

Applying this formula to the result of part (e), we get

$$e_n \le \left(\frac{K_1^n - 1}{K_1 - 1} \right) K_2.$$

(g) Since $K_1 = 1 + M_2 \Delta t$ and $K_2 = M_2 (\Delta t)^2 / 2$, we have

$$e_n \le \left(\frac{(1 + M_2 \Delta t)^n - 1}{M_2 \Delta t} \right) \left(M_1 \frac{(\Delta t)^2}{2} \right)$$

$$= \frac{M_1}{2M_2} \left((1 + M_2 \Delta t)^n - 1 \right) \Delta t.$$

(h) If we let $\alpha = M_2 \Delta t$ in Exercise 10, then

$$\left((1 + M_2 \Delta t)^n - 1 \right) \le \left(\left(e^{(M_2 \Delta t)} \right)^n - 1 \right)$$

$$= \left(e^{(M_2 \Delta t)n} - 1 \right).$$

Consequently, we can conclude that

$$e_n \le \frac{M_1}{2M_2} \left(e^{(M_2 \Delta t)n} - 1 \right) \Delta t.$$

(i) Since $\Delta t = (t_n - t_0)/n$, the product $(M_2 \Delta t)n$ is equal to $M_2 (t_n - t_0)$. Therefore, we can conclude that

$$e_n \le \frac{M_1}{2M_2} \left(e^{M_2 (t_n - t_0)} - 1 \right) \Delta t.$$

(j) Note that the quantities M_1 and M_2 are determined by the right-hand side of the differential equation and the rectangle R in the ty-plane. Also, the quantity $(t_n - t_0)$ is precisely the length of the interval over which we are approximating the solution. Therefore, all of the terms in the expression

$$\frac{M_1}{2M_2} \left(e^{M_2 (t_n - t_0)} - 1 \right)$$

do not depend on the number of steps involved in the application of Euler's method. In other words, when considering the effectiveness of Euler's method, we can treat this expression as a constant C determined only by the right-hand side of the differential equation and the rectangle R under consideration.

This expression includes all of the right-hand side of part (i) with the exception of the Δt factor. Consequently, we can consider this long and involved derivation as a justification of the simple inequality

$$e_n \le C \cdot \Delta t.$$

(k) This estimate is a rigorous one. In other words, given the hypotheses stated at the beginning of the exercise, we can be certain that the error is indeed bounded by the quantities specified. The "estimates" calculated in Exercises 7–9 are not as certain. The logic that justifies their calculation is valid, but we cannot be certain that these quantities always give us a true indication of the accuracy of our approximations.

EXERCISES FOR SECTION 7.2

1. Table 7.12 includes the approximate values y_k obtained using improved Euler's method. In addition to the results of improved Euler's method, we graph the results of Euler's method and the results obtained when we used a built-in numerical solver.

Table 7.12
Results of improved Euler's method

k	t_k	y_k
0	0.0	3.0000
1	0.5	8.2500
2	1.0	21.3750
3	1.5	54.1875
4	2.0	136.2187

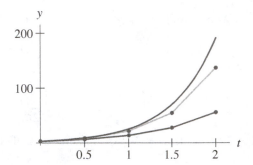

3. Table 7.13 includes the approximate values y_k obtained using improved Euler's method. In addition to the results of improved Euler's method, we graph the results of Euler's method and the results obtained when we used a built-in numerical solver.

Table 7.13
Results of improved Euler's method

k	t_k	y_k
0	0.00	0.5
1	0.25	0.445801
2	0.50	0.103176
3	0.75	−0.501073
4	1.00	−1.16818
5	1.25	−1.70535
6	1.50	−2.09616
7	1.75	−2.39212
8	2.00	−2.63277

5. Table 7.14 includes the approximate values w_k obtained using improved Euler's method. In addition to the results of improved Euler's method, we graph the results of Euler's method and the results obtained when we used a built-in numerical solver.

Table 7.14

Results of improved Euler's method

k	t_k	y_k
0	0.	0.
1	0.5	1.6875
2	1.	2.06728
3	1.5	2.22283
4	2.	2.31738
5	2.5	2.38333
6	3.	2.43292
7	3.5	2.47205
8	4.	2.50398
9	4.5	2.53071
10	5.	2.55351

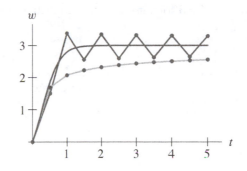

7. Table 7.15 includes the approximate values y_k obtained using improved Euler's method. Compare the results of this calculation with the results obtained in Exercise 6. In addition to the results of improved Euler's method, we graph the results of Euler's method and the results obtained when we used a built-in numerical solver.

Table 7.15

Results of improved Euler's method

k	t_k	y_k
0	1.	2.
1	1.5	3.13301
2	2.	4.01452
3	2.5	4.80396
4	3.	5.54124

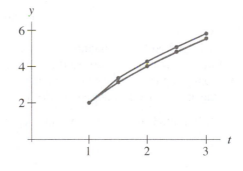

9. **(a)** The differential equation is both separable and linear. Therefore, one way to obtain the solution to the initial-value problem is to integrate

$$\int \frac{1}{1-y}\, dy = \int 1\, dt.$$

We obtain

$$-\ln|1 - y| = t + c,$$

which yields

$$y = 1 - ke^{-t}.$$

We determine the value of k using the initial condition $y(0) = 0$. Hence, $k = 1$, and the solution to the given initial-value problem is $y(t) = 1 - e^{-t}$.

(b) Table 7.16 contains the steps involved in applying improved Euler's method to this initial-value problem.

Table 7.16
Results of improved Euler's method

t_k	y_k
0.	0.
0.25	0.21875
0.5	0.389648
0.75	0.523163
1.	0.627471

Using the analytic solution, we know that the actual value of $y(1)$ is $1 - 1/e$. Therefore, we can compute the error

$$e_4 = |y(1) - y_4| = 0.00464959.$$

(c) If we want an approximation that is accurate to 0.0001, we need an improvement by a factor of

$$\frac{0.00464959}{0.0001} = 46.4959.$$

Since improved Euler's method is a second-order numerical scheme, we expect to get that kind of improvement if we increase the number of steps by a factor of $\sqrt{46.4959}$. In other words, we compute the smallest integer larger than $4\sqrt{46.4959} = 27.2752$. Using $n = 28$ steps, we get the approximate value $y_{28} = 0.63204$ and, consequently, an error $e_{28} \approx 0.00008$.

11. (a) The differential equation is separable. Therefore, we integrate

$$\int \frac{1}{y^2}\, dy = \int -1\, dt.$$

We obtain

$$-\frac{1}{y} = -t + c$$

$$y = \frac{1}{t + k}.$$

We determine the value of k using the initial condition $y(0) = 1/2$. Hence, $k = 2$, and the solution to the given initial-value problem is $y(t) = 1/(t + 2)$.

(b) Table 7.17 contains the steps involved in applying improved Euler's method to this initial-value problem.

Table 7.17
Results of improved Euler's method

t_k	y_k
0.	0.5
0.5	0.402344
1.	0.336049
1.5	0.288275
2.	0.252281

Using the analytic solution, we know that the actual value of $y(2)$ is $1/4$. Therefore, we can compute the error

$$e_4 = |y(2) - y_4| = 0.002281.$$

(c) If we want an approximation that is accurate to 0.0001, we need an improvement by a factor of

$$\frac{0.002281}{0.0001} = 22.81.$$

Since improved Euler's method is a second-order numerical scheme, we expect to get that kind of improvement if we increase the number of steps by a factor of $\sqrt{22.81}$. In other words, we compute the smallest integer larger than $4\sqrt{22.81} = 19.103$. Using $n = 20$ steps, we get the approximate value $y_{20} = 0.250081$ and, consequently, an error $e_{20} \approx 0.00008$.

13. **(a)** If we want an approximation that is accurate to 0.0001, we need an improvement by a factor of

$$\frac{0.000695}{0.0001} = 6.95.$$

Since improved Euler's method is a second-order numerical scheme, we expect to get that kind of improvement if we increase the number of steps by a factor of $\sqrt{6.95}$. In other words, we compute the smallest integer larger than $20\sqrt{6.95} = 52.7257$.

(b) Using $n = 53$ steps, we get the approximate value $y_{53} = 0.200095$.

(c) Consequently, the error e_{53} is the difference between the actual value $y(2) = 0.2$ and y_{53}. We get $e_{53} = 0.000095$.

15.

17.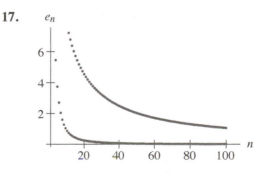

EXERCISES FOR SECTION 7.3

1. Runge-Kutta applied to this initial-value problem yields the points given in Table 7.18. The graph illustrates the results of Runge-Kutta as compared to those of Euler's method, improved Euler's method, and a built-in solver.

Table 7.18
Results of Runge-Kutta

t_k	y_k
0.0	3.000
0.5	8.979
1.0	25.173
1.5	69.030
2.0	187.811

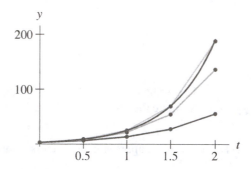

3. Runge-Kutta applied to this initial-value problem yields the points given in Table 7.19. The graph illustrates the results of Runge-Kutta as compared to those of Euler's method, improved Euler's method, and a built-in solver.

Table 7.19
Results of Runge-Kutta

t_k	y_k	t_k	y_k
0.0	0.00000	3.0	2.99645
0.5	1.82290	3.5	2.99882
1.0	2.70058	4.0	2.99961
1.5	2.90368	4.5	2.99987
2.0	2.96803	5.0	2.99996
2.5	2.98935		

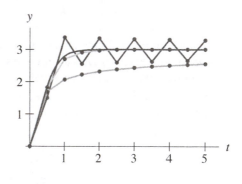

5. Note the relationship between this exercise and Exercise 4.

Runge-Kutta applied to this initial-value problem yields the points given in Table 7.20. The graph illustrates the results of Runge-Kutta as compared to those of Euler's method, improved Euler's method, and a built-in solver.

Table 7.20
Results of Runge-Kutta

t_k	y_k
1.0	2.00000
1.5	3.10456
2.0	3.98546
2.5	4.77554
3.0	5.51352

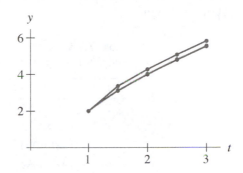

7. **(a)** The results of Runge-Kutta applied to the predator-prey system are given in Table 7.21 and the figure illustrates this computation in the phase plane.

Table 7.21
Results of Runge-Kutta for the Predator-Prey System

t_k	R_k	F_k
0	1.	1.
1	1.50412	1.91806
2	0.641301	2.48192
3	0.416812	1.62154
4	0.636774	1.08434

(b)

(c)

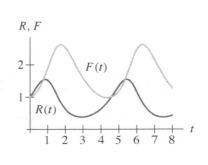

EXERCISES FOR SECTION 7.4

1. Table 7.22 contains the results of approximating the value $y(1)$ of the solution using y_k, where y_k is obtained by applying Runge-Kutta over the interval $0 \leq t \leq 1$ using k steps with single precision arithmetic. Note that we obtain the approximation $y(1) \approx 0.941274$, but we cannot be confident about the next digit.

Table 7.22
Runge-Kutta approximations

k	y_k	k	y_k
2	0.94188476	64	0.94127458
4	0.92484933	128	0.94127458
8	0.94101000	256	0.94127434
16	0.94127089	512	0.94127452
32	0.94127452	1024	0.94127500

3. Table 7.23 contains the results of approximating the value $y(2)$ of the solution using y_k, where y_k is obtained by applying Runge-Kutta over the interval $0 \leq t \leq 2$ using k steps with single precision arithmetic. Note that we obtain the approximation $y(2) \approx 1.25938$, but we cannot be confident about the next digit.

Table 7.23
Runge-Kutta approximations

k	y_k	k	y_k
2	1.33185744	64	1.25938189
4	1.25684679	128	1.25938177
8	1.25911093	256	1.25938165
16	1.25936496	512	1.25938165
32	1.25938094	1024	1.25938201

REVIEW EXERCISES FOR CHAPTER 7

1. True. In a first-order method, the error is, at worst, proportional to the first power of the step size. In other words, if e_n is the error at the nth step, there is a constant C such that

$$e_n \leq C \cdot (\Delta t)^1.$$

Assuming that this inequality is basically an equality for large values of n, we see that halving Δt would halve e_n.

3. False. In a fourth-order method, the error is, at worst, proportional to the fourth power of the step size. In other words, if e_n is the error at the nth step, there is a constant C such that

$$e_n \leq C \cdot (\Delta t)^4.$$

Assuming that this inequality is basically an equality for large values of n, we see that halving Δt would lower e_n by a factor of $(1/2)^4 = 1/16$.

5. False. Runge-Kutta is a fourth-order method. See the discussion in the text on page 659.

7. **(a)** Euler's method:

Table 7.24
Results of Euler's method

t_k	x_k	y_k	t_k	x_k	y_k
0.0	3.	3.	⋮	⋮	⋮
0.1	1.8	1.2	⋮	⋮	⋮
0.2	1.62	0.984	4.7	1.87797	0.0737875
0.3	1.52215	0.863558	4.8	1.88703	0.0676651
0.4	1.46344	0.785159	4.9	1.89558	0.0619696
0.5	1.42706	0.729253	5.0	1.90363	0.0566828

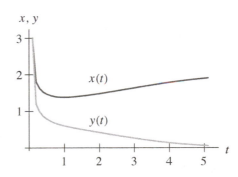

(b) improved Euler's method:

Table 7.25
Results of improved Euler's method

t_k	x_k	y_k	t_k	x_k	y_k
0.0	3.	3.	⋮	⋮	⋮
0.1	2.31	1.992	⋮	⋮	⋮
0.2	1.9507	1.51588	4.7	1.80751	0.124714
0.3	1.7399	1.24768	4.8	1.81953	0.115741
0.4	1.60693	1.07962	4.9	1.83107	0.107255
0.5	1.5191	0.96594	5.0	1.84212	0.0992498

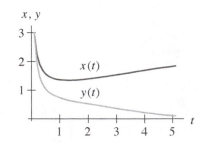

(c) Runge Kutta:

Table 7.26
Results of Runge Kutta

t_k	x_k	y_k	t_k	x_k	y_k
0.0	3.	3.	⋮	⋮	⋮
0.1	2.23985	1.91269	⋮	⋮	⋮
0.2	1.88951	1.45316	4.7	1.80448	0.126944
0.3	1.69272	1.2038	4.8	1.81662	0.117848
0.4	1.57072	1.04912	4.9	1.82828	0.10924
0.5	1.49087	0.944565	5.0	1.83946	0.101114

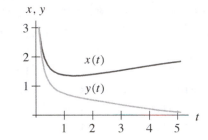

CHAPTER 8

Discrete Dynamical Systems

EXERCISES FOR SECTION 8.1

1. $x_0 = 0, x_1 = -2, x_2 = 2$, and $x_3 = 2$. The orbit is eventually fixed.

3. $x_0 = 0, x_1 = 1, x_2 = e$, and $x_3 = e^e$. The orbit tends to infinity.

5. $x_0 = 0, x_1 = 2, x_2 = 0$, and $x_3 = 2$. The orbit is periodic of period 2.

7. $x_0 = 0, x_1 = 1, x_2 = 1$, and $x_3 = 1$. The orbit is eventually fixed.

9. To find the fixed points, we solve $F(x) = x$ or

$$-x + 2 = x.$$

The solution is $x = 1$. For periodic points of period 2, we solve $F^2(x) = x$ or

$$-(-x + 2) + 2 = x.$$

This equation reduces to the identity $x = x$. Therefore, all real numbers (except the fixed point $x = 1$) are periodic points of period 2.

11. To find the fixed points, we solve $F(x) = x$ or

$$x^2 + 1 = x.$$

This equation can be written as $x^2 - x + 1 = 0$ which has imaginary solutions. Alternatively, since

$$F(x) - x = x^2 - x + 1 = (x - \frac{1}{2})^2 + \frac{3}{4} > 0,$$

we see that $F(x) > x$. Thus, the graph of $F(x)$ does not intersect the diagonal and there are no fixed points. For periodic points of period 2, we first compute that

$$F^2(x) = (x^2 + 1)^2 + 1 = x^4 + 2x^2 + 2.$$

Then we solve $F^2(x) = x$ or

$$x^4 + 2x^2 - x + 2 = 0.$$

We already know two solutions of this equation. They are the two imaginary solutions to

$$x^2 - x + 1 = 0.$$

(Any x-value satisfying $F(x) = x$ also satisfies $F^2(x) = x$.) Using long division, we have

$$x^4 + 2x^2 - x + 2 = (x^2 - x + 1)(x^2 + x + 2).$$

Since the roots of the quadratic factor $x^2 + x + 2$ are imaginary also, we have no period 2 points. We also could have used the fact that

$$F^2(x) - x = (x^2 + 1)^2 + 1 - x = x^4 + 2(x - \frac{1}{4})^2 + \frac{15}{8} > 0.$$

Therefore, $F^2(x) > x$ and the graph of $F^2(x)$ does not intersect the diagonal.

13. To find the fixed points, we solve $F(x) = x$ or

$$\sin x = x.$$

Since the graph of $F(x) = \sin x$ only crosses the diagonal at the origin, $x = 0$ is the only fixed point. For periodic points of period 2, consider the function

$$G(x) = F^2(x) - x = \sin(\sin x) - x.$$

Any period 2 point will be a root of $G(x)$. Differentiation yields

$$G'(x) = (\cos(\sin x))\cos x - 1 < 0 \text{ for } x \neq 0,$$

so that $G(x)$ is strictly decreasing. Since $G(0) = 0$, we have $G(x) > 0$ for $x < 0$ and $G(x) < 0$ for $x > 0$. This means that $x = 0$ is the only root of $G(x)$ and thus there are no periodic points of period 2. Another method would be to use the fact that $|\sin x| < |x|$ for $x \neq 0$. This implies that

$$|\sin(\sin x)| < |\sin x| < |x| \text{ for } x \neq 0.$$

15. For fixed points,
$$F(x) = -2x - x^2 = x,$$

or $x = 0, -3$. For periodic points of period 2,

$$F^2(x) = -x^4 - 4x^3 - 2x^2 + 4x = x,$$

or

$$x(x + 3)(-x^2 - x + 1) = 0.$$

Therefore, the periodic points of period 2 are $x = (-1 \pm \sqrt{5})/2$.

17. From the graph of $F(x) = -e^x$ and $f(x) = x$, there is one intersection and, therefore, there is one fixed point. For periodic points of period 2, consider

$$g(x) = -e^{-e^x} - x.$$

By the graph of g shown to the right (or by computing g'), $g(x)$ is zero only once, so this point must be the fixed point and there are no periodic points of period 2.

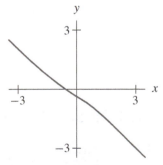

19. The graph of $y = -x$ meets $y = x$ only at $x = 0$, so 0 is a fixed point. Since $F^2(x) = x$, it follows that all other real numbers lie on periodic orbits of period 2.

21. There is a fixed point at $x = 2$; all other points are sent to $x = 2$ by F so they are eventually fixed. Therefore there are no other periodic points.

23. $x_0 = x, x_1 = -x + 4, x_2 = x, x_3 = -x + 4$, and $x_4 = x$. The orbit of any real number is a periodic orbit of period 2, except for $x = 2$, which is fixed.

25. The orbit is $1/6, 1/3, 2/3, 2/3, \ldots$, so $1/6$ is eventually fixed.

27. The orbit is $2/7, 4/7, 6/7, 2/7, \ldots$, so $2/7$ is periodic with period 3.

29. The orbit is $1/8, 1/4, 1/2, 1, 0, 0, 0, \ldots$, so $1/8$ is eventually fixed.

31. The orbit of $6/11$ is $6/11, 10/11, 2/11, 4/11, 8/11, 6/11, \ldots$, so $6/11$ is periodic with period 5.

33. The orbit is $0, 0, 0, \ldots$, so it is a fixed point.

35. The orbit is $1/2, 1, 0, 0, \ldots$, so it is eventually fixed.

37. For fixed points, we must solve
$$F_c(x) = x^2 + c = x$$
or
$$x^2 - x + c = 0.$$
From the quadratic formula we find
$$x = (1 \pm \sqrt{1 - 4c})/2.$$
Thus we need $1 - 4c \geq 0$ or $c \leq \frac{1}{4}$. For $c = 1/4$, we have $x = 1/2$ and this is the only fixed point. For $c < 1/4$, there are two roots and thus there are two fixed points. For $c > 1/4$, there are no fixed points.

39. For periodic points of period 2, we have
$$F_c^2(x) = (x^2 + c)^2 + c = x,$$
or
$$x^4 + 2cx^2 - x + c^2 + c = 0.$$
Since we have fixed points when $x^2 + c = x$, the left-hand side can be factored and one obtains
$$(x^2 - x + c)(x^2 + x + c + 1) = 0.$$
This is zero when $x^2 - x + c = 0$ or when $x^2 + x + c + 1 = 0$. The first factor corresponds to the fixed points, which are present when $c \leq 1/4$. The second factor corresponds to the period two points and has two solutions when $c < -3/4$. At $c = -3/4$ there are two fixed points (roots of the second factor coincide with one of the fixed points).

EXERCISES FOR SECTION 8.2

1. For the fixed points,
$$F(x) = x^2 - 2x = x,$$
or
$$x^2 - 3x = 0.$$
Then, $x = 0, 3$ are fixed points. Differentiation yields
$$F'(x) = 2x - 2.$$
Then, $F'(0) = -2$ and $F'(3) = 4$. Therefore, both $x = 0$ and $x = 3$ are repelling fixed points.

3. For the fixed points,

$$F(x) = \sin x = x,$$

or $x = 0$. Differentiation yields $F'(0) = 1$. For $x > 0$, $x > \sin x$ and for $x < 0$, $x < \sin x$. Therefore, $x = 0$ is attracting by graphical analysis.

5. For the fixed points,

$$F(x) = \arctan x = x.$$

Suppose $f(x) = \arctan x - x$ and differentiation yields

$$f'(x) = \frac{-x^2}{1 + x^2} - 1.$$

Then, $f'(x) < 0$ for $x \neq 0$ and $f'(x) = 0$ only for $x = 0$. $f(x)$ is decreasing for $x \neq 0$. Since $f(0) = 0$, $f(x)$ is tangent to x-axis at $x = 0$. Therefore, $x = 0$ is the only fixed point. Also, $f(x) = F(x) - x > 0$ for $x < 0$ and $f(x) = F(x) - x < 0$ for $x > 0$. Therefore, $x = 0$ is attracting.

7. For the fixed points,

$$F(x) = \frac{\pi}{2} \sin x = x.$$

Then, $x = 0, \pm \pi/2$ are fixed points. Differentiation yields

$$F'(x) = \frac{\pi}{2} \cos x.$$

Then, $F'(0) = \pi/2$ and $F'(\pm\pi/2) = 0$. Therefore, $x = 0$ is repelling, and $x = \pm\pi/2$ are attracting.

9. For the fixed points,

$$F(x) = \frac{1}{x} = x,$$

or

$$x^2 = 1.$$

Then, $x = \pm 1$ are fixed points. Differentiation yields

$$F'(x) = -\frac{1}{x^2}.$$

Then, $F'(\pm 1) = -1$. Since $F^2(x) = x$ for any non-zero x, $x = \pm 1$ are neutral.

11. For the fixed points,

$$F(x) = e^x = x.$$

Suppose $f(x) = e^x - x$ and differentiation yields

$$f'(x) = e^x - 1.$$

Therefore, $f(x)$ is increasing for $x > 0$, decreasing for $x < 0$ and equal to 1 for $x = 0$. $f(x) \neq 0$ for any x and there are no fixed points.

13. $F(0) = 1$, and $F^2(0) = 0$. The period is 2 and since $(F^2)'(0) = 0$, it is attracting.

15. $F(0) = 1$, $F(1) = 2$, $F(2) = 3$, $F(3) = 4$, and $F(4) = 0$. The period is 5 and since $(F^5)'(0) = 2$, the cycle is repelling.

17. $F(0) = 1$ and $F^2(0) = 0$. The period is 2. For $x < 2$ close to zero, $F(x) = 1 - x$. Therefore, any x close to zero is periodic point and the cycle of $x = 0$ is neutral.

19. For fixed points,

$$F(x) = \frac{1}{x} = x,$$

or $x = \pm 1$. Since $F^2(x) = x$ for all other x, these fixed points are neutral. We can also see this by graphical analysis.

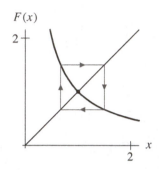

21. For fixed points,

$$F(x) = \tan x = x,$$

or $x = 0$. Since $(\tan x)' = 1/\cos^2 x = 1$ if and only if $x = 0, \pi$, $x = 0$ is the only fixed point with $F'(0) = 1$. By graphical analysis, the fixed point is repelling.

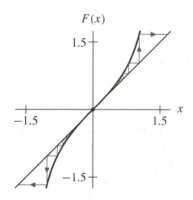

23. For fixed points,

$$F(x) = -x - x^3 = x,$$

or $x = 0$. By graphical analysis, the fixed point is repelling.

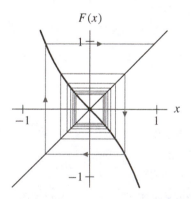

25. For fixed points,

$$F(x) = -e \cdot e^x = x,$$

or $x = -1$. By graphical analysis, the fixed point is attracting.

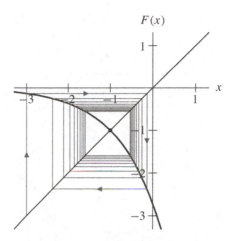

F(x)

27. For $F(x) = \tan x = x$, there are infinitely many fixed points because $\tan x$ is periodic, and $|\tan x|$ tends to infinity as x tends to $n\pi/2$ for odd integers n. As we saw in Exercise 21, $x = 0$ is repelling. First, note that for $x = n\pi$ for $n \neq 0$,

$$F(n\pi) = \tan n\pi = 0 \neq n\pi.$$

Therefore, $x = n\pi$ for $n \neq 0$ are not fixed points. At the fixed points,

$$F'(x) = \frac{1}{\cos^2 x} > 1$$

because $|\cos x| < 1$. Therefore, all other fixed points are repelling.

29. There is a cycle for T. Try $x = 16/21$. Suppose T has n-cycle, $x_0, x_1, \cdots, x_n = x_0$. Then,

$$|(T^n)'(x_0)| = |T'(x_0) \cdot T'(x_1) \cdots T'(x_{n-1})|$$
$$= 4^n,$$

where n is integer larger than zero. Therefore, the cycle is repelling.

EXERCISES FOR SECTION 8.3

1. For fixed points, we must have

$$F_\alpha(x) = x + x^2 + \alpha = x$$

or $x^2 + \alpha = 0$. Therefore, for $\alpha > 0$, there are no fixed points; for $\alpha = 0$, there is one fixed point; and for $\alpha < 0$, there are two fixed points at $x = \pm\sqrt{-\alpha}$. Differentiation yields

$$F_\alpha'(x) = 1 + 2x$$

and

$$F_\alpha'(\pm\sqrt{-\alpha}) = 1 \pm 2\sqrt{-\alpha}.$$

Therefore, for $-\alpha$ small, $0 < 1 - 2\sqrt{-\alpha} < 1$ and $x = -\sqrt{-\alpha}$ is attracting. Since $1 + 2\sqrt{-\alpha} > 1$, $x = \sqrt{-\alpha}$ is repelling. For $\alpha = 0$, $F_\alpha(x)$ is tangent to $y = x$ from above and, therefore, $x = 0$ is neutral. The bifurcation is a tangent bifurcation.

3. For α slightly smaller than 1, the origin is the only fixed point and it is attracting. For $\alpha = 1$, F_1 is tangent to $y = x$ and 0 is attracting. For $\alpha > 1$, two more fixed points appear and they are attracting for α slightly larger than 1. The origin becomes a repelling fixed point. This is a pitchfork bifurcation.

5. For fixed points,

$$\alpha - x^2 = x,$$

and $x = (-1 \pm \sqrt{1 + 4\alpha})/2$. Therefore, there are no fixed point for $\alpha < -1/4$, there is one fixed point for $\alpha = -1/4$, and there are two fixed points for α slightly larger than $-1/4$. This is a tangent bifurcation. Also, for α slightly larger than $-1/4$,

$$\left| F'_\alpha \left(\frac{-1 + \sqrt{1 + 4\alpha}}{2} \right) \right| = |-1 + \sqrt{1 + 4\alpha}| < 1$$

and

$$\left| F'_\alpha \left(\frac{-1 - \sqrt{1 + 4\alpha}}{2} \right) \right| = |-1 - \sqrt{1 + 4\alpha}| > 1.$$

Therefore, $x = (-1 + \sqrt{1 + 4\alpha})/2$ is attracting and $x = (-1 - \sqrt{1 + 4\alpha})/2$ is repelling. For $\alpha = -1/4$, $F'_{-1/4}(-1/2) = 1$ and, therefore, $F_{-1/4}$ is tangent to $y = x$ from below at $x = -1/2$. By graphical analysis, $x = -1/2$ is neutral. Nearby orbits on the right of the fixed point tend to the fixed point and nearby orbits on the left tend away from it.

7. This bifurcation is none of the above. When $\alpha = 0$, we have $F_\alpha(x) = 0$ for all x. So all nonzero points are eventually fixed. For $0 < |\alpha| < 1$, the origin is an attracting fixed point and there is a second (repelling) fixed point at $x = (1 - \alpha)/\alpha$.

9. The function T_μ has a single fixed point at the origin when $0 < \mu < 1$. This fixed point is attracting. If $\mu = 1$, all $x \le 1/2$ are fixed points, and each is neutral. If $\mu > 1$, T_μ has two fixed points, at $x = 0$ and $x = \mu/(\mu + 1)$, and both are repelling.

11. The graph of the function $F_c(x)$ is a parabola with minimum at $x = 0$ and $F_c(0) = c$. The function $F_c^2(x)$ is a quartic with local maximum at $x = 0$. Consider the small portion of the graph of F_c^2 defined over the interval $[p_c, -p_c]$, where p_c is the leftmost fixed point of F_c (not F_c^2). This piece of the graph of F_c^2 resembles that of F_c, only upside down. Note that, as c decreases, this piece of graph behaves similarly to that of F_c relative to the diagonal. Thus we expect F_c^2 to undergo a period-doubling bifurcation in this interval, much the same as F_c did.

EXERCISES FOR SECTION 8.4

1.

Histogram for seed .3.   Histogram for seed .3001.

Time series for seed .3.   Time series for seed .3001.

3. The image under T of a point of the form $p/2^n$ where p is an integer is of the form $q/2^{n-1}$ where q is an integer. This is because T multiplies by 2 and then subtracts one if necessary to keep the image in the interval $[0, 1)$ (where 1 is not included). Hence, after n iterates, $T^n(p/2^n)$ must be the fixed point 0.

5. The graph of T^n crosses the diagonal in each interval $[i/2^n, (i+1)/2^n)$ for $i = 0, 1, \ldots, 2^n - 2$. Hence, each of these intervals contains a point of period n (n may not be the least period). We may take n as large as we like, so periodic points of T are dense.

7. The eventually fixed points of T are those whose orbits hit 0. These are precisely the points which have finite binary expansion, that is, points of the form

$$\frac{a_1}{2} + \frac{a_2}{2^2} + \frac{a_3}{2^3} + \ldots + \frac{a_n}{2^n}$$

where $a_1, a_2, \ldots, a_n$ equal zero or one. Using 2^n as a common denominator, we can express such a point in the form $p/2^n$ for an integer p between zero and $2^n - 1$.

9. The points which are fixed point T^n are those whose binary expansion repeats every n summands, that is, those of the form

$$x = \frac{a_1}{2} + \frac{a_2}{2^2} + \ldots + \frac{a_n}{2^n} + \frac{a_1}{2^{n+1}} + \frac{a_2}{2^{n+2}} + \ldots + \frac{a_n}{2^{2n}} + \frac{a_1}{2^{2n+1}} + \ldots.$$

We may choose $a_1, a_2, \ldots$, however we like provided they are not all one (since this corresponds to $x = 1$ which is not in the domain). Hence, T^n has $2^n - 1$ fixed points (and T has $2^n - 1$ points of period n). This can also be seen by looking at the graph of T^n.

11. First form a list of all the possible finite strings of zeros and ones,

$$0, 1, 00, 01, 10, 11, 000, 001, 010, 011, 100, 101, 110, 111, 0000, \ldots$$

then concatenate this list into the binary expansion of a number x_0,

$$x_0 = .01000110110000010100111001011110111\ldots.$$

Since every finite string appears in this list and application of T shifts the decimal expansion to the left by one place, the orbit of this point will eventually come close to every point in $[0, 1)$.

13. The graph of T^n on the segment $[i/2^n, (i+1)/2^n]$ is a straight line from zero to one or one to zero depending on whether i is even or odd. In either case, the graph of T^n crosses the diagonal in each such interval. Hence, T^n has a fixed point and T has a periodic point in each such interval. Since n can be taken as large as we like, periodic points are dense.

REVIEW EXERCISES FOR CHAPTER 8

1. We solve $x^3 - x/2 = x$ and obtain $x = 0$ or $x = \pm\sqrt{3/2}$.

To determine the type of each fixed point, we compute $F'(x) = 3x^2 - 1/2$. So $F'(0) = -1/2$, and $x = 0$ is attracting. On the other hand, $F'(\pm\sqrt{3/2}) = 4$, so both of these fixed points are repelling.

3. The graph of the second iterate of T consists of four line segments.

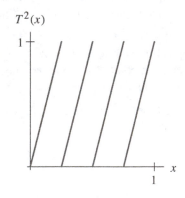

The points $x = 0$ and $x = 1$ are fixed points. To find the cycle of period two, we note that $T^2(x) = 4x - 1$ for $1/4 \leq x < 1/2$. Solving $4x - 1 = x$ produces the period-two point $x = 1/3$. Note that $T(1/3) = 2/3$ and $T(2/3) = 1/3$.

5. The bifurcation at $c = 1$ is a pitchfork bifurcation. The attracting fixed point at $x = 0$ for $-1 < c < 1$ becomes repelling for $c > 1$, and two new attracting fixed points appear, one on each side of $x = 0$.

7. For every x, $-1 \leq F(x) \leq 1$. So the first iterate of every point lies in this interval. Moreover, in this interval, it is easy to see that $|F(x)| < |x|$ for $x \neq 0$. Since the only fixed point is $x = 0$, all orbits tend to this fixed point.

9. For every x, $-1 \le F(x) \le 1$. For all x between -1 and 1, the orbit of x tends to the attracting fixed point at $0.73908\dots$ (see Exercise 8). Hence, every orbit tends to this attracting fixed point.

11. False. If $c > 1/4$, the equation $x^2 + c = x$ has no solutions.

13. False. Note that $F(1) = -1$ and $F(-1) = 1$, so $x = 1$ does lie on a cycle of period two. However, using the fact that $F'(x) = -3x^2$, we see that $F'(\pm 1) = -3$, so $(F^2)'(\pm 1) = 9$. Therefore, $x = 1$ lies on a repelling cycle of period two.

15. False. The bifurcation that occurs at $a = 1$ is a pitchfork bifurcation. For $a < 1$, there is one fixed point (at $x = 0$). For $a > 1$, there are three fixed points, $x = 0$ and one on each side of $x = 0$.

17. All points of the form $x = .abababab\dots$ where $a \ne b$ are digits from 0 to 9 are periodic points of period two. Note that the points of the form $.aaaa\dots$ are fixed points, not points of period two.

19. The graph of $F(x)$ shows that each point in the interval $-2 < x \le 2$ is the image of exactly two points in -2 to 2. The same observation holds for the logistic map $L_4(x)$ on the interval $[0, 1]$, so there are infinitely many periodic points for F.

Appendices

EXERCISES FOR APPENDIX A

1. We rewrite the equation as

$$\frac{dy}{dt} = (y - 4t) + (y - 4t)^2 + 4,$$

use that

$$\frac{dy}{dt} = \frac{du}{dt} + 4,$$

and substitute to obtain

$$\frac{du}{dt} + 4 = u + u^2 + 4.$$

This equation simplifies to

$$\frac{du}{dt} = u^2 + u,$$

which is nonlinear, autonomous, and separable.

3. Rewrite the equation as

$$\frac{dy}{dt} = ty + (ty)^2 + \cos(ty).$$

Using that $y = u/t$, we have

$$\frac{dy}{dt} = \frac{1}{t}\frac{du}{dt} - \frac{1}{t^2}u.$$

We substitute to obtain

$$\frac{1}{t}\frac{du}{dt} - \frac{1}{t^2}u = u + u^2 + \cos u,$$

which simplifies to

$$\frac{du}{dt} = \frac{u}{t} + t(u + u^2 + \cos u).$$

This equation is nonlinear and nonautonomous.

This is a good example of a change of variables that looks like it is going to greatly simplify the equation but does not because the term that replaces dy/dt is complicated.

5. Let $u = y - t$. Then

$$\frac{du}{dt} = \frac{dy}{dt} - 1,$$

and the differential equation becomes

$$\frac{du}{dt} + 1 = u^2 - u - 1,$$

which simplifies to

$$\frac{du}{dt} = u^2 - u - 2 = (u - 2)(u + 1).$$

The equilibrium points are $u = 2$ (a source) and $u = -1$ (a sink). These equilibria correspond to the solutions $y_1(t) = 2 + t$ and $y_2(t) = -1 + t$.

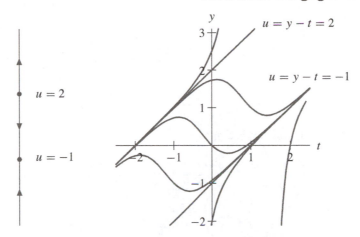

7. First we clear the denominator, so the equation becomes

$$t\frac{dy}{dt} = ty(\cos ty) - y.$$

Then we change variables using $u = ty$. We have

$$\frac{du}{dt} = t\frac{dy}{dt} + y,$$

and the differential equation becomes

$$\frac{du}{dt} - \frac{u}{t} = u(\cos u) - \frac{u}{t}.$$

Thus we get

$$\frac{du}{dt} = u\cos u.$$

The equilibrium point $u = 0$ corresponds to the solution $y(t)$ that is constantly zero for all t, and the equilibrium points $u = \pi/2 \pm n\pi$ for $n = 0, \pm1, \pm2, \ldots$ correspond to solutions $y(t) = (\pi/2 \pm n\pi)/t$, for $n = 0, \pm1, \pm2, \ldots$. Note that the change of variables "blows up" at $t = 0$.

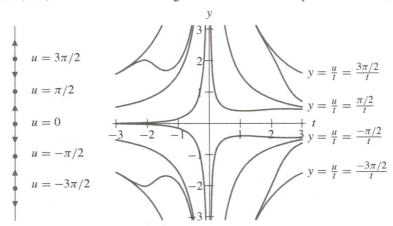

9. If $u = y/(1 + t)$, we have

$$\frac{du}{dt} = \frac{1}{1 + t}\frac{dy}{dt} - \frac{y}{(1 + t)^2}.$$

Then

$$\frac{dy}{dt} = (1 + t)\frac{du}{dt} + u,$$

and the differential equation becomes

$$(1 + t)\frac{du}{dt} + u = u - \frac{u(1 + t)}{t} + t^2(t + 1),$$

which reduces to

$$\frac{du}{dt} = -\frac{u}{t} + t^2.$$

This differential equation is linear, and we rewrite it as

$$\frac{du}{dt} + \frac{u}{t} = t^2.$$

Its integrating factor is

$$\mu(t) = e^{\int (1/t)\, dt} = e^{\ln t} = t.$$

Multiplying both sides of the differential equation by $\mu(t)$, we obtain

$$t\frac{du}{dt} + u = t^3,$$

which is equivalent to

$$\frac{d(tu)}{dt} = t^3.$$

Integrating both sides, we have

$$tu = \frac{t^4}{4} + c,$$

where c is an arbitrary constant. Therefore,

$$u(t) = \frac{t^3}{4} + \frac{c}{t}.$$

To determine the general solution for $y(t)$, we have $y = u(1 + t)$, and therefore

$$y(t) = \frac{t^3(1 + t)}{4} + \frac{c(1 + t)}{t}$$

$$= \frac{t^3(1 + t)}{4} + \frac{c}{t} + c.$$

11. We know that

$$\frac{du}{dt} = \frac{dy}{dt} - 1.$$

Substituting y and dy/dt into the equation $du/dt = (1 - u)u$, we get

$$\frac{dy}{dt} - 1 = (1 - (y - t))(y - t),$$

which simplifies to

$$\frac{dy}{dt} = -y^2 + 2yt + y - t^2 - t + 1.$$

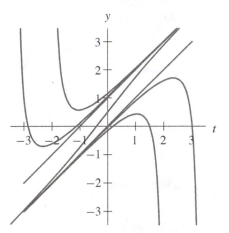

Graphs of solutions $y(t)$ are obtained from the graphs of the solutions to $du/dt = (1 - u)u$ by "tilting up" the plane so that horizontal lines in the tu-plane become lines of slope 1 in the ty-plane. As a result, the graphs of the equilibrium solutions in the tu-plane become lines with slope 1 in the ty-plane.

13. If $y = u^2$, then $u = \sqrt{y}$ and

$$\frac{du}{dt} = \frac{1}{2\sqrt{y}}\frac{dy}{dt}.$$

Note that the change of variables $u = \sqrt{y}$ is defined (real valued) only if $y \geq 0$ and results in $u \geq 0$. Substituting y and dy/dt into $du/dt = (1 - u)u$, we have

$$\frac{1}{2\sqrt{y}}\frac{dy}{dt} = (1 - \sqrt{y})\sqrt{y},$$

which simplifies to

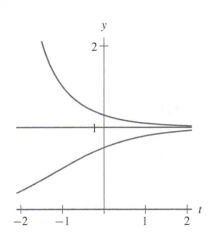

$$\frac{dy}{dt} = 2y(1 - \sqrt{y}).$$

Note that the change of variables fixes the equilibrium points $u = 0$ and $u = 1$. In other words, $y = 0$ and $y = 1$ are also equilibrium points.

15. **(a)** Since 1 gallon per minute of salt water containing 2 pounds of salt per gallon and 5 gallons per minute of salt water containing 0.2 pounds of salt per gallon enter the vat, 3 pounds of salt per minute enters the vat. The concentration of the salt water at time t is $S(t)/(10 + 3t)$. Since 3 gallons per minute of salt water leaves the vat, $3S(t)/(10 + 3t)$ pounds of salt is removed. The rate of change of the amount of salt in the vat is

$$\frac{dS}{dt} = 3 - \frac{3S}{10 + 3t}.$$

(b) The concentration of the salt in the vat at time t is $C(t) = S(t)/(10 + 3t)$. Differentiating, we obtain

$$\frac{dC}{dt} = \frac{1}{10 + 3t}\frac{dS}{dt} - \frac{3C}{10 + 3t},$$

and thus

$$\frac{dS}{dt} = (10 + 3t)\frac{dC}{dt} + 3C.$$

Since $dS/dt = 3 - 3(S/(10 + 3t)) = 3 - 3C$, we obtain

$$(10 + 3t)\frac{dC}{dt} + 3C = 3 - 3C,$$

which yields

$$\frac{dC}{dt} = \frac{3 - 6C}{10 + 3t}.$$

(c) From the right-hand side of the differential equation, note that $C(t) = 1/2$ is an equilibrium solution. Moreover, for solutions whose initial conditions satisfy $C(0) < 1/2$, we see that $dC/dt > 0$. In our case, $C(0) = 0$, and we know that $C(t) \to 1/2$ as $t \to \infty$.

(d) From the differential equation in part (b), we separate variables and obtain

$$\frac{dC}{3 - 6C} = \frac{dt}{10 + 3t}.$$

Integration yields

$$C = \frac{1}{2} - \frac{k}{(10 + 3t)^2}.$$

At $t = 0$, $C(t) = 0$. Therefore, we have $0 = 1/2 - k/100$. Hence $k = 50$, and

$$C(t) = \frac{1}{2} - \frac{50}{(10 + 3t)^2}.$$

Evaluating this expression at $t = 5$ yields $C(5) = 0.42$.

17.   **(a)** To find the equilibrium points, we solve $dy/dt = 0$, which is equivalent to

$$10y^3 = 1.$$

We get $y = \sqrt[-3]{10}$.

(b) Let $u = y - \sqrt[-3]{10}$. Then, $du/dt = dy/dt$, and

$$10y^3 - 1 = 10\left(u + \sqrt[-3]{10}\right)^3 - 1$$

$$= 10\left(u^3 + 3\sqrt[-3]{10}\,u^2 + 3\sqrt[-3]{10^2}\,u + \frac{1}{10}\right) - 1$$

$$= 10u^3 + 3 \cdot 10^{2/3}u^2 + 3\sqrt[3]{10}\,u.$$

The linear approximation is

$$\frac{du}{dt} = 3\sqrt[3]{10}\,u.$$

Since $3\sqrt[3]{10} > 0$, the origin is a source.

(c) Solving the linear system, we get

$$u(t) = u_0 e^{3\sqrt[3]{10}\,t}$$

near the origin. In other words, solutions move away from the equilibrium solution exponentially fast with exponent $3\sqrt[3]{10}\,t$. In particular, to derive the amount of time that it takes a solution to double its distance from the equilibrium, we solve

$$2u_0 = u_0 e^{3\sqrt[3]{10}\,t}$$

and obtain $t = (\ln 2)/(3\sqrt[3]{10})$.

19. **(a)** To find the equilibria, we solve $dy/dt = 0$ and obtain $y = -1$ and $y = 3$.

(b) For $y = -1$, let $u = y + 1$. Then, $du/dt = dy/dt$ and

$$\frac{du}{dt} = \frac{dy}{dt} = (y+1)(3-y) = u(4-u).$$

The linear approximation is

$$\frac{du}{dt} = 4u,$$

so $u = 0$ is a source.

Similarly, for $y = 3$, we let $v = y - 3$ and obtain

$$\frac{dv}{dt} = -v(v+4) = -4v - v^2.$$

The linear approximation is

$$\frac{dv}{dt} = -4v.$$

so $v = 0$ is a sink.

(c) The linearizations at $y = -1$ and $y = 3$ have general solutions $u = u_0 e^{4t}$ and $v = v_0 e^{-4t}$, respectively. Hence, the time necessary to halve or double the distance to the equilibrium point starting nearby is $t = (\ln 2)/4$.

21. **(a)** Since $y = u + y_0$, $du/dt = dy/dt$, and we can replace y by $u + y_0$ in the right-hand side of the differential equation. We get

$$\frac{du}{dt} = f(u + y_0).$$

(b) Using the Taylor series

$$f(y) = f(y_0) + f'(y_0)(y - y_0) + \frac{f''(y_0)}{2!}(y - y_0)^2 + \dots$$

about $y = y_0$, we have

$$\frac{du}{dt} = f(y_0) + f'(y_0)\,u + \frac{f''(y_0)}{2!}\,u^2 + \dots$$

because $u = y - y_0$. Since y_0 is an equilibrium point, $f(y_0) = 0$. Thus, when we truncate higher-order terms (order ≥ 2) in u, we get the linearized equation

$$\frac{du}{dt} = f'(y_0)\,u.$$

23. This equation is separable, and we can solve it by separating variables. However, treating it as a Bernoulli equation and changing variables is easier.

First, note that the equilibrium solution $y(t) = 0$ for all t is a solution to this equation. So we assume $y \neq 0$ when we change variables.

We change variables using $z = y^{-2}$. We get

$$\frac{dz}{dt} = -2y^{-3}\left(\frac{dy}{dt}\right)$$

$$= -2y^{-3}(y + y^3)$$

$$= -2y^{-2} - 2$$

$$= -2z - 2 = -2(z + 1).$$

The differential equation for $z(t)$ is autonomous and linear. One particular solution is the equilibrium solution $z(t) = -1$. Also, the general solution of the associated homogeneous equation is ke^{-2t}. Hence, the general solution of the differential equation for $z(t)$ is $z(t) = ke^{-2t} - 1$.

In our case, $z = y^{-2}$. Therefore, $z(t)$ must be positive, and we discard those solutions for which $k \leq 0$. Converting back to y, using $y = z^{-1/2}$, we have

$$y(t) = \frac{\pm 1}{\sqrt{ke^{-2t} - 1}},$$

where $k > 0$.

25. Note that this differential equation is not defined at $t = 0$. Therefore, we assume that $t \neq 0$ when we derive the general solution. Also, the equilibrium solution $y(t) = 0$ for all $t \neq 0$ is a solution. So we also assume that $y \neq 0$.

This equation is a Bernoulli equation, so we change variables using $z = y^{-3}$. We obtain

$$\frac{dz}{dt} = -3y^{-4}\left(\frac{dy}{dt}\right)$$

$$= 3y^{-4}\left(\frac{1}{t}y - y^4\right)$$

$$= -\frac{3}{t}z + 3.$$

This equation is linear, but it does not have constant coefficients. Therefore, we solve it using an integrating factor (see Section 1.9). We rewrite the equation as

$$\frac{dz}{dt} + \frac{3}{t}z = 3,$$

and observe that the integrating factor is $e^{\int (3/t)\,dt} = e^{3\ln t} = t^3$. Multiplying both sides of the differential equation for z by t^3, we have

$$t^3\frac{dz}{dt} + 3t^2 z = 3t^3.$$

$$t^3 z = \tfrac{3}{4} t^4 + c$$

$$z = \tfrac{3}{4} t + ct^{-3}.$$

So

$$y(t) = \left(\tfrac{3}{4} t + ct^{-3} \right)^{-1/3}.$$

27. **(a)** Substituting y_1 into the right-hand side of the equation, we get

$$\left(2t + \frac{1}{t} \right) t - t^2 - t^2 = 1,$$

which agrees with dy_1/dt. Hence, $y_1(t) = t$ is a solution.

(b) This differential equation is a Riccati equation, and we know one solution. Therefore, we change variables using $w = y - t$. We get

$$\frac{dw}{dt} = \frac{dy}{dt} - 1$$

$$= \left(2t + \frac{1}{t} \right) y - y^2 - t^2 - 1$$

$$= \left(2t + \frac{1}{t} \right) (w + t) - (w + t)^2 - t^2 - 1$$

$$= \left(2t + \frac{1}{t} \right) w + 2t^2 + 1 - \left(w^2 + 2tw + t^2 \right) - t^2 - 1$$

$$= \frac{1}{t} w - w^2.$$

This differential equation for w is a Bernoulli equation. So change variables once more using $z = w^{-1}$. Hence,

$$\frac{dz}{dt} = -w^{-2} \left(\frac{dw}{dt} \right)$$

$$= -w^{-2} \left(\frac{1}{t} w - w^2 \right)$$

$$= -\frac{1}{t} w^{-1} + 1$$

$$= -\frac{1}{t} z + 1.$$

This differential equation for z is linear, but it does not have constant coefficients. Therefore, we use an integrating factor to solve it (see Section 1.9). We rewrite the equation as

$$\frac{dz}{dt} + \frac{1}{t} z = 1,$$

and derive the integrating factor $\mu = e^{\int (1/t)dt} = e^{\ln t} = t$.
Multiplying both sides by t and integrating, we obtain

$$t\frac{dz}{dt} + z = t$$

$$\frac{d(tz)}{dt} = t$$

$$tz = \frac{t^2}{2} + c,$$

where c is the constant of integration. Therefore, the solution is

$$z(t) = \frac{t}{2} + \frac{c}{t}.$$

Hence, the solution to the equation in w is

$$w(t) = \frac{2t}{t^2 + k},$$

where $k = 2c$ is an arbitrary constant. The solution of the original differential equation is

$$y(t) = w(t) + t = \frac{(2+k)t + t^3}{t^2 + k}.$$

29. **(a)** Substituting y_1 into the right-hand side of the equation, we get

$$2\sin t + \cos t + t^2 \sin^2 t - 2(1 + t^2 \sin t)\sin t + t^2 \sin^2 t = \cos t,$$

which equals dy_1/dt. Hence, $y_1(t) = \sin t$ is a solution.

(b) This differential equation is a Riccati equation, and we know one solution. Therefore, we change variables using $w = y - \sin t$. We get

$$\frac{dw}{dt} = \frac{dy}{dt} - \cos t$$

$$= 2\sin t + t^2 \sin^2 t - 2(1 + t^2 \sin t)y + t^2 y^2$$

$$= 2\sin t + t^2 \sin^2 t - 2(1 + t^2 \sin t)(w + \sin t) + t^2 (w + \sin t)^2$$

$$= -2w + t^2 w^2.$$

This differential equation for w is a Bernoulli equation. So change variables once more using $z = w^{-1}$. Hence,

$$\frac{dz}{dt} = -w^{-2}\left(\frac{dw}{dt}\right)$$

$$= -w^{-2}\left(-2w + t^2 w^2\right)$$

$$= 2z - t^2.$$

This differential equation for z is linear. We solve it using the Extended Linearity Principle and a guessing technique. The general solution of the associated homogeneous equation is ke^{2t}. To find one particular solution of the nonhomogeneous equation, we rewrite the equation in the form

$$\frac{dz}{dt} - 2z = -t^2,$$

and we guess $z_p(t) = at^2 + bt + c$. Substituting this guess into the equation yields

$$(2at + b) - 2(at^2 + bt + c) = -t^2,$$

which simplifies to $(-2a)t^2 + (2a - 2b)t + (b - 2c) = -t^2$. The function $z_p(t)$ is a solution if $a = 1/2, b = 1/2$, and $c = 1/4$. Therefore,

$$z(t) = \tfrac{1}{2}t^2 + \tfrac{1}{2}t + \tfrac{1}{4} + ke^{2t}$$

is the general solution to the linear equation for z.

Hence, the functions

$$w(t) = \frac{4}{2t^2 + 2t + 1 + Ke^{2t}}$$

where K is an arbitrary constant are solutions to the the the Bernoulli equation for w. Note also that the equilibrium solution $w(t) = 0$ for all t is a solution.

Solutions to the original Riccati equation consist of function of the form

$$y(t) = \frac{4}{2t^2 + 2t + 1 + Ke^{2t}} + \sin t$$

as well as the solution $y(t) = \sin t$.

31. Let $u = (y - y_1)^{-1}$. Then

$$\begin{aligned}
\frac{du}{dt} &= -(y - y_1)^{-2}\left(\frac{dy}{dt} - \frac{dy_1}{dt}\right)\\
&= -(y - y_1)^{-2}\left(r(t) + a(t)y + b(t)y^2 - r(t) - a(t)y_1 - b(t)y_1^2\right)\\
&= -(y - y_1)^{-2}\left(a(t)(y - y_1) + b(t)(y^2 - y_1^2)\right)\\
&= -(y - y_1)^{-2}\left(a(t)(y - y_1) + b(t)(y - y_1)(y + y_1)\right)\\
&= -(y - y_1)^{-2}\left(a(t)(y - y_1) + b(t)(y - y_1)(y - y_1 + 2y_1)\right)\\
&= -u^2\left(a(t)u^{-1} + b(t)u^{-1}\left(u^{-1} + 2y_1\right)\right)\\
&= -a(t)u - b(t) - 2b(t)y_1(t)u\\
&= -(a(t) + 2b(t)y_1(t))u - b(t),
\end{aligned}$$

which is a linear differential equation for $u(t)$.

33. We begin with $w = y - y_1$, so $w = y - (t^2 + 2)$. Equivalently, $y = w + t^2 + 2$. We have

$$\frac{dy}{dt} = \frac{dw}{dt} + 2t$$

$$
\begin{aligned}
&= w + t^2 w^2 + 2t \\
&= (y - (t^2 + 2)) + t^2 (y - (t^2 + 2))^2 + 2t \\
&= y - t^2 - 2 + t^2 y^2 - 2t^4 y - 4t^2 y + t^6 + 4t^4 + 4t^2 + 2t \\
&= (t^6 + 4t^4 + 3t^2 + 2t - 2) + (1 - 4t^2 - 2t^4)y + t^2 y^2.
\end{aligned}
$$

Hence, our Riccati equation is

$$
\frac{dy}{dt} = (t^6 + 4t^4 + 3t^2 + 2t - 2) + (1 - 4t^2 - 2t^4)y + t^2 y^2.
$$

EXERCISES FOR APPENDIX B

1. We guess the power series $y(t) = a_0 + a_1 t + a_2 t^2 + a_3 t^3 + a_4 t^4 + \dots$. Since the exercise only requests the coefficients $a_0, a_1, \dots, a_4$, we stop at the $a_4 t^4$ term. Substituting the power series for $y(t)$ and dy/dt into the differential equation, we get

$$
a_1 + 2a_2 t + 3a_3 t^2 + 4a_4 t^3 + \dots = a_0 + a_1 t + a_2 t^2 + a_3 t^3 + a_4 t^4 + \dots.
$$

Equating coefficients for each power of t yields

$$
\begin{aligned}
a_1 &= a_0 \\
2a_2 &= a_1 \\
3a_3 &= a_2 \\
4a_4 &= a_3 \\
&\;\;\vdots
\end{aligned}
$$

Expressing all of the coefficients in terms of a_0, we get

$$
a_1 = a_0, \quad a_2 = \frac{a_0}{2}, \quad a_3 = \frac{a_0}{2 \cdot 3}, \quad a_4 = \frac{a_0}{2 \cdot 3 \cdot 4}, \quad \dots.
$$

The power series solution is

$$
y(t) = a_0 \left(1 + t + \frac{t^2}{2!} + \frac{t^3}{3!} + \frac{t^4}{4!} + \dots \right).
$$

As expected, this series is the power series for $a_0 e^t$, the general solution of the differential equation $dy/dt = y$.

3. We guess the power series $y(t) = a_0 + a_1 t + a_2 t^2 + a_3 t^3 + a_4 t^4 + \dots$. Since the exercise only requests the coefficients $a_0, a_1, \dots, a_4$, we stop at the $a_4 t^4$ term. Substituting the power series for $y(t)$ and dy/dt into the differential equation, we get

$$
a_1 + 2a_2 t + 3a_3 t^2 + 4a_4 t^3 + \dots = -2t(a_0 + a_1 t + a_2 t^2 + a_3 t^3 + a_4 t^4 + \dots).
$$

Equating coefficients for each power of t yields

$$a_1 = 0$$
$$2a_2 = -2a_0$$
$$3a_3 = -2a_1$$
$$4a_4 = -2a_2$$
$$\vdots$$

Expressing all of the coefficients in terms of a_0, we get

$$a_1 = 0, \quad a_2 = -a_0, \quad a_3 = 0, \quad a_4 = \frac{a_0}{2}, \quad \ldots$$

The power series solution is

$$y(t) = a_0 \left(1 - t^2 + \frac{t^4}{2} \mp \ldots\right).$$

As expected, this series is the power series for $a_0 e^{-t^2}$, the general solution of the differential equation $dy/dt = -2ty$.

5. We guess the power series $y(t) = a_0 + a_1 t + a_2 t^2 + a_3 t^3 + a_4 t^4 + \ldots$. Since the exercise only requests the coefficients $a_0, a_1, \ldots, a_4$, we stop at the $a_4 t^4$ term. Using the power series for e^u with $u = 2t$ (see Appendix C), we have

$$e^{2t} = 1 + 2t + \frac{4t^2}{2!} + \frac{8t^3}{3!} + \frac{16t^4}{4!} + \ldots.$$

Substituting the power series for $y(t)$, dy/dt, and e^{2t} into the differential equation, we get

$$a_1 + 2a_2 t + 3a_3 t^2 + 4a_4 t^3 + \ldots = -\left(a_0 + a_1 t + a_2 t^2 + a_3 t^3 + a_4 t^4 + \ldots\right) +$$

$$\left(1 + 2t + \frac{4t^2}{2!} + \frac{8t^3}{3!} + \frac{16t^4}{4!} + \ldots\right).$$

Equating coefficients for each power of t yields

$$a_1 = -a_0 + 1$$
$$2a_2 = -a_1 + 2$$
$$3a_3 = -a_2 + \frac{4}{2!}$$
$$4a_4 = -a_3 + \frac{8}{3!}$$
$$\vdots$$

Expressing all of the coefficients in terms of a_0, we get

$$a_1 = -a_0 + 1, \quad a_2 = \frac{a_0 + 1}{2!}, \quad a_3 = \frac{-a_0 + 3}{3!}, \quad a_4 = \frac{a_0 + 5}{4!}, \quad \ldots$$

The power series solution is

$$y(t) = a_0 + (-a_0 + 1)t + \left(\frac{a_0 + 1}{2!}\right)t^2 + \left(\frac{-a_0 + 3}{3!}\right)t^3 + \left(\frac{a_0 + 5}{4!}\right)t^4 + \dots.$$

7. We guess the power series $y(t) = a_0 + a_1 t + a_2 t^2 + a_3 t^3 + a_4 t^4 + \dots$. Since the exercise only requests the coefficients $a_0, a_1, \dots, a_4$, we stop at the $a_4 t^4$ term. The power series for $\cos t$ is

$$1 - \frac{t^2}{2!} + \frac{t^4}{4!} - \frac{t^6}{6!} \pm \dots$$

(see Appendix C). Substituting the power series for $y(t)$, $d^2 y/dt^2$, and $\cos t$ into the differential equation, we get

$$(2a_2 + (2 \cdot 3)a_3 t + (3 \cdot 4)a_4 t^2 + \dots) + 2(a_0 + a_1 t + a_2 t^2 + a_3 t^3 + a_4 t^4 + \dots)$$
$$= 1 - \frac{t^2}{2!} + \frac{t^4}{4!} \mp \dots.$$

Equating coefficients for each power of t yields

$$2a_2 + 2a_0 = 1$$
$$6a_3 + 2a_1 = 0$$
$$12a_4 + 2a_2 = -\tfrac{1}{2}$$
$$\vdots$$

Expressing all of the coefficients in terms of a_0 and a_1, we get

$$a_2 = \tfrac{1}{2} - a_0, \quad a_3 = -\tfrac{1}{3}a_1, \quad a_4 = \tfrac{1}{6}a_0 - \tfrac{1}{8}, \quad \dots.$$

The power series solution is

$$y(t) = a_0 + a_1 t + \left(\tfrac{1}{2} - a_0\right)t^2 - \tfrac{1}{3}a_1 t^3 + \left(\tfrac{1}{6}a_0 - \tfrac{1}{8}\right)t^4 + \dots.$$

9. First we verify that $y(t) = \tan t$ is a solution of the differential equation. We note that

$$\frac{dy}{dt} = \sec^2 t$$
$$= 1 + \tan^2 t$$
$$= 1 + y^2.$$

We guess the power series $y(t) = a_0 + a_1 t + a_2 t^2 + a_3 t^3 + a_4 t^4 + \dots$. Since the exercise only requests the coefficients $a_0, a_1, \dots, a_6$, we stop at the $a_6 t^6$ term. We also need the coefficients of the power series for y^2 up to degree five. We compute

$$\left(a_0 + a_1 t + a_2 t^2 + a_3 t^3 + a_4 t^4 + a_5 t^5 + \dots\right)^2 = a_0^2 + (2a_0 a_1)t + (2a_0 a_2 + a_1^2)t^2 +$$
$$(2a_0 a_3 + 2a_1 a_2)t^3 +$$
$$(2a_0 a_4 + 2a_1 a_3 + a_2^2)t^4 +$$
$$(2a_0 a_5 + 2a_1 a_4 + 2a_2 a_3)t^5 + \dots.$$

Substituting the power series for dy/dt and y^2 into the differential equation, we get

$$a_1 + 2a_2t + 3a_3t^2 + 4a_4t^3 + 5a_5t^4 + 6a_6t^5 + \ldots =$$
$$1 + a_0^2 + (2a_0a_1)\,t + (2a_0a_2 + a_1^2)\,t^2 + (2a_0a_3 + 2a_1a_2)\,t^3 +$$
$$(2a_0a_4 + 2a_1a_3 + a_2^2)\,t^4 + (2a_0a_5 + 2a_1a_4 + 2a_2a_3)\,t^5 + \ldots.$$

Equating coefficients for each power of t, we obtain the equations

$$a_1 = 1 + a_0^2$$
$$2a_2 = 2a_0a_1$$
$$3a_3 = 2a_0a_2 + a_1^2$$
$$4a_4 = 2a_0a_3 + 2a_1a_2$$
$$5a_5 = 2a_0a_4 + 2a_1a_3 + a_2^2$$
$$6a_6 = 2a_0a_5 + 2a_1a_4 + 2a_2a_3$$
$$\vdots$$

The function $y(t) = \tan t$ satisfies the initial condition $y(0) = 0$, so we know that $a_0 = 0$. We have

$$a_1 = 1, \quad a_2 = 0, \quad a_3 = \tfrac{1}{3}, \quad a_4 = 0, \quad a_5 = \tfrac{2}{15}, \quad a_6 = 0, \quad \ldots.$$

The power series for $\tan t$ is

$$t + \tfrac{1}{3}t^3 + \tfrac{2}{15}t^5 + \ldots.$$

11. We guess the power series $y(t) = a_0 + a_1t + a_2t^2 + a_3t^3 + a_4t^4 + \ldots$. Since the exercise only requests the coefficients $a_0, a_1, \ldots, a_6$, we stop at the a_6t^6 term. Substituting the power series for $y(t), dy/dt$, and d^2y/dt^2 into the differential equation, we get

$$(2a_2 + 6a_3t + 12a_4t^2 + 20a_5t^3 + 30a_6t^4 + \ldots)$$
$$+ (a_1 + 2a_2t + 3a_3t^2 + 4a_4t^3 + 5a_5t^4 + 6a_6t^5 + \ldots)$$
$$+ (a_0 + a_1t + a_2t^2 + a_3t^3 + a_4t^4 + a_5t^5 + a_6t^6 + \ldots) = 0.$$

Equating coefficients for each power of t yields

$$2a_2 + a_1 + a_0 = 0$$
$$6a_3 + 2a_2 + a_1 = 0$$
$$12a_4 + 3a_3 + a_2 = 0$$
$$20a_5 + 4a_4 + a_3 = 0$$
$$30a_6 + 5a_5 + a_4 = 0$$
$$\vdots$$

Expressing all of the coefficients in terms of a_0 and a_1, we get

$$a_2 = -\tfrac{1}{2}(a_0 + a_1), \quad a_3 = \tfrac{1}{6}a_0, \quad a_4 = \tfrac{1}{24}a_1, \quad a_5 = -\tfrac{1}{120}(a_0 + a_1), \quad a_6 = \tfrac{1}{720}a_0, \quad \ldots.$$

The power series solution is

$$y(t) = a_0 + a_1 t - \tfrac{1}{2}(a_0 + a_1)t^2 + \tfrac{1}{6}a_0 t^3 + \tfrac{1}{24}a_1 t^4 - \tfrac{1}{120}(a_0 + a_1)t^5 + \tfrac{1}{720}a_0 t^6 + \dots.$$

13. We guess the power series $y(t) = a_0 + a_1 t + a_2 t^2 + a_3 t^3 + a_4 t^4 + \dots.$ Since the exercise only requests the coefficients $a_0, a_1, \dots, a_6$, we stop at the $a_6 t^6$ term. Using the power series for e^u with $u = -2t$ (see Appendix C), we have

$$e^{-2t} = 1 - 2t + \frac{4t^2}{2!} - \frac{8t^3}{3!} + \frac{16t^4}{4!} - \frac{32t^5}{5!} \pm \dots.$$

Substituting the power series for $y(t), dy/dt, d^2y/dt^2$, and e^{-2t} into the differential equation, we get

$$(2a_2 + 6a_3 t + 12a_4 t^2 + 20a_5 t^3 + 30a_6 t^4 + \dots)$$
$$+ t\,(a_1 + 2a_2 t + 3a_3 t^2 + 4a_4 t^3 + 5a_5 t^4 + 6a_6 t^5 + \dots)$$
$$+ (a_0 + a_1 t + a_2 t^2 + a_3 t^3 + a_4 t^4 + a_5 t^5 + a_6 t^6 + \dots) =$$
$$1 - 2t + 2t^2 - \tfrac{4}{3}t^3 + \tfrac{2}{3}t^4 + \dots.$$

Equating coefficients for each power of t yields

$$2a_2 + a_0 = 1$$
$$6a_3 + 2a_1 = -2$$
$$12a_4 + 3a_2 = 2$$
$$20a_5 + 4a_3 = -\tfrac{4}{3}$$
$$30a_6 + 5a_4 = \tfrac{2}{3}$$
$$\vdots$$

Expressing all of the coefficients in terms of a_0 and a_1, we get

$$a_2 = \tfrac{1}{2}(1 - a_0), \quad a_3 = -\tfrac{1}{3}(a_1 + 1), \quad a_4 = \tfrac{1}{24}(3a_0 + 1), \quad a_5 = \tfrac{1}{15}a_1, \quad a_6 = \tfrac{1}{720}(11 - 15a_0), \quad \dots.$$

The power series solution is

$$y(t) = a_0 + a_1 t + \tfrac{1}{2}(1 - a_0)t^2 - \tfrac{1}{3}(a_1 + 1)t^3 + \tfrac{1}{24}(3a_0 + 1)t^4 + \tfrac{1}{15}a_1 t^5 + \tfrac{1}{720}(11 - 15a_0)t^6 + \dots.$$

15. (a) We guess the power series $y(t) = a_0 + a_1 t + a_2 t^2 + a_3 t^3 + a_4 t^4 + \dots.$ To do this part of the exercise, we need only consider terms up to degree four. Substituting the power series for $y(t)$, dy/dt, and d^2y/dt^2 into the differential equation, we get

$$(1 - t^2)(2a_2 + 6a_3 t + 12a_4 t^2 + \dots)$$
$$- 2t(a_1 + 2a_2 t + 3a_3 t^2 + 4a_4 t^3 + \dots)$$
$$+ v(v + 1)(a_0 + a_1 t + a_2 t^2 + a_3 t^3 + a_4 t^4 + \dots) = 0.$$

Collecting the coefficients of the constant terms, linear terms, and quadratic terms, we see that

$$2a_2 + v(v+1)a_0 = 0$$
$$6a_3 - 2a_1 + v(v+1)a_1 = 0$$
$$12a_4 - 6a_2 + v(v+1)a_2 = 0.$$

Therefore,

$$a_2 = \frac{-v(v+1)}{2}a_0$$

$$a_3 = \frac{2 - v(v+1)}{6}a_1$$

$$a_4 = \left(\frac{6 - v(v+1)}{12}\right)\left(\frac{-v(v+1)}{2}\right)a_0.$$

(b) If we substitute the power series for $y(t)$, dy/dt, and d^2y/dt^2 into the differential equation and collect all of the t^n terms ($n \geq 2$), we get

$$(n+1)(n+2)a_{n+2}t^n - n(n-1)a_n t^n - 2na_n t^n + v(v+1)a_n t^n.$$

Consequently, $y(t)$ is a solution only if

$$(n+1)(n+2)a_{n+2} - n(n-1)a_n - 2na_n + v(v+1)a_n = 0$$

for all $n \geq 2$. This equality is satisfied if

$$a_{n+2} = \frac{n(n+1) - v(v+1)}{(n+1)(n+2)}a_n.$$

Note that this relationship between a_n and a_{n+2} generalizes the results obtained in part (a).

An important consequence of this relationship is that, if any coefficient, $a_1, a_3, a_5, \ldots$, of an odd power of t is zero, then all coefficients of higher odd powers are also zero. For example, if $a_7 = 0$, then $a_9 = 0$, $a_{11} = 0$, $\ldots$. Moreover, the same can be said for coefficients of even powers of t.

To produce polynomial solutions if v is a positive integer, we consider two cases. If v is even, we use the initial condition $(y(0), y'(0)) = (1, 0)$. Then $a_1 = 0$, and all of the coefficients of odd powers are zero. Moreover, the formula that relates a_{n+2} to a_n implies that $a_{v+2} = 0$. Therefore, all of the coefficients of higher even powers are also zero, and the solution is a polynomial (with no odd powers of t).

If v is odd, we use the initial condition $(y(0), y'(0)) = (0, 1)$ and obtain a polynomial solution with no even powers of t.

(c) To compute $P_0(t)$, take $v = 0$ and $(y(0), y'(0)) = (1, 0)$. Then $a_0 = 1$ and $a_1 = 0$. From part (b), $a_3 = a_5 = a_7 = \ldots = 0$. Also, from the formula in part (a), we get $a_2 = 0$, and consequently, $a_4 = a_6 = \ldots = 0$ as well. Hence, the solution is

$$P_0(t) = 1.$$

Similarly, taking $v = 1$ and $(y(0), y'(0)) = (0, 1)$, we obtain $a_0 = a_2 = \ldots = 0$, and since $a_3 = 0$, $a_5 = a_7 = \ldots = 0$. Hence, the solution is

$$P_1(t) = t.$$

To compute $P_2(t)$, take $\nu = 2$ and $(y(0), y'(0)) = (1, 0)$. We have $a_1 = a_3 = \ldots = 0$. Also, $a_0 = 1$ implies that $a_2 = -3$ and $a_4 = a_6 = a_8 = \ldots = 0$. We have the solution

$$P_2(t) = 1 - 3t^2.$$

(d) To compute $P_3(t)$, take $\nu = 3$ and $(y(0), y'(0)) = (0, 1)$. Then $a_3 = -5/3$ and

$$P_3(t) = t - \tfrac{5}{3}t^3.$$

To compute $P_4(t)$, take $\nu = 4$ and $(y(0), y'(0)) = (1, 0)$. Then $a_2 = -10, a_4 = 35/3$, and

$$P_4(t) = 1 - 10t^2 + \tfrac{35}{3}t^4.$$

To compute $P_5(t)$, take $\nu = 5$ and $(y(0), y'(0)) = (0, 1)$. Then $a_3 = -14/3, a_5 = 21/5$, and

$$P_5(t) = t - \tfrac{14}{3}t^3 + \tfrac{21}{5}t^5.$$

To compute $P_6(t)$, take $\nu = 6$ and $(y(0), y'(0)) = (1, 0)$. Then $a_2 = -21, a_4 = 63, a_6 = -231/5$, and

$$P_6(t) = 1 - 21t^2 + 63t^4 - \tfrac{231}{5}t^6.$$

(e) Legendre's equation is a homogeneous linear equation. The Linearity Principle implies that $kP_\nu(t)$ is a solution for any constant k if $P_\nu(t)$ is a solution. (However, the initial conditions change by a factor of k.)

17. We guess the power series $y(t) = a_0 + a_1 t + a_2 t^2 + a_3 t^3 + a_4 t^4 + \ldots$. Using the power series for e^u with $u = -t$ (see Appendix C), we have

$$e^{-t} = 1 - t + \frac{t^2}{2!} - \frac{t^3}{3!} \pm \ldots.$$

Substituting the power series for $y(t)$ and dy/dt into the differential equation, we get

$$a_1 + 2a_2 t + 3a_3 t^2 + 4a_4 t^3 + \ldots =$$
$$-\left(a_0 + a_1 t + a_2 t^2 + a_3 t^3 + \ldots \right) + \left(1 - t + \frac{t^2}{2!} - \frac{t^3}{3!} \pm \ldots \right).$$

Equating coefficients for each power of t yields

$$a_1 = -a_0 + 1$$
$$2a_2 = -a_1 - 1$$
$$3a_3 = -a_2 + \tfrac{1}{2}$$
$$4a_4 = -a_3 - \tfrac{1}{6}$$
$$\vdots$$

In fact, by comparing coefficients of t^n, we get

$$(n+1)a_{n+1} = -a_n + (-1)^n \frac{1}{n!}.$$

Using the initial condition $y(0) = a_0 = 0$, we observe that

$$a_0 = 0, \quad a_1 = 1, \quad a_2 = -1, \quad a_3 = \tfrac{1}{2}, \quad a_4 = -\tfrac{1}{6}, \quad \ldots$$

Using induction, we can show that

$$a_n = (-1)^{n-1} \frac{1}{(n-1)!}$$

for $n \geq 2$. The power series solution is

$$y(t) = t - t^2 + \frac{t^3}{2!} - \frac{t^4}{3!} \pm \cdots$$

$$= t \left(1 - t + \frac{t^2}{2!} - \frac{t^3}{3!} \pm \cdots \right),$$

which is the power series for te^{-t}.